Pedestrians 2018

AA005604

Transportation Research Record: Journal of the Transportation Research Board

Volume 2672, Issue 35

ISBN: 978-1-5108-7836-5

Printed from e-media with permission by:

Curran Associates, Inc.
57 Morehouse Lane
Red Hook, NY 12571

Some format issues inherent in the e-media version may also appear in this print version.

Published by Sage Publications USA

ISBN (Print) 978-1-5108-7836-5
ISBN 2018 Printed Set (All Issues) 978-1-5108-7735-1

Printed by Curran Associates, Inc. (2019)

Additional copies of this publication are available from:

Curran Associates, Inc.
57 Morehouse Lane
Red Hook, NY 12571 USA
Phone: 845-758-0400
Fax: 845-758-2633
Email: curran@proceedings.com
Web: www.proceedings.com

Volume 2672 Number 35 December 2018

TRR
JOURNAL OF THE TRANSPORTATION RESEARCH BOARD

Contents

Pedestrians 2018

Articles

TRB

TRANSPORTATION RESEARCH BOARD

CORE SPONSORS

Transportation Departments of the 50 States and the District of Columbia

Federal Government

U.S. Department of Transportation
- Federal Aviation Administration (FAA)
- Federal Highway Administration
- Federal Motor Carrier Safety Administration (FMCSA)
- Federal Railroad Administration (FRA)
- Federal Transit Administration
- National Highway Traffic Safety Administration (NHTSA)
- Office of the Assistant Secretary of Transportation for Research and Technology (OST-R)

U.S. Air Force Civil Engineer Center (AFCEC)
U.S. Coast Guard (USCG)
U.S. Department of Energy
U.S. Department of the Interior (DOI)
U.S. Environmental Protection Agency (EPA)

Nongovernmental Organizations

American Public Transportation Association (APTA)
Association of American Railroads (AAR)
California Air Resources Board

The National Academies of
SCIENCES · ENGINEERING · MEDICINE

The **National Academy of Sciences** was established in 1863 by an Act of Congress, signed by President Lincoln, as a private, nongovernmental institution to advise the nation on issues related to science and technology. Members are elected by their peers for outstanding contributions to research. Dr. Marcia McNutt is president.

The **National Academy of Engineering** was established in 1964 under the charter of the National Academy of Sciences to bring the practices of engineering to advising the nation. Members are elected by their peers for extraordinary contributions to engineering. Dr. C. D. Mote, Jr., is president.

The **National Academy of Medicine** (formerly the Institute of Medicine) was established in 1970 under the charter of the National Academy of Sciences to advise the nation on medical and health issues. Members are elected by their peers for distinguished contributions to medicine and health. Dr. Victor J. Dzau is president.

The three Academies work together as the National Academies of Sciences, Engineering, and Medicine to provide independent, objective analysis and advice to the nation and conduct other activities to solve complex problems and inform public policy decisions. The Academies also encourage education and research, recognize outstanding contributions to knowledge, and increase public understanding in matters of science, engineering, and medicine.

Learn more about the National Academies of Sciences, Engineering, and Medicine at **www.national-academies.org.**

The **Transportation Research Board** is one of seven major programs of the National Academies of Sciences, Engineering, and Medicine. The mission of the Transportation Research Board is to increase the benefits that transportation contributes to society by providing leadership in transportation innovation and progress through research and information exchange, conducted within a setting that is objective, interdisciplinary, and multimodal. The Board's varied activities annually engage about 7,000 engineers, scientists, and other transportation researchers and practitioners from the public and private sectors and academia, all of whom contribute their expertise in the public interest. The program is supported by state transportation departments, federal agencies including the component administrations of the U.S. Department of Transportation, and other organizations and individuals interested in the development of transportation.

Learn more about the Transportation Research Board at **www.TRB.org**.

Article

TRR
JOURNAL OF THE TRANSPORTATION RESEARCH BOARD

Transportation Research Record
2018, Vol. 2672(35) 1–12
© National Academy of Sciences:
Transportation Research Board 2018
Article reuse guidelines:
sagepub.com/journals-permissions
DOI: 10.1177/0361198118759949
journals.sagepub.com/home/trr
SAGE

Pedestrian Behavior at Signalized Intersection Crosswalks: Observational Study of Factors Associated with Distracted Walking, Pedestrian Violations, and Walking Speed

Brendan J. Russo[1], **Emmanuel James**[1], **Cristopher Y. Aguilar**[1], **and Edward J. Smaglik**[1]

Abstract

In the past two decades, cell phone and smartphone use in the United States has increased substantially. Although mobile phones provide a convenient way for people to communicate, the distraction caused by the use of these devices has led to unintended traffic safety and operational consequences. Although it is well recognized that distracted driving is extremely dangerous for all road users (including pedestrians), the potential impacts of distracted walking have not been as comprehensively studied. Although practitioners should design facilities with the safety, efficiency, and comfort of pedestrians in mind, it is still important to investigate certain pedestrian behaviors at existing facilities to minimize the risk of pedestrian–vehicle crashes, and to reduce behaviors that may unnecessarily increase delay at signalized intersections. To gain new insights into factors associated with distracted walking, pedestrian violations, and walking speed, 3,038 pedestrians were observed across four signalized intersections in New York and Arizona using high-definition video cameras. The video data were reduced and summarized, and an ordinary least squares (OLS) regression model was estimated to analyze factors affecting walking speeds. In addition, binary logit models were estimated to analyze both pedestrian distraction and pedestrian violations. Ultimately, several site- and pedestrian-specific variables were found to be significantly associated with pedestrian distraction, violation behavior, and walking speeds. The results provide important information for researchers, practitioners, and legislators, and may be useful in planning strategies to reduce or mitigate the impacts of pedestrian behavior that may be considered unsafe or potentially inefficient.

Vision Zero is a strategy to eliminate all traffic-related fatalities, at the same time increasing safe mobility for all road users, including pedestrians. First implemented in Sweden in the 1990s, Vision Zero has been gaining momentum across the United States in recent years (*1*), with its needed focus on improving conditions for vulnerable road users. However, the behavior of all road users affects the ability to eliminate fatalities across the network, including pedestrians who are crossing the street legally at a signalized intersection. Whereas some may perceive this as victim blaming, the concept of permitting vehicles to turn across a crosswalk at the same time as pedestrians are crossing with the signal requires all parties to be cognizant of their surroundings.

In the past two decades, cell phone use in the United States has increased substantially. As of 2016, 95% of adults in the United States own a cell phone (up from 87% in 2011) and 77% of US adults own smartphones (up from only 35% in 2011) (*2*). Although mobile phones provide a convenient way for people to communicate, the increased use of these devices by road users on public roadways has led to unintended traffic safety and operational consequences. Recent research has found conclusively that drivers using mobile

[1]Department of Civil Engineering, Construction Management, and Environmental Engineering, Northern Arizona University, Flagstaff, AZ

Corresponding Author:
Address correspondence to Brendan J. Russo: brendan.russo@nau.edu

phones exhibit reduced reaction times and have an increased risk of crashing (*3*). In fact, a meta-analysis of 33 studies showed mobile phone use to increase mean reaction time by 0.25 seconds (*4*). It is clear that distracted drivers are extremely dangerous to pedestrians (as well as to all other road users); however, as users operating in a shared space, pedestrians displaying distracted walking may also have an impact on pedestrian safety and operations at some locations.

Contrary to mobile phone use by vehicle drivers, the way in which pedestrians are affected by using their mobile phones has not been studied as extensively, and data related to crashes involving pedestrians using mobile phones are often not as readily available in crash databases as they are for vehicle driver mobile phone use. The use of mobile phones has the potential to affect pedestrian behavior with regard to delay and walking speed (primarily at signalized intersection crossings), and with regard to unsafe crossing behavior at signalized crosswalks (i.e., violating pedestrian signals and crossing outside of crosswalks). There is currently a need to investigate how mobile phone use may affect pedestrian behavior so that engineers, planners, and law enforcement agencies can address any potential issues that may affect pedestrian safety. A recent report published by the National Highway Traffic Safety Administration concluded that "a very limited number of studies have investigated the effect of electronic device use by pedestrians and drivers on pedestrian safety" and that "there is a need to conduct naturalistic observations of the effect of electronic device use on pedestrian distraction and safety" (*5*). The general public are becoming aware of this issue, with one municipality recently passing legislation introducing fines for pedestrians looking at their mobile phones while crossing the street (*6*). Although there have been a limited number of studies related to pedestrian mobile phone use in the United States, several rely on simulator and survey data (*7, 8*), which may potentially introduce bias. There has been one study using observational data focused on signalized intersections (*9*). In all of these studies, mobile phone use was found to affect pedestrian behavior to some degree; however, more research using observed field data is warranted.

In addition to distracted walking, pedestrian signal or crosswalk violations (i.e., disregarding pedestrian signals or crossing outside of marked crosswalks), as well as walking speed, can have potential impacts on operations and safety at signalized intersections. For example, vehicle drivers may not expect to encounter pedestrians outside of a crosswalk as they turn, and pedestrians violating the pedestrian signal (i.e., starting to cross during the "flashing don't walk" or "don't walk" phases may be more likely to be struck by a vehicle, and may increase overall delay at the intersection if vehicles must wait to let them cross against the signal. This potential increased delay for vehicle drivers may also increase driver frustration and lead to riskier maneuvers around pedestrians. Although practitioners should design facilities with the safety, efficiency, and comfort of all road users in mind, including pedestrians, it is still important to investigate certain pedestrian behaviors at existing facilities to minimize the risk of pedestrian–vehicle crashes, and to reduce behaviors that may unnecessarily increase delay at signalized intersections.

To gain new insights into factors associated with distracted walking, pedestrian walking speed, and pedestrian signal or crosswalk violations at signalized intersections, this study utilized observational field-recorded data from one intersection in New York, and three intersections in Flagstaff, Arizona. An ordinary least squares (OLS) regression model was estimated to examine factors affecting pedestrian walking speed and binary logit models were estimated to examine factors associated with pedestrian distraction and signal or crosswalk violation behavior. Ultimately, several site- and pedestrian-specific variables were found to be significantly associated with pedestrian distraction, violation behavior, and walking speeds.

Literature Review

In 2015, 5,376 pedestrians were killed and 70,000 were injured in traffic crashes in the United States, representing 15% of total traffic fatalities and 3% of total injuries (*10*). As noted previously, distracted driving has received much attention because of its greater impact on traffic crashes; however, research on pedestrian distraction, which is often attributed to behaviors such as talking on a mobile phone, texting or reading, using headphones, and eating or drinking, among others, is much more limited. Research that has been conducted in this area can be divided into two general areas: observational studies and simulation-based studies.

In the area of observational studies, researchers have examined the behavior of pedestrians at signalized and unsignalized crossings. Basch et al. studied pedestrian behavior at five busy signalized intersections in Manhattan, New York. Their findings revealed that headphone use was the most common observed distraction. Of all the pedestrians crossing, 27.8% crossing on a "walk" signal and 42% crossing on a "don't walk" signal were distracted (wearing headphones, talking on a mobile phone, or looking down at an electronic device) (*11*). Hatfield and Murphy conducted a study of observed pedestrian behavior at signalized and unsignalized intersections and found that females who were talking on a mobile phone crossed more slowly and were: less likely to look at traffic before starting to cross; wait for traffic to stop; or look at traffic while crossing as compared with other pedestrians (*12*). Thompson et al. observed pedestrians at 20 high-risk intersections in Seattle, Washington in 2012 (*13*). Their findings revealed that crossing times for pedestrians using mobile devices were slower, and texting was associated with an 18% increase in crossing times. They also found that 20% of the pedestrians engaged in distracted behavior while crossing the intersection, and that females were less likely to demonstrate optimal crossing behavior compared with males.

Cooper et al. studied pedestrian, cyclist, and driver behavior at 12 signalized intersections near transit stations in San Francisco (*14*). Their findings revealed 3% to 18% of pedestrians used mobile devices while crossing at various sites and females were more likely to talk on mobile devices while crossing a street than males; however, males were more likely to violate the signal display while walking and cycling. Gillette et al. conducted an observational study to understand the impacts of distraction on pedestrian start-up time and crossing behaviors (*9*). Researchers observed 760 pedestrians at three intersections in College Station, Texas in real time and found that pedestrians who texted and those who talked on their mobile phones had 21% and 31% higher start-up times, respectively. Pedestrians in mixed age groups also had a 14% higher start-up time. The findings also showed that pedestrians who were texting and talking on their mobile phones were less likely to glance around before entering the crosswalk.

In the area of simulation, Schwebel et al. investigated the impacts of behavioral intervention on crossing behavior, both in a virtual environment and in a real-world setting (*15*). Their results revealed that individuals who took the test in the virtual environment reported changes to their intended behavior while crossing the street, but there was no evident change in observed behavior before and after the intervention compared with a control site. Lin and Huang studied the impact of using a smartphone on a pedestrian's awareness of roadside events in a semi-virtual environment (*16*). Their findings revealed that pedestrians took longer to detect roadside events while reading and texting as compared with dragging pictures. Schwebel et al. studied the impacts of talking on the phone, texting, and listening to music on pedestrian safety in a virtual environment (*7*). They found that participants who were distracted by music or texting were more likely to be hit by a vehicle in the virtual environment than were undistracted pedestrians.

In summary, previous studies have found that certain demographic characteristics affect pedestrian behaviors, including distracted walking. This study will focus on pedestrians at signalized crosswalks, and investigate person- and site-specific factors associated with pedestrian distraction, pedestrian violations, and pedestrian walking speeds.

Data Description

Data for this study were obtained from one signalized intersection in New York, and three signalized intersections in Flagstaff, Arizona. The intersection in New York was chosen based on existing video data and high pedestrian volumes, and the three Flagstaff locations were chosen based on their high pedestrian activity. The primary goal of this study was to examine pedestrian behavior at signalized intersection crossings and, to this end, data were discreetly recorded using high-definition, field-mounted video cameras. This type of video data collection is advantageous as it is relatively inconspicuous, allowing for observation of natural road user behavior, and it allows for very accurate data collection as recordings can be observed frame by frame and stopped and rewound during data extraction. All video data were recorded during the spring and summer of 2017 during daylight hours and clear weather conditions. It should be noted that cameras were set up such that one entire crosswalk was visible, along with the face of the pedestrian signals for that crosswalk. For all study sites, a total of 12 hours of video were collected and analyzed, resulting in 1,398 pedestrian observations for the New York site and 1,640 pedestrian observations across all three Arizona sites, giving a total of 3,038 pedestrian observations. Site characteristics were obtained for each site, including crossing length (in feet) for each study crosswalk, "walk" and "flashing don't walk" pedestrian signal phase times (in seconds), cycle length (in seconds), posted speed limit on the road being crossed for each crosswalk studied, and whether the intersection was equipped with a pedestrian push button. A summary of study site characteristics is presented in Table 1, and Figure 1 shows overhead satellite views of all four crosswalks studied.

The pedestrian behavior data utilized for this study were extracted from the field-collected videos by data collectors in an office setting. Data collectors were trained to accurately record the desired pedestrian behavior, and test videos were used to ensure consistency between data collectors. For each pedestrian observed at the study crosswalks, data collectors extracted the following data:

- Pedestrian distraction, recorded as one of five discrete categories: no distraction, talking on mobile phone, texting on mobile phone, listening to headphones, or other. The 'other' distraction category includes behaviors besides mobile phone or headphone use that may otherwise distract the pedestrian from focusing on the task of safely crossing the street, such as looking in a purse, reading a newspaper, tending to a young child, etc. Example video screen shots of undistracted and distracted pedestrians are shown in Figure 2.
- Pedestrian age estimate using four discrete age group categories: age 0 to 16, age 16 to 29, age 30 to 59, and age 60 or older.
- Pedestrian gender: male or female.
- Group size at time of crossing the street.
- Number of pedestrians crossing in the opposing direction during crossing.
- Time stamp of pedestrian arrival at crossing corner.
- Time stamp of "walk" pedestrian signal.
- Time stamp of the start of pedestrian crossing (i.e., the moment the pedestrian stepped off the sidewalk on to the street). The difference between the time stamp of pedestrian arrival and crossing start time stamps was subsequently used to determine waiting time.

Table 1. Summary of Study Site Characteristics

Site characteristic	Study site			
	Site 1	Site 2	Site 3	Site 4
Intersection	6th Avenue/14th Street	Beulah/McConnell	Beaver/Route 66	San Francisco/Route 66
Location	New York	Flagstaff, Arizona	Flagstaff, Arizona	Flagstaff, Arizona
"Walk" time (s)	32	4	7	7
"Flashing don't walk" time (s)	14	24	18	20
Cycle length (s)	90	60	110	110
Crossing length (ft)	52.3	91.3	63.4	63.5
Posted speed limit (mph)	25	35	30	30
Pedestrian push button equipped?	No	Yes	Yes	Yes
Total pedestrian observations	1,398	120	530	990

Figure 1. Overhead satellite views of study crosswalks for (*a*) site 1, (*b*) site 2, (*c*) site 3, and (*d*) site 4.

Figure 2. Video screenshot examples of (*a*) undistracted pedestrian, (*b*) pedestrian with headphones, (*c*) pedestrian talking on a mobile phone, and (*d*) pedestrian texting on a mobile phone.

- Time stamp of the end of pedestrian crossing (i.e., the moment the pedestrian stepped off the street on to the sidewalk.
- Pedestrian signal at the start and end of crossing: "walk," "flashing don't walk," or "don't walk."
- Whether the pedestrian crossed outside of the marked crosswalk during their crossing.
- Whether the pedestrian was crossing with or against adjacent roadway traffic.
- Whether the pedestrian used the pedestrian push button (if available).

Because data collectors were able to examine video recordings frame by frame with the ability to pause and rewind, the time stamps of pedestrian start and end crossing times and the calculation of average walking speed in feet per second (ft/s) were very precise. Table 2 presents summary statistics for pedestrian observations. It should be noted that walking speed statistics are presented as both a continuous variable, and as a series of binary indicator variables for different speed ranges.

Analysis Methodology

The objective of this study was to analyze factors associated with distracted walking, pedestrian signal or crosswalk violations, and walking speed at signalized intersections. As these characteristics consist of different types of data (i.e., walking speed is a continuous variable and distraction and violation are binary discrete variables), two separate types of statistical models were utilized for this analysis. Pedestrian walking speed was modeled using OLS regression, and distraction and violation behavior were modeled using binary logistic regression.

For this study, an OLS regression (i.e., multiple linear regression) model was estimated using the field-observed pedestrian walking speed data. This is an appropriate statistical framework given the continuous nature of the walking

Table 2. Summary Statistics for Pedestrian Observations

Variable	Study site: Site 1	Site 2	Site 3	Site 4	Total
Pedestrian walking speed (ft/s)					
Mean (SD)	4.76 (1.1)	5.19 (1.4)	5.11 (1.7)	4.65 (1.1)	4.80 (1.2)
Speed ≤3.5	85 (6.1%)	0 (0.0%)	8 (1.5%)	37 (3.7%)	130 (4.3%)
Speed >3.5–4.5	486 (34.8%)	31 (25.8%)	191 (36.0%)	450 (45.5%)	1,158 (38.1%)
Speed >4.5–5.5	595 (42.6%)	54 (45.0%)	228 (43.0%)	402 (40.6%)	1,279 (42.1%)
Speed >5.5	232 (16.6%)	35 (29.2%)	103 (19.4%)	101 (10.2%)	471 (15.5%)
Distraction					
No distraction	1,169 (83.6%)	95 (79.2%)	458 (86.4%)	906 (91.5%)	2,628 (86.5%)
Talking on mobile phone	59 (4.2%)	1 (0.8%)	11 (2.1%)	17 (1.7%)	88 (2.9%)
Texting on mobile phone	130 (9.3%)	6 (5.0%)	16 (3.0%)	22 (2.2%)	174 (5.7%)
Headphones	35 (2.5%)	14 (11.7%)	39 (7.4%)	25 (2.5%)	113 (3.7%)
Other	5 (0.4%)	4 (3.3%)	6 (1.1%)	20 (2.0%)	35 (1.2%)
Adjacent traffic direction					
Cross with traffic	579 (41.4%)	65 (54.2%)	244 (46.0%)	502 (50.7%)	1,390 (45.8%)
Cross against traffic	819 (58.6%)	55 (45.8%)	286 (54.0%)	488 (49.3%)	1,648 (54.2%)
Pedestrian gender					
Male	689 (49.3%)	72 (60.0%)	303 (57.2%)	479 (48.4%)	1,543 (50.8%)
Female	709 (50.7%)	48 (40.0%)	227 (42.8%)	511 (51.6%)	1,495 (49.2%)
Pedestrian age					
Age 16 or younger	16 (1.1%)	7 (5.8%)	35 (6.6%)	54 (5.5%)	112 (3.7%)
Age 16–29	687 (49.1%)	68 (56.7%)	222 (41.9%)	417 (42.1%)	1,394 (45.9%)
Age 30–59	587 (42.0%)	43 (35.8%)	264 (49.8%)	498 (50.3%)	1,392 (45.8%)
Age 60 or older	108 (7.7%)	2 (1.7%)	9 (1.7%)	21 (2.1%)	140 (4.6%)
Pedestrian signal at start					
"Walk"	1,086 (77.7%)	95 (79.2%)	415 (78.3%)	756 (76.4%)	2,352 (77.4%)
"Flashing don't walk"	194 (13.9%)	19 (15.8%)	74 (14.0%)	204 (20.6%)	491 (16.2%)
"Don't walk"	118 (8.4%)	6 (5.0%)	41 (7.7%)	30 (3.0%)	195 (6.4%)
Crosswalk violation					
Crossed within crosswalk	1,285 (91.9%)	83 (69.2%)	380 (71.7%)	805 (81.3%)	2,553 (84.0%)
Crossed outside of crosswalk	113 (8.1%)	37 (30.8%)	150 (28.3%)	185 (18.7%)	485 (16.0%)
Pushed button (NA for New York)					
Pedestrian pushed button	0 (0%)	77 (64.2%)	173 (32.6%)	249 (25.2%)	499 (16.4%)
Pedestrian did not push button	1,398 (100.0%)	43 (35.8%)	357 (67.4%)	741 (74.8%)	2,539 (83.6%)
Group size					
Single pedestrian	675 (48.3%)	58 (48.3%)	177 (33.4%)	335 (33.8%)	1,245 (41.0%)
2 pedestrians	290 (20.7%)	43 (35.8%)	85 (16.0%)	191 (19.3%)	609 (20.0%)
3–4 pedestrians	171 (12.2%)	13 (10.8%)	142 (26.8%)	229 (23.1%)	555 (18.3%)
5 or more pedestrians	262 (18.7%)	6 (5.0%)	126 (23.8%)	235 (23.7%)	629 (20.7%)
Opposing pedestrians					
No opposing pedestrians	247 (17.7%)	106 (88.3%)	248 (46.8%)	325 (32.8%)	926 (30.5%)
1 opposing pedestrian	233 (16.7%)	4 (3.3%)	112 (21.1%)	156 (15.8%)	505 (16.6%)
2 or more opposing pedestrians	918 (65.7%)	10 (8.3%)	170 (32.1%)	509 (51.4%)	1,607 (52.9%)
Waiting time					
0 s (no wait time)	497 (35.6%)	97 (80.8%)	384 (72.5%)	691 (69.8%)	1,669 (54.9%)
1 s or more	901 (64.5%)	23 (19.2%)	146 (27.5%)	299 (30.2%)	1,369 (45.1%)
Grand total	1,398 (100%)	120 (100%)	530 (100%)	990 (100%)	3,038 (100%)

Note: SD = standard deviation; ft/s = feet per second; na = not applicable.

speed observations. The OLS equation takes the following form (*17*):

$$S_i = \beta_0 + \beta_i X + \varepsilon_i \tag{1}$$

where S_i is the walking speed (in ft/s) for pedestrian *i*, *X* is a vector of study site and pedestrian characteristics (i.e., site location, pedestrian distraction, age, gender, etc.), β are vectors of estimable parameters, and εs are disturbance terms.

In contrast to pedestrian walking speed, pedestrian distraction and violation are binary discrete variables with only two possible outcomes each: distraction versus no distraction, and violation versus no violation. As such, binary logistic (logit) models were estimated to analyze these behaviors. In the binary logit model, the probability of pedestrian distraction or violation is estimated as (*17*):

$$P_i = \frac{\mathrm{EXP}\begin{bmatrix}\beta_o + \beta_1 X_{1,i} + \beta_2 X_{2,i} \\ + \ldots + \beta_K X_{K,i}\end{bmatrix}}{1 + \mathrm{EXP}\begin{bmatrix}\beta_o + \beta_1 X_{1,i} \\ + \beta_2 X_{2,i} + \ldots + \beta_K X_{K,i}\end{bmatrix}} \tag{2}$$

where P_i is the probability of pedestrian *i* being distracted or committing a violation, β_0 is the model constant, and $\beta_I, \ldots, \beta_K$ are estimable parameters corresponding with explanatory site and pedestrian variables X_K. In binary logistic regression, the β are estimated using maximum likelihood methods. To improve interpretability of results, all explanatory variables included in the logit models were binary indicator variables (taking the value of either 0 or 1), and marginal effects for each variable were provided with the model results. Marginal effects are used to determine the direction and magnitude of the effects of each variable on the probability of pedestrian distraction or violation. Details with regard to marginal effect determination are provided elsewhere (*17*). In the context of this analysis, the marginal effect for each variable represents the change in probability of pedestrian distraction or violation when an indicator variable is changed from 0 to 1.

Results and Discussion

Ultimately, separate models were estimated for walking speed and each type of distraction and violation, resulting in one OLS walking speed model, and five total binary logit models: talking distraction, texting distraction, headphone distraction, pedestrian signal violation (i.e., starting the cross during the "flashing don't walk" or "don't walk" phase), and crosswalk violation (i.e., crossing outside of the marked crosswalk). Previous research has shown pedestrian gender and age to be associated with certain crossing behaviors (*9, 13*); therefore, these variables were included in all the models estimated as part of this study. In addition, indicator variables for each site were included (with Site 4 left out as the reference category), to account for potential site-by-site differences not explicitly captured by other variables included in the models. In addition to the aforementioned demographic and site variables, pedestrian distraction indicators were included in the walking speed and violation models, and additional variables were included if they exhibited statistical significance at a 90% or greater confidence level (i.e., *p*-values ≤0.10).

Pedestrian Walking Speed Analysis

Table 3 presents the results of the pedestrian walking speed analysis. The dependent variable for this model is walking speed in ft/s, and the constant represents the "baseline" walking speed, with the β estimate for each variable representing the expected change (in ft/s) from the baseline speed if that variable is changed from 0 to 1. With regard to the site indicator variables, Site 4 is excluded as the reference category, and walking speeds at this site tended to be slower than the other three sites (as evidenced by the positive parameter estimates for the three site indicators included in the model). This may be because of the downtown Flagstaff location of Site 4, which tends to experience significant tourist pedestrian traffic during the summer months. Turning to the distraction variables, talking or texting did not have a statistically significant impact on walking speed (the parameter estimates for these variables are negative; however, they are not statistically significant as evidenced by their high *p*-values). A previous observational study conducted in 2012 found that talking or texting on a mobile phone tended to decrease walking speeds (*13*); however cell phone and, in particular, smartphone use has increased substantially since that study (*2*). It is possible that pedestrians are now more accustomed to walking while talking or texting and do not significantly slow their walking speeds while engaging in these activities.

Pedestrians using headphones tended to exhibit faster walking speeds than undistracted pedestrians, a result that is similar to previous studies (*13*) and may be in part because these pedestrians walk or jog faster than normal walking speed for exercise. Pedestrians in the "other distraction" group tended to exhibit slower walking speeds than undistracted pedestrians, an expected result also consistent with past research (*13*). With respect to age, pedestrians aged 60 or over (the reference category) exhibited the slowest walking speeds, whereas pedestrians aged between 16 and 29 exhibited the fastest walking speeds. These were both expected results similar to prior research in this area (*13*). With respect to gender, males tended to exhibit slightly faster walking speeds than females.

Pedestrians crossing by themselves or with one other person exhibited faster walking speeds compared with those crossing in groups of three or more pedestrians, indicating locations with high numbers of pedestrians during each cycle may experience overall slower pedestrian walking speeds.

Table 3. Results of the OLS Pedestrian Walking Speed Model

Variable	β	Standard error	*p*-value
Constant**	3.715	0.112	<0.001
Site 1 (New York)**	0.134	0.053	0.012
Site 2 (Arizona)*	0.194	0.116	0.093
Site 3 (Arizona)**	0.352	0.062	<0.001
Talking	−0.136	0.123	0.268
Texting	−0.023	0.090	0.802
Headphones**	0.578	0.111	<0.001
Other distraction**	−0.420	0.192	0.029
Age less than 16**	0.450	0.146	0.002
Age 16–29**	0.643	0.101	<0.001
Age 30–59**	0.484	0.101	<0.001
Male pedestrian**	0.206	0.042	<0.001
Group: single pedestrian**	0.328	0.050	<0.001
Group: 2 pedestrians*	0.111	0.057	0.052
Opposing pedestrians: 0**	0.178	0.049	<0.001
Cross with traffic**	−0.128	0.042	0.002
Start during "flashing don't walk"**	0.382	0.061	<0.001
Start during "don't walk"**	0.690	0.086	<0.001
Crosswalk violation*	0.113	0.058	0.050
Pushed button**	0.195	0.064	0.002
R^2		0.132	

Note: **Denotes variable significant at 95% confidence level; *denotes variable significant at 90% confidence level.

Additionally, pedestrians crossing where there were no opposing pedestrians exhibited faster walking speeds. Pedestrians crossing in the same direction as adjacent roadway traffic exhibited slower walking speeds than those crossing against adjacent traffic. This may be a result of pedestrians being more cautious with regard to potential turning vehicles coming from behind them not in their immediate field of view.

Pedestrians who started their crossing during the "flashing don't walk" or "don't walk" pedestrian signal phases (i.e., they violated the pedestrian signal), exhibited significantly faster walking speeds than those who started their crossing during the "walk" signal, probably because of their rushed behavior knowing they had violated the pedestrian signal. Similarly, pedestrians who crossed outside of the crosswalk for at least a portion of their crossing (i.e., crosswalk violation) also exhibited faster walking speeds. Finally, pedestrians who used a pedestrian push button before their crossing also exhibited faster walking speeds.

Pedestrian Distraction Analysis

Table 4 presents the results of the binary logit models used to examine factors associated with pedestrian distraction. Overall, 13.5% of pedestrians observed in this study were deemed to have been distracted by some type of behavior as they crossed (talking, texting, headphones, or other). As stated previously, three separate models were estimated: one for talking on a mobile phone, one for texting on a mobile phone, and one for headphone use. With regard to the results of the "talking on phone" model, pedestrians observed at the New York site were more likely to talk on their mobile phones while walking, which may be because of general differences in the population with respect to mobile phone use behaviors. None of the age groups were statistically significant predictors of talking on a mobile phone; however, males were significantly less likely than females to talk on their mobile phones while walking. Finally, pedestrians who crossed alone (i.e., not in a group) and those who crossed outside of the crosswalk (crosswalk violation) were more likely to be talking on a mobile phone while walking.

Turning to the results of the "texting on phone" model, pedestrians observed at the New York site were more likely to text on their mobile phones while walking, a similar finding to the "talking on phone" model. With respect to age, pedestrians aged between 16 and 29 were more likely than any other age group to be observed texting while walking, a result that may be associated with the general cell phone and smartphone use behaviors of this demographic as compared with others. Similar to the "talking on phone" model results, pedestrians who were walking alone were also more likely to text while walking. Finally, pedestrians who did not have to wait at the intersection for a "walk" signal (i.e., wait time = 0 s) were less likely to text while walking compared with those who had to wait 1 or more seconds.

The results of the "headphone distraction" model indicated that pedestrians were more likely to be observed wearing headphones at the Arizona study sites compared with the New York site. Pedestrians aged between 16 and 29, male pedestrians, and pedestrians walking alone were all significantly more likely to be wearing headphones while they crossed the street. In addition, pedestrians who did not have to wait for a "walk" signal, and those who walked faster than 4.5 ft/s were more likely to be wearing headphones. Ultimately, the results of the pedestrian distraction models provide new insights into both demographic and site-related factors associated with distracted walking for researchers, practitioners, and legislators. The results may be valuable in creating and planning targeted educational or enforcement campaigns aimed at mitigating potential negative impacts of distracted walking.

Pedestrian Violation Analysis

Table 5 presents the results of the binary logit models used to examine factors associated with pedestrian violations. As stated previously, two separate models were estimated to examine pedestrian violations: one for pedestrian signal violation (i.e., starting the cross during the "flashing don't walk" or "don't walk" phase), and one for crosswalk violation (i.e., crossing outside of the marked crosswalk). Based on the

Table 4. Results of the Binary Logit Pedestrian Distraction Models

Variable	β	Standard error	*p*-value	Marginal effect
Talking on Mobile Phone Distraction Model				
Constant**	−5.202	0.773	<0.001	na
Site 1 (New York)**	0.912	0.288	0.002	0.025
Site 2 (Arizona)	−0.843	1.036	0.416	−0.016
Site 3 (Arizona)	0.183	0.394	0.643	0.005
Age less than 16	0.195	1.244	0.876	0.006
Age 16–29	0.977	0.732	0.182	0.030
Age 30–59	0.975	0.735	0.185	0.031
Male pedestrian**	−0.508	0.225	0.024	−0.014
Group: single pedestrian**	0.745	0.226	0.001	0.021
Crosswalk violation*	0.561	0.289	0.053	0.019
Restricted LL			−398.375	
LL at convergence			−378.853	
Texting on Phone Distraction Model				
Constant**	−5.050	0.635	<0.001	na
Site 1 (New York)**	1.805	0.244	<0.001	0.093
Site 2 (Arizona)	0.417	0.482	0.388	0.024
Site 3 (Arizona)	0.300	0.336	0.372	0.017
Age less than 16	1.183	0.842	0.160	0.091
Age 16–29**	1.889	0.595	0.002	0.096
Age 30–59	0.588	0.612	0.337	0.033
Male pedestrian	0.006	0.163	0.973	0.000
Group: single pedestrian**	0.532	0.188	0.005	0.028
Wait time = 0 s**	−1.212	0.200	<0.001	−0.062
Restricted LL			−666.542	
LL at convergence			−582.699	
Headphones Distraction Model				
Constant**	−5.323	0.353	<0.001	na
Site 1 (New York)	−0.006	0.278	0.984	0.000
Site 2 (Arizona)**	0.907	0.375	0.016	0.042
Site 3 (Arizona)**	1.020	0.270	0.000	0.043
Age 16–29**	1.474	0.240	<0.001	0.045
Male pedestrian*	0.373	0.206	0.070	0.012
Group: single pedestrian**	0.683	0.230	0.003	0.024
Walking speed 4.5–5.5 ft/s**	0.727	0.267	0.007	0.025
Walking speed > 5.5 ft/s**	1.181	0.298	<0.001	0.054
Wait time = 0 s**	−0.832	0.250	0.001	−0.027
Restricted LL			−482.819	
LL at convergence			−415.255	

Note: **Denotes variable significant at 95% confidence level; *denotes variable significant at 90% confidence level. na = not applicable; LL = log likelihood; ft/s = feet per second.

results of both the signal and crosswalk violation models, pedestrians observed at the New York site were less likely to commit these violations than those at the Arizona sites. For the signal violation model, none of the distraction, age, or gender variables were statistically significant in predicting pedestrian signal violation behavior. Pedestrians in groups of five or more were more likely to commit a signal violation, indicating this behavior is deemed more acceptable when it is done with a group. Pedestrians crossing with no other or only one opposing pedestrian(s) were more likely to commit signal violations, as were pedestrians walking faster than 5.5 ft/s. Finally, pedestrians who used the pedestrian push button were less likely to commit a signal violation, whereas those who had a wait time of 0 s were more likely to commit a signal violation. These were both expected results.

Turning to the results of the crosswalk violation model, pedestrians who were texting while walking were more likely to commit a crosswalk violation, which is probably because their vision is focused on their mobile phone and not the crosswalk markings as they cross. Male pedestrians were more likely to commit crosswalk violations, as were groups of three or more pedestrians, which may be a result of crowding in the

Table 5. Results of the Binary Logit Pedestrian Violation Models

Variable	β	Standard error	*p*-value	Marginal effect
Signal Violation Model				
Constant**	−2.143	0.345	<0.001	na
Site 1 (New York)**	−1.324	0.141	<0.001	−0.153
Site 2 (Arizona)	−0.217	0.329	0.509	−0.025
Site 3 (Arizona)**	−0.426	0.176	0.015	−0.047
Talking	−0.103	0.303	0.734	−0.012
Texting	−0.196	0.272	0.470	−0.022
Headphones	0.289	0.305	0.343	0.035
Other distraction	0.172	0.479	0.720	0.020
Age less than 16	0.077	0.365	0.833	0.009
Age 16–29	−0.261	0.264	0.323	−0.030
Age 30–59	−0.314	0.263	0.232	−0.036
Male pedestrian	0.055	0.109	0.611	0.006
Group: 5 or more pedestrians**	0.550	0.177	0.002	0.066
Opposing pedestrians: 0**	1.343	0.126	<0.001	0.170
Opposing pedestrians: 1**	0.501	0.148	<0.001	0.060
Walking speed <3.5 ft/s**	−1.484	0.319	<0.001	−0.138
Walking speed 3.5–4.5 ft/s**	−1.210	0.163	<0.001	−0.137
Walking Speed 4.5–5.5 ft/s**	−0.732	0.149	<0.001	−0.083
Pushed button**	−0.797	0.217	<0.001	−0.085
Wait time = 0 s**	3.308	0.165	<0.001	0.423
Restricted LL			−1,622.776	
LL at convergence			−1,110.659	
Crosswalk Violation Model				
Constant**	−1.614	0.331	<0.001	na
Site 1 (New York)**	−1.161	0.140	<0.001	−0.137
Site 2 (Arizona)**	1.054	0.236	<0.001	0.168
Site 3 (Arizona)**	0.634	0.132	<0.001	0.088
Talking*	0.533	0.291	0.067	0.076
Texting	−0.101	0.267	0.705	−0.012
Headphones	0.158	0.249	0.524	0.021
Other distraction	0.270	0.426	0.527	0.036
Age less than 16	−0.443	0.416	0.287	−0.049
Age 16–29	0.220	0.313	0.482	0.028
Age 30–59	0.118	0.312	0.706	0.015
Male pedestrian**	0.280	0.106	0.008	0.035
Group = single pedestrian**	−0.391	0.140	0.005	−0.047
Group: 2 pedestrians*	−0.266	0.149	0.074	−0.032
Opposing pedestrians: 0**	−0.603	0.129	<0.001	−0.070
Opposing pedestrians: 1**	−0.318	0.150	0.034	−0.037
Walking speed <3.5 ft/s**	−1.108	0.429	0.010	−0.099
Wait time = 0 s**	0.511	0.133	<0.001	0.065
Cross with traffic**	0.228	0.105	0.029	0.028
Restricted LL			−1,333.925	
LL at convergence			−1,225.325	

Note: **Denotes variable significant at 95% confidence level; *denotes variable significant at 90% confidence level. na = not applicable; LL = log likelihood; ft/s = feet per second.

crosswalk when larger groups are crossing. Pedestrians crossing with no other or only one opposing pedestrian(s) were less likely to commit a crosswalk violation as compared with those encountering two or more opposing pedestrians, again possibly a result of the crosswalk crowding issue. Finally, pedestrians walking less than 3.5 ft/s were less likely to commit crosswalk violation, whereas those with a wait time of 0 s and those who crossed in the same direction as adjacent traffic were more likely to commit crosswalk violations. The results of these violation models provide insights that can be used in planning strategies to reduce such potentially unsafe behavior, targeted enforcement, for example.

Conclusions

Given the increasing use of cell phones and smartphones, it is important to consider the impacts of distracted walking on the safety and efficiency of transportation systems. Although the dangers of distracted driving have been well established, little research has investigated the potential impact of distracted walking on safety. To fill this research gap and gain new insights into factors associated with distracted walking, pedestrian walking speed, and pedestrian signal or crosswalk violations at signalized intersections, this study utilized observational field-recorded data from one intersection in New York and three intersections in Flagstaff, Arizona. An OLS model was estimated to examine factors affecting pedestrian walking speed, and binary logit models were estimated to examine factors associated with pedestrian distraction and signal or crosswalk violation behavior.

The results of the OLS walking speed model showed that talking and texting while walking were not statistically significantly associated with walking speed, indicating that pedestrians may be accustomed to walking while talking or texting and do not significantly slow (or increase) their walking speeds while engaging in these activities. This is important, as engineers may not necessarily need to reduce design walking speeds to accommodate distracted pedestrians when determining pedestrian clearance intervals as cell phone and smartphone use continues to increase. The results of the OLS model also showed that pedestrians wearing headphones exhibited faster walking speeds, and that pedestrians walking in groups of two or more exhibited slower walking speeds, among other results. These findings provide information that may be useful in planning or designing pedestrian facilities at signalized intersections, including the design walking speed for pedestrian signal clearance intervals.

The results of the binary logit pedestrian distraction models showed, among other findings, that male pedestrians were less likely to talk on a mobile phone while walking as compared with females, and that pedestrians walking alone were more likely to exhibit all three types of distraction (talking, texting, or wearing headphones). Additionally, pedestrians aged between 16 and 29 were more likely to text on a phone while walking or wear headphones as compared with all other age groups. Finally, pedestrians who did not have to wait at the intersection for a "walk" signal (i.e., wait time = 0 s) were less likely to text while walking or to wear headphones as compared with those who had to wait 1 or more seconds. The pedestrian violation models indicated that males were more likely to commit a crosswalk violation, whereas those who used a pedestrian push button were less likely to commit a pedestrian signal violation, among other findings. Future research into the differences in pedestrian behavior at actuated versus non-actuated pedestrian signals may be warranted.

Ultimately, the results of these models provide new insights into those factors associated with distracted walking and pedestrian violation behavior for researchers, practitioners, and legislators. The results may be useful in planning strategies to reduce or mitigate the impacts of such potentially unsafe behavior. These strategies may include educational or enforcement campaigns aimed at certain demographics, or engineering solutions such as restricting right-turns-on-red at locations that may be more prone to experiencing pedestrian distraction or violation behavior. Several such campaigns have already been enacted across the United States (*18*), and the results of this study provide further demographic and site-specific information to help properly design these efforts.

Future research in this area could include a larger sample of intersection locations, as statistically significant differences in certain pedestrian behaviors between sites were observed in this study. Additionally, traffic volumes may be incorporated into future analyses on pedestrian violation behavior, as sites with lower vehicle volumes may be more prone to pedestrian violations. Furthermore, future research in this area may be extended to examine the impact of distracted walking on safety at unsignalized crosswalks, and pedestrian–vehicle conflict data could be utilized to gain further insights into this emerging issue. Finally, educational or enforcement campaigns aimed at distracted walking (either existing or new campaigns) should be evaluated to determine their effectiveness in affecting pedestrian behavior.

References

1. Vision Network. What is Vision Zero? http://visionzeronetwork.org/about/what-is-vision-zero/. Accessed July 27, 2017.
2. Pew Research Center. Mobile Fact Sheet. http://www.pewinternet.org/fact-sheet/mobile/. Accessed July 25, 2017.
3. Horrey, W. J., and C. D. Wickens. The Impact of Cell Phone Conversations on Driving Using Meta-Analytic Techniques. *Human Factors*, Vol. 48, No. 1, 2006, pp. 196–205.
4. Caird, J. K., C. R. Willness, P. Steel, and C. Scialfa. A Meta-Analysis of the Effects of Cell Phones on Driver Performance. *Accident Analysis and Prevention*, Vol. 40, No. 4, 2008, pp. 1282–1293.
5. Scopatz, R. A., and Y. Zhou. *Effect of Electronic Device Use on Pedestrian Safety: A Literature Review*. Report No. DOT HS 812 256. National Highway Traffic Safety Administration, 2016.
6. Ellis, R. Look Left, Look Right—But Not at Your Cell Phone in Honolulu Crosswalks. CNN, July 29, 2017. http://www.cnn.com/2017/07/29/us/smart-phones-crosswalks-hawaii-illegal/index.html. Accessed July 31, 2017.
7. Schwebel, D. C., D. Stavrinos, K. W. Byington, T. Davis, E. E. O'Neal, and D. de Jong. Distraction and Pedestrian Safety: How Talking on the Phone, Texting, and Listening to Music Impact Crossing the Street. *Accident Analysis and Prevention*, Vol. 45, 2012, pp. 266–271.

8. White, J., J. L. Mwakalonge, and S. Siuhi. Distracted Walking: What We Know and Data Needs. Presented at 93rd Annual Meeting of the Transportation Research Board, Washington, D. C., 2014.
9. Gillette, G., K. Fitzpatrick, S. Chrylser, and R. Avelar. The Effect of Distractions on a Pedestrian's Waiting Behavior at Traffic Signals: Observational Study. *Transportation Record: Journal of the Transportation Research Board*, 2016. 2586: 111–119.
10. National Highway Traffic Safety Administration. *2015 Motor Vehicle Crashes: Overview*. U.S. Department of Transportation, 2016.
11. Basch, C., D. Ethan, P. Zybert, and C. E. Basch. Pedestrian Behavior at Five Dangerous and Busy Manhattan Intersections. *Journal of Community Health*, Vol. 40, No. 4, 2015, pp. 789–792.
12. Hatfield, J., and S. Murphy. The Effects of Mobile Phone Use on Pedestrian Crossing Behavior at Signalized and Unsignalized Intersections. *Accident Analysis and Prevention*, Vol. 39, No. 1, 2007, pp. 197–205.
13. Thompson, L., F. Rivara, R. Ayyagari, and B. Ebel. Impact of Social and Technological Distraction on Pedestrian Crossing Behavior: An Observational Study. *Injury Prevention*, Vol. 19, No. 4, 2012, pp. 232–237.
14. Cooper, J., R. Schneider, S. Ryan, and S. Co. Documenting Targeted Behaviors Associated with Pedestrian Safety. *Transportation Research Record, Journal of the Transportation Research Board*, 2012. 2299: 1–10.
15. Schwebel, D., L. McClure, and B. Porter. Experiential Exposure to Texting and Walking in Virtual Reality: A Randomized Trial to Reduce Distracted Pedestrian Behavior. *Accident Analysis and Prevention*, Vol. 102, 2017, pp. 116–122.
16. Lin, M., and Y. Huang. The Impact of Walking While Using a SmartPhone on Pedestrians' Awareness of Roadside Events. *Accident Analysis and Prevention*, Vol. 101, 2017, pp. 87–96.
17. Washington, S., M. Karlaftis, and F. Mannering. Statistical and Econometric Methods for Transportation Data Analysis. 2nd ed. CRC Press, Boca Raton, Fla., 2011.
18. Mwakalonge, J., S. Saidi Siuhi, and J. White. Distracted Walking: Examining the Extent to Pedestrian Safety Problems. *Journal of Transportation and Traffic Engineering*, Vol. 2, Issue No. 5, 2015, pp. 327–337.

The Standing Committee on Pedestrians (ANF10) peer-reviewed this paper (18-03541).

Article

TRR
JOURNAL OF THE TRANSPORTATION RESEARCH BOARD

Transportation Research Record
2018, Vol. 2672(35) 13–20

DOI: 10.1177/0361198118776480
journals.sagepub.com/home/trr
SAGE

Influence of Advanced Placement of the In-Street Sign Gateway on Distance of Yielding from the Crosswalk

Jonathan Hochmuth[1] **and Ron Van Houten**[1]

Abstract

A multiple-threat accident occurs when a pedestrian attempts to traverse a crosswalk on a multilane road, and the vehicle nearest to the pedestrian yields close to the crosswalk, obscuring the pedestrian's view of the next lane, and the motorist's view in the adjacent lane of the pedestrian. This set of circumstances may lead to pedestrians being struck at higher speeds by a vehicle in the adjacent oncoming lane, often resulting in serious injury or a fatality. The present study compared advanced placement of a gateway in-street sign configuration with a gateway deployed at the crosswalk, on two multilane roads. Data were collected using a reversal design, capturing alternating series of data points in each condition, and counter-balancing the order of those conditions. The advanced placement of the sign produced comparable levels of yielding to the crosswalk placement, but produced a large increase in the proportion of yielding 50 ft or further from the crosswalk. On multilane roads, increasing the distance at which vehicles yield from the crosswalk mitigates the sight distance issues that produce multiple-threat accidents, thereby reducing the likelihood of pedestrian fatalities or serious injuries. Advanced placement of the gateway configuration may also take the signs out of the turning radii of larger vehicles at a driveway or intersection, which may improve the longevity of the in-street signs.

According the National Highway Traffic Safety Association (NHTSA) in 2015 there were 5,376 pedestrian fatalities in the United States, and an additional 70,000 injuries. There has been an increasing trend in pedestrian deaths since 2009 with 4,109 fatalities, as well as an increase in the proportion of pedestrian fatalities out of all traffic-related fatalities from 12 to 15% (*1*). NHTSA also reports that over one quarter of those fatalities involved a failure to yield right-of-way to the pedestrian (*1*).

One countermeasure that has been demonstrated to be effective in increasing the percentage of motorists yielding right-of-way to pedestrians at uncontrolled crosswalks is the gateway configuration of the R1-6 in-street sign. The R1-6 sign has a yellow-green reflective backing, with the message "YIELD TO PEDESTRIANS IN CROSSWALK" on its face (*2*). When deployed in a gateway configuration, a sign is placed on both edges of the roadway and on each lane line. Bennett, Manal, and Van Houten (*3*) found that the R1-6 gateway produced a substantial increase in yielding over both baseline and a single centerline sign placement. They also found that the gateway produced comparable levels of yielding to that of the more expensive rectangular rapid flashing beacon (RRFB) and the pedestrian hybrid beacon (PHB). Moreover, when the signs were used in conjunction with the RRFB and PHB, yielding increased beyond the RRFB and PHB alone conditions (*3*). Additionally, Bennett and Van Houten (*4*) found that it was not merely the presence of a between-lane obstacle producing the increase in yielding. When the authors replaced fluorescent yellow-green blank signs of the same size and shape in a gateway configuration, average yielding fell from 78 to 33%. A photograph of the Gateway treatment is shown in Figure 1.

In Michigan, the Uniform Traffic Code (R 28.1703 Rule 703 Part 1) states "When any vehicle is stopped at a marked crosswalk or at any unmarked crosswalk at an intersection to permit a pedestrian to cross the roadway, the driver of any other vehicle approaching from the rear shall not overtake and pass the stopped vehicle" (*5*), and similar laws exist in many other states. Failure of a vehicle to abide by these laws results in an incident, which can be particularly deadly at crosswalks on multilane

[1]Psychology Department, Western Michigan University, Kalamazoo, MI

Corresponding Author:
Address correspondence to Ron Van Houten: Ron.vanhouten@wmich.edu

Figure 1. Gateway configuration of the R1-6 in-street sign showing the two lanes in the east-bound direction.

roads, described as a multiple-threat crash type. A 2005 Federal Highway Administration (FHWA) report describes this type of crash as involving a motorist yielding close to the crosswalk in one lane of a multilane road permitting the pedestrian to cross, and an oncoming vehicle in the next vehicle lane (traveling in the same direction) striking the pedestrian who is screened by the vehicle stopped in the first lane the pedestrian crossed (*6*). To reduce the likelihood of this sort of crash, the authors suggest that drivers should reduce speed and scan the area for pedestrians when a vehicle has stopped in an adjacent lane, and pedestrians should stop at the edge of the stopped car and check the next oncoming lane before stepping into it (*6*). Fortunately, the gateway sign configuration may already be addressing the issue of speed. Van Houten and Hochmuth (*7*) reported preliminary data showing an average reduction of up to 10 mph at the crosswalk when no pedestrian was present (i.e., passive speed reductions), though additional data are needed to confirm the generality of this effect.

Outside of reducing speeds, multiple-threat accidents are fundamentally a sight distance or visibility issue; motorists cannot see the pedestrian, and pedestrians cannot see the oncoming vehicle, until it is too late. Increasing yielding right-of-way to pedestrians is a beneficial outcome, but depending on where those vehicles yield, it may actually produce some of the factors implicated in multiple-threat crashes on multilane roadways (i.e., the near lane vehicle yields close to the crosswalk). Advanced yield pavement markings and sign prompts have been demonstrated to be effective in reducing motorist–pedestrian conflicts (i.e., near collisions), marginally increasing yielding, and increasing yielding distances (*8*, *9*). A recent study by Zegeer and colleagues (*10*) estimated Crash Modification Factors (CMF) for several pedestrian crossing treatments, including advanced yield or stop markings and signs (AS). The AS treatment produced reductions in total crashes (11%), rear-end and sideswipe crashes (20%), and pedestrian crashes (36%), with an estimated CMF for pedestrian crashes of 0.863 (*10*). It is worth noting that the cost of such treatments is comparable to the in-street gateway (*11*), however the gateway produces substantially higher levels of motorist yielding (*3*, *8*, *9*).

In the effort to reduce multiple-threat accidents, increasing yielding alone is not sufficient, rather yielding distances in both lanes (particularly the lane nearest to the pedestrian) must occur to ameliorate the visibility issue. Therefore, it is of interest when examining the gateway configuration, if deployment of the signs in advance of the crosswalk can produce similar effects as the AS treatment. The present study examined whether a gateway configuration placed in advance of the crosswalk would 1) produce comparable levels of yielding to the traditional placement at the crosswalk, and 2) increase the distance from the crosswalk at which vehicles yield.

Method

Participants

The participants were motorists using the road when pedestrians (i.e., research assistants) were crossing in the crosswalk.

Site Selection

Data were collected at two locations in Grand Rapids, Michigan. The first site (Site 1) was a midblock crossing with a median island, with two travel lanes in each direction, a speed limit of 25 mph and an average daily traffic (ADT) of 9,400. At this site, only data on the two east-bound lanes were collected. The second site (Site 2) was also a midblock crossing; however, it was on a one-way road, with three travel lanes, and a parking lane on one side, with a speed limit of 25 mph, and ADT of 8,200.

Dependent Variables

At each location, brightly colored duct tape was used to mark the curb at 10 ft, 30 ft, and 50 ft. Another tape line was used to demarcate the dilemma zone (DZ), which is the distance that a motorist traveling the speed limit should be able to appropriately respond to a change in roadway conditions, including yielding right-of-way to a pedestrian in a crosswalk. This distance is calculated using the ITE traffic signal timing formula (*12*), and only vehicles beyond this distance were considered when calculating the number of motorists failing to yield right-of-way to the pedestrian.

All data reported in the present study resulted from staged crossings by trained research assistants who followed a safe crossing protocol, and data were collected with a single pedestrian (i.e., the research assistant) present. Research assistants initiated a trial by approaching the crosswalk, placing their foot into the crosswalk, and turning to look toward oncoming traffic. Trained observers began counting vehicles which did not yield, starting with vehicles beyond the DZ at the time the research assistant's foot entered the crosswalk. The research assistant held position until traffic yielded or cleared, then moved to the edge of the first lane, stopped to look, and waited for traffic to yield or clear before continuing. Upon completing crossing of the entire crosswalk, yielding and non-yielding, as well as vehicle yielding distance were recorded for each lane traversed. Data were collected on "natural" pedestrian crossings, but samples were not sufficient in size for presentation and analysis. Using data from staged crossings has two separate, but related, advantages. More aggressive crossing methods, as sometimes witnessed in the natural pedestrian population, may "force" motorists to yield, potentially skewing results, and also increasing the risk to pedestrians and motorists of an accident. As such, the staged crossing protocol increases experimental control by reducing the variability in pedestrian behavior, and reduces risk to the research team by ensuring that crossings adhere to both legal standards as well as recommended safety practices.

Yielding Behavior

The number of motorists who did and did not yield to pedestrians in crosswalks was measured in the same way as reported in the original study (*6*). Yielding was considered to occur when a vehicle stopped or slowed sufficiently to create a gap large enough for the pedestrian to cross safely. Driver yielding was measured in relationship to an objective DZ (a location beyond which a driver can easily yield if a pedestrian enters the crosswalk). This distance was calculated using the ITE signal timing formula that takes into account driver reaction time, safe deceleration rate, the posted speed, and the grade of the road (*10*). Motorists who had not passed the outer boundary of the DZ when a pedestrian (i.e., a research assistant) entered the crosswalk were scored as yielding or not yielding because they had sufficient time and space to stop safely for the pedestrian. Motorists who entered the DZ before the pedestrian placed a foot in the crosswalk could be scored as yielding, but could not be scored as failing to yield because the motorist did not have adequate distance to yield based on the calculated distance. However, the signal timing formula is relatively lenient; hence, many vehicles that passed the DZ could yield safely, particularly those traveling below the speed limit.

At the road with two-way traffic (Site 1), drivers in the first two travel lanes nearest the pedestrian were scored for yielding after the pedestrian had entered the crosswalk. This procedure was used because it conforms to the obligations of motorists specified in the Michigan Uniform Traffic Code R 28.1702 Rule 702 and local ordinance in Grand Rapids with regard to who has the right-of-way at what time. Although research assistants completed the entire crossing, data were only collected on the east-bound leg of the road, and each travel direction was separated by a median island reducing the influence of oncoming traffic on observed motorist behavior. At the site with one-way traffic (Site 2) drivers were scored for yielding in all three travel lanes after the pedestrian placed a foot in the crosswalk. Vehicles were scored in reference to the DZ in the same manner as described for Site 1. Each data sheet (a data point) consisted of 20 pedestrian crossings when drivers were present who could yield or fail to yield to the pedestrian.

Inter-observer agreement was calculated in the field for at least 30% of all observations in this experiment, and data were collected independently during each condition of each experiment to validate the observational data. Each event that was scored the same by both observers was counted as an agreement, and each event that was scored differently by each observer was scored as a disagreement. Inter-observer agreement was calculated by dividing the number of agreements during each session by the sum of agreements plus disagreements for that session. Inter-observer agreement on driver yielding right-of-way to pedestrians across both locations averaged 95.8%, and a range of 90 to 100%.

Yielding Distance

Yielding distance was recorded as the distance at which a vehicle yielded from the crosswalk: less than 10 ft from the crosswalk, between 10 and 30 ft from the crosswalk, between 30 and 50 ft, and 50 ft or more in advance of the crosswalk. These data show the proportion of vehicles that yielded in a given range, and were calculated as the sum of the yielding that occurred for each range of distances, divided by the total number of vehicles yielding. Inter-observer agreement was also calculated on yielding distance range category, and averaged 89%, with a range of 82.6 to 100%.

Independent Variable

The *crosswalk placement* utilized a gateway configuration of the in-street sign placed at the leading edge of the crosswalk, whereas the *advanced placement* deployed a gateway configuration of the in-street sign 50 ft in advance of the leading edge of the crosswalk (Figure 2). Baseline data were collected using the same crossing procedures, but without the presence of the R1-6 in-street gateway signs.

Experimental Design

A replication logic, reversal design was used, which involves collecting multiple daily points of baseline data, followed by multiple data points of each experimental condition, and a return to the baseline condition. The experimental conditions are then tested again, and the order counterbalanced against the previous order. This design allows changes in driver behavior to be confirmed through multiple replications, and extended across time. This longitudinal type of design allows one to see variation over the duration of the study, and includes possible trends in the data, as well as the amount of overlap between conditions.

Data Analysis and Results

Across both sites, a total of 1,702 motorist reactions were measured, constituting 880 independent, staged crossing trials. Table 1 shows the overall yielding percentages, as well as yielding percentage by lane for both sites. At Site 1, both the crosswalk and advanced placements of the in-street gateway produced a large increase in driver yielding, with no trend in baseline, and no overlap between baseline and the treatments (Figure 3). Additionally, in baseline, only 11.9% of motorists yielded 50 ft or further from the crosswalk, compared with 32.6% in the crosswalk placement, and 75.3% in the advanced placement conditions (Figure 4). For Site 1, correlated-samples *t*-tests were conducted on the proportions of yielding in the 50 ft or greater interval for both treatment conditions, comparing the overall proportions between conditions, and near and far lane yielding within each condition. There was a significant difference ($p < .01$) found between crosswalk placement and advanced placement, however a significant difference was not detected when comparing the near and far lane yielding in the crosswalk placement and advanced placement conditions. These data suggest that at Site 1 the advanced placement produced a significant increase in the number of vehicles yielding 50 ft or further from the crosswalk compared with the crosswalk placement of

Figure 2. Site 1 crosswalk placement (*a*), and advanced placement (*b*); Site 2 crosswalk placement (*c*), and advanced placement (*d*).

Table 1. Overall and Individual Lane Yielding Percentage at Sites 1 and 2 by Condition; *n* Represents the Number of Motorists Observed in the Condition

Percent yielding						
	Condition	*n*	Average	Near lane	Mid lane	Far lane
Site 1	Baseline	517	11.9	14.5	n/a	9.7
	Crosswalk	265	85.0	84.5	n/a	85.6
	Advanced	291	80.3	82.5	n/a	78.0
Site 2	Baseline	244	48.1	61.3	43.6	42.7
	Crosswalk	180	91.1	96.7	93.5	88.3
	Advanced	205	85.8	94.5	85.2	81.1

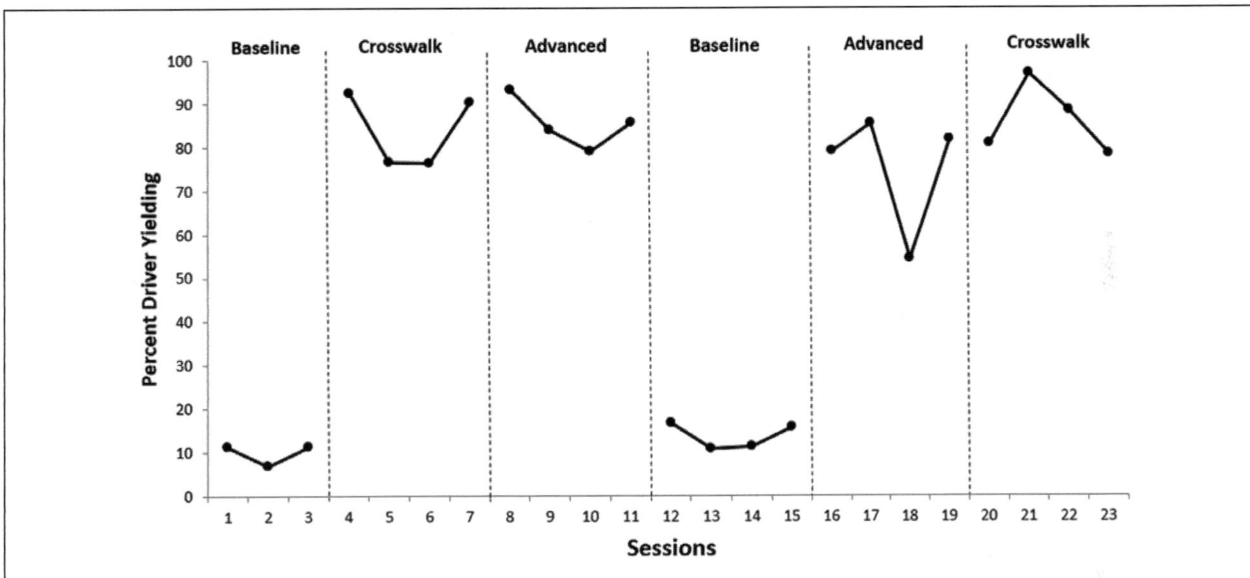

Figure 3. The overall percentage of drivers yielding to pedestrians at Site 1.

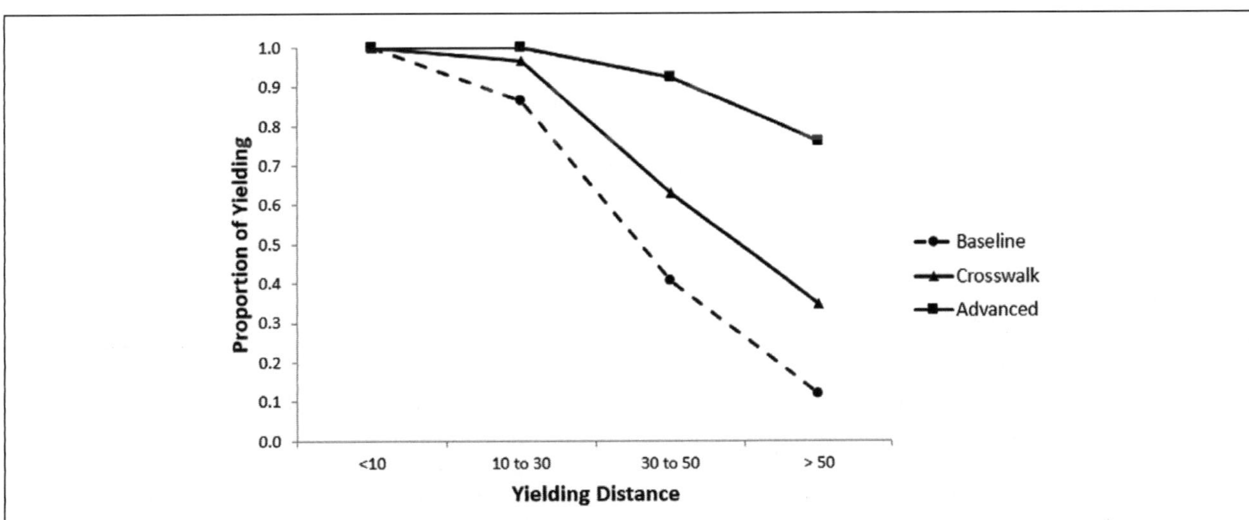

Figure 4. Occurrence of yielding distances at Site 1.

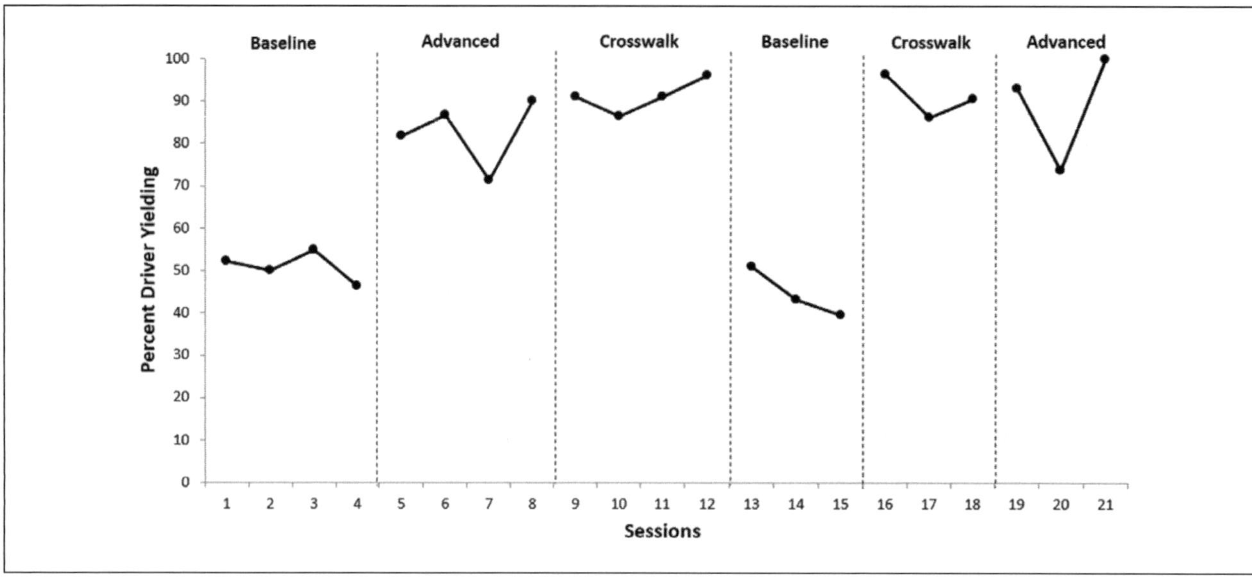

Figure 5. Percentage of drivers yielding to pedestrians at Site 2.

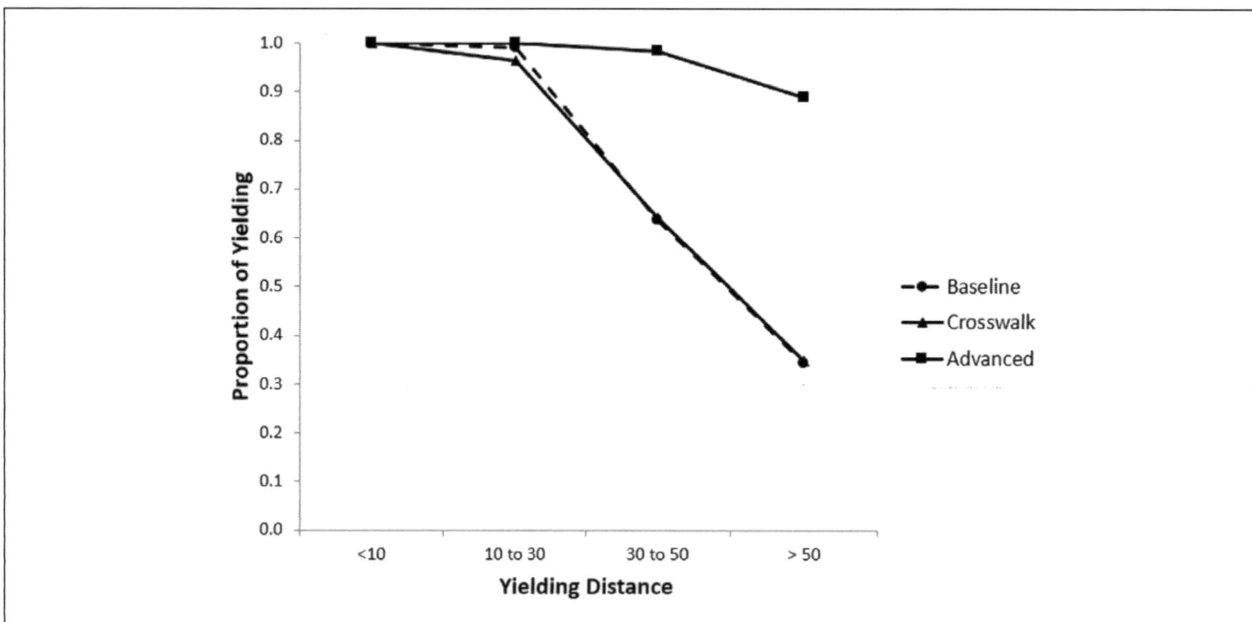

Figure 6. Occurrence of yielding distances at Site 2.

the signs, but that there was not a significant difference in yielding distances by lane between the treatments.

At Site 2, both gateway treatments also produced a large, convincing change in driver yielding behavior, with no overlap between treatment and baseline, and no trend in baseline (Figure 5). Once again, however, the advanced placement condition produced substantially higher yielding 50 ft and greater in advance of the crosswalk compared with the crosswalk placement condition (Figure 6). The advanced placement resulted in 88.9% of all yielding to occur at 50 ft or greater, compared with 34.8% in the crosswalk placement, and 34.3% in the baseline condition. It is also interesting to note that, compared with Site 1, the decrease in yielding distance in both the baseline and crosswalk placements are nearly identical to each other (Figure 6). The same statistical analyses described above for Site 1 were also conducted for Site 2. There was a significant difference ($p < .01$) found between crosswalk placement and advanced placement. A significant difference ($p < .05$) was also found in the crosswalk placement condition, between the proportions of near and far lane yielding distances. A significant difference was not detected between near and far lane yielding in the advanced placement condition. Unlike Site 1, these data suggest that at this location there may also be a disparity in the likelihood of near and far lane yielding distance when the signs are placed at the crosswalk, and the advanced placement of a gateway appears to mitigate this disparity.

Limitations and Recommendations

The results of the present study replicate previous research demonstrating the efficacy of the gateway configuration in increasing yielding right-of-way to pedestrians. The data also show that *advanced placement* of the signs produces a significant increase in the likelihood of motorists yielding in advance of the crosswalk, when compared with *crosswalk placement* of the signs, though producing comparable levels of overall yielding.

One limitation of this investigation is that yielding distance was only measured up to 50 ft in advance of the crosswalk and, as a result, yielding distance more than 50 ft "stacked" into the last category. Anecdotally, in the *advanced placement* condition, some vehicles would begin to yield noticeably before the signs. Future studies should consider extending the range of distance values that are observed and recorded, allowing the shift in yielding distance distributions to be fully documented. Moreover, the similar results obtained for multilane roads with a one-way street and a two-way street add to the generality of these data, however future research should replicate these effects with additional road configurations and with roads with a higher speed limit.

Data should also be collected on the effects of gateway *advanced placement* on passive speed reduction. If the *advanced placement* of the in-street gateway increases the distance at which motorists begin to decelerate it may also reduce the likelihood of a billiard ball crash (*8*), wherein a vehicle strikes a vehicle that has yielded, causing it to strike a pedestrian, and more generally may reduce the probability of a rear-end collision.

Perhaps most importantly, future research should work to determine if a crash modification factor is associated with the in-street sign gateway. This, however, requires wide-scale implementation of the treatment, and would ideally be sufficiently large to compare several gateway variations, across a range of site characteristics. Given that the in-street gateway has been repeatedly demonstrated as an effective and low-cost treatment for increasing motorists yielding right-of-way, comparable to more expensive interventions (*3*) that do have CMFs associated with them (*10*), the outlook for such a study seems promising. Finally, although several studies have been conducted that intensively examined the variables influencing the efficacy and longevity of the gateway treatment at a small number of sites, future research could demonstrate whether the advanced placement of the gateway might reduce the number of multiple-threat conflicts and crashes.

It should be noted that the *Manual on Uniform Traffic Control Devices* (MUTCD), Section 2B.12 paragraph 5, currently prohibits the advanced placement of in-street signs in order "to educate" motorists prior to reaching a crosswalk. This recommendation was in place to prevent their use well in advance of the crosswalk. Based on our data showing the efficacy of placing signs close to the crosswalk in advance of the crosswalk the full national committee has recommended a change in the MUTCD to allow placement near rather than at the crosswalk. FHWA has not yet issued an interim approval, but one can use engineering judgment if the location has a high risk of a multiple-threat crash.

Acknowledgments

This research was supported by a grant from the Western Michigan University Traffic Research Center. The authors thank the City of Grand Rapids, and the city's Traffic Safety Manager, Chris Zull, for their cooperation in identifying research sites, as well as providing permits for those locations.

Author Contributions

The authors confirm contribution to the paper as follows: study conception and design: J.H. and R.V.H.; data collection: J.H.; analysis and interpretation of results: J.H. and R.V.H.; draft manuscript preparation: J.H. and R.V.H. All authors reviewed the results and approved the final version of the manuscript.

References

1. National Highway Traffic Safety Association. *Traffic Safety Facts 2015*. NHTSA Publication DOT HS-812-384. NHTSA, U.S. Department of Transportation, Washington, D.C., 2017. https://crashstats.nhtsa.dot.gov/Api/Public/ViewPublication/812384. Accessed July 10, 2017.
2. Federal Highway Administration. *Manual on Uniform Traffic Control Devices for Streets and Highways*. FHWA, U.S. Department of Transportation, Washington, D.C., 2009.
3. Bennett, M. K., H. Manal, and R. Van Houten. A Comparison of Gateway In-Street Sign Configuration to Other Driver Prompts to Increase Yielding to Pedestrians at Crosswalks. *Journal of Applied Behavior Analysis*, Vol. 47, No. 1, 2014, pp. 3–15. doi:10.1002/jaba.103
4. Bennett, M. K., and R. Van Houten. Variables Influencing Efficacy of Gateway In-Street Sign Configuration on Yielding at Crosswalks. *Transportation Research Record:*

Journal of the Transportation Research Board, 2016. 2586: 100–105.

5. Michigan Department of State Police. *Uniform Traffic Code for Cities, Townships, and Villages*. Lansing, MI, 2003. http://www.michigan.gov/documents/UTC_50201_7.pdf. Accessed November 9, 2017.
6. Federal Highway Administration. *Safety Effects of Marked Versus Unmarked Crosswalks at Uncontrolled Locations: Final Report and Recommended Guidelines*. FHWA Publication HRT-04-100. FHWA, U.S. Department of Transportation, Washington, D.C., 2005 https://www.fhwa.dot.gov/publications/research/safety/04100/04100.pdf. Accessed July 10, 2017.
7. Van Houten, R., and J. Hochmuth. *Evaluation of R1-6 Gateway Treatment Alternatives for Pedestrian Crossings*. Publication RC-1638. Michigan Department of Transportation, 2016 http://www.michigan.gov/documents/mdot/SPR-1638_552736_7.pdf. Accessed July 10, 2017.
8. Van Houten, R. The Effects of Advance Stop Lines and Sign Prompts on Pedestrian Safety in a Crosswalk on a Multilane Highway. *Journal of Applied Behavior Analysis*, Vol. 21, 1988, pp. 245–251. doi:10.1901/jaba.1988.21-245
9. Huybers, S., R. Van Houten, and J. E. L. Malenfant. Reducing Conflicts Between Motor Vehicles and Pedestrians: The Separate and Combined Effects of Pavement Markings and a Sign Prompt. *Journal of Applied Behavior Analysis*, Vol. 37, 2004, pp. 445–456. doi: 10.1901/jaba.2004.37-445
10. Zegeer, C., C. Lyon, R. Srinivasan, B. Persaud, B. Lan, S. Smith, D. Carter, N. J. Thirsk, K. Zegeer, E. Ferguson, R. Van Houten, and C. Sundstrom. Development of Crash Modification Factors for Uncontrolled Pedestrian Crossing Treatments. *Transportation Research Record: Journal of the Transportation Research Board*, 2017. 2636: 1–8. doi:10.3141/2636-01
11. Bushell, M. A., B. W. Poole, C. V. Zegeer, and D. A. Rodriguez. *Costs for Pedestrian and Bicyclist Infrastructure Improvements: A Resource for Researchers, Engineers, Planners, and the General Public*. Federal Highway Administration, Chapel Hill, NC, 2013. http://www.pedbikeinfo.org/cms/downloads/Countermeasure%20Costs_Report_Nov2013.pdf. Accessed July 10, 2017.
12. Institute of Transportation Engineers. Determining Vehicle Change Intervals: Proposed Recommended Practice. *ITE Journal*, Vol. 32, 1989, pp. 126–129.

The Pedestrians and Cycles Section (ANF00) peer-reviewed this paper (18-01080).

Article

TRR
JOURNAL OF THE TRANSPORTATION RESEARCH BOARD

Exploratory Analysis of Driver Yielding at Low-Speed, Uncontrolled Crosswalks in Milwaukee, Wisconsin

Transportation Research Record
2018, Vol. 2672(35) 21–32

DOI: 10.1177/0361198118782251
journals.sagepub.com/home/trr
SAGE

Robert J. Schneider[1], Aida Sanatizadeh[1], Mohammad Razaur Rahman Shaon[2], Zhaoxiang He[2], and Xiao Qin[2]

Abstract

One of the most common circumstances contributing to pedestrian crashes is drivers failing to yield to pedestrians in crosswalks. A better understanding of driver yielding behavior can help identify optimal safety treatments to improve driver yielding and prevent pedestrian injuries and fatalities. Recognizing this need, this study observed driver yielding behavior at 20 uncontrolled intersections along two-lane arterial and collector roadways with posted speed limits of 25 or 30 miles per hour in Milwaukee, Wisconsin during weekday afternoon peak travel periods in fall 2016. The naturalistic observations showed that drivers yielded 60 times out of 364 opportunities when the pedestrian wished to cross (16% driver yielding rate). Yielding rates differed between intersections, ranging from a high of 60% to a low of 0%. A binary logistic model showed that drivers were more likely to yield to pedestrians when the major roadway had a lower speed limit or less traffic; when the intersection had a shorter crossing distance or a bus stop; and when the pedestrian was White, standing in the street, or acting assertively. Finally, all else equal, intersections with no reported pedestrian crashes in the last 5 years had higher driver yielding rates than intersections with at least two reported pedestrian crashes. While this exploratory study is based on a small sample of observations, it supports several engineering, education, and enforcement strategies and provides suggestions for future studies of driver yielding behavior.

United States pedestrian fatalities have increased from approximately 4,100 in 2009 to approximately 6,000 in 2016 (a 46% increase) (*1*, *2*). The current number represents the highest level of annual fatalities in nearly three decades (*3*). Reversing this trend is critical, and it is supported by federal (*4*), state (*5*), and local (*6*) goals to move toward zero traffic fatalities.

Drivers failing to yield to pedestrians in crosswalks is a common factor contributing to pedestrian crashes. It was the top factor in California and Wisconsin pedestrian crashes (*7*) and was listed as the top common contributing factor (besides "hit-and-run," which is a post-crash characteristic) in a study of six states (*8*). Increasing driver yielding through education, enforcement, and engineering strategies may help improve pedestrian safety (*9*, *10*).

This study conducted naturalistic field observations in Milwaukee, Wisconsin to explore two questions related to driver yielding behavior: (1) How often do drivers yield to pedestrians in specific locations? (2) What characteristics of drivers, pedestrians, and crosswalk locations are associated with whether or not drivers yield to pedestrians?

This study focuses on pedestrians crossing at uncontrolled crosswalks—locations without a traffic signal or stop sign for the driver. These locations are particularly important because they often involve interactions between pedestrians crossing the roadway and drivers traveling straight. Vehicles going straight tend to travel at higher speeds than turning vehicles, leading to more severe pedestrian injuries when collisions occur. A sample of Wisconsin pedestrian crashes from 2011 to 2013 showed that 49% of non-severe crashes involved a vehicle going straight but 77% of fatal crashes involved a vehicle going straight (*11*).

This study is exploratory. Field data were collected in one community at 20 intersections with a relatively narrow range of characteristics. Still, the results are theoretically consistent with previous studies on driver yielding; support several specific education, enforcement, and engineering pedestrian safety strategies; and provide suggestions for future studies.

[1]Department of Urban Planning, University of Wisconsin-Milwaukee, Milwaukee, WI
[2]Department of Civil and Environmental Engineering, University of Wisconsin-Milwaukee, Milwaukee, WI

Corresponding Author:
Address correspondence to Robert J. Schneider: rjschnei@uwm.edu

Literature Review

Previous research suggests that a variety of factors influence whether or not drivers yield to pedestrians in crosswalks. For example, Schneider and Sanders developed a broad conceptual framework to explain driver yielding behavior (*7*). According to this framework, whether or not a driver yields to a pedestrian at a specific crosswalk depends on roadway design, land-use patterns, pedestrian safety education and enforcement, pedestrian and driver sociodemographic characteristics, and how drivers and pedestrians typically interact throughout a community.

Yet, with several exceptions (*12–15*), there are few driver yielding field studies that examine more than one set of factors. This exploratory study contributes to pedestrian safety research by examining the association between driver yielding and a broad set of roadway design and pedestrian and driver characteristics.

Two of the most common categories of driver yielding factors studied in the literature and evaluated in this study are roadway design characteristics and individual pedestrian and driver characteristics.

Roadway Design Characteristics

A survey of nearly 400 pedestrian safety professionals throughout North America suggested that lower speed limits and fewer roadway lanes were associated with higher driver yielding rates (*7*). These perceptions are supported by field data: lower posted speed limits were associated with higher driver yielding rates for several pedestrian crossing treatments (*16*), and lower actual motor vehicle approach speeds were associated with higher rates of driver yielding in Massachusetts (*17*) and at roundabouts in six states (*12*). Further, shorter crossing distances were associated with higher yielding rates in a multi-state study (*18*). In general, these studies suggest that roadways designed primarily to move large amounts of motor vehicle traffic quickly are less conducive to yielding than narrower, lower-speed roadways.

Many studies have tested how drivers respond to design treatments at specific crosswalk sites. Driver yielding rates tend to be higher at crossings with MUTCD R1-6 in-street pedestrian crossing signs (*13*, *19*), gateway treatments (i.e., MUTCD R1-6 in-street signs at the center of the roadway and on the curbs near either end of the crosswalk) (*20*), advance yield markings (*21*), median islands (*18*), rapid flashing beacons (*22–24*), pedestrian hybrid beacons (*25*), and leading pedestrian intervals at signalized intersections (*26*).

Pedestrian and Driver Demographic and Behavior Characteristics

Previous research has found pedestrian and driver characteristics to be associated with driver yielding. Drivers tend to yield more often to pedestrians holding a cane (*12*, *27*) and pedestrians wearing brighter clothing (*28*). Drivers may also yield more often to pedestrians who are White than to pedestrians of color (*29*, *30*). Drivers who are older (*31*), male (*31*), and drive less expensive automobiles (*32*) may yield more often to pedestrians. Yet, few studies have collected these demographic and socioeconomic characteristics, especially in combination with roadway characteristics.

Studies also suggest that there is a relationship between pedestrian behavior and driver yielding. More assertive pedestrians tend to produce higher driver yielding rates (*7*, *14*, *15*, *28*, *33*). However, researchers do not use a consistent measure of "assertiveness." For example, pedestrian assertiveness has been defined as waiting in the street (rather than on the curb) (*12*, *28*), walking quickly toward the crossing (*14*, *15*), or raising a hand or extending an arm in the direction of crossing (*33*). Overall, higher pedestrian volumes may be associated with higher rates of driver yielding (*13*, *31*). This may be attributable to heightened driver awareness or greater pedestrian assertiveness in areas with high pedestrian activity.

Driver Yielding Data Collection Approaches

Most field studies observe how drivers react to staged pedestrians (i.e., members of the research team) as they approach a crosswalk (*10*, *12*, *13*, *17*, *18*, *20*, *24*, *29*, *30*, *33*). This method has several advantages. First, researchers can collect a large sample of driver yielding opportunities relatively quickly. Second, researchers can control pedestrian demographic and assertiveness characteristics. In particular, pedestrians can be directed to behave similarly for each approaching vehicle (e.g., set one foot into the roadway and look at the approaching driver). Some staged pedestrian studies have also observed unstaged crossings of pedestrians who happened to pass through the study site (*10*).

This study focuses exclusively on unstaged crossings, following the approach used by Schroeder (*14*) and Schroeder and Rouphail (*15*) in Raleigh, NC. This approach was chosen to document naturally occurring interactions between drivers and pedestrians in the Milwaukee community.

Data Collection

Driver yielding data were collected at 20 uncontrolled intersections along two-lane arterial and collector roadways in Milwaukee during fall 2016. Ten of the intersections had experienced at least two reported daytime pedestrian crashes during a 5-year period (2010–2014). The other ten had similar characteristics but experienced one or zero reported crashes during this period (Figure 1). Data were collected in central Milwaukee because it tends to have moderate to high pedestrian activity as a result of its relatively high development density, neighborhood commercial streets, proximity to the central business district, and high-frequency bus lines. Regular pedestrian activity allowed observation of a

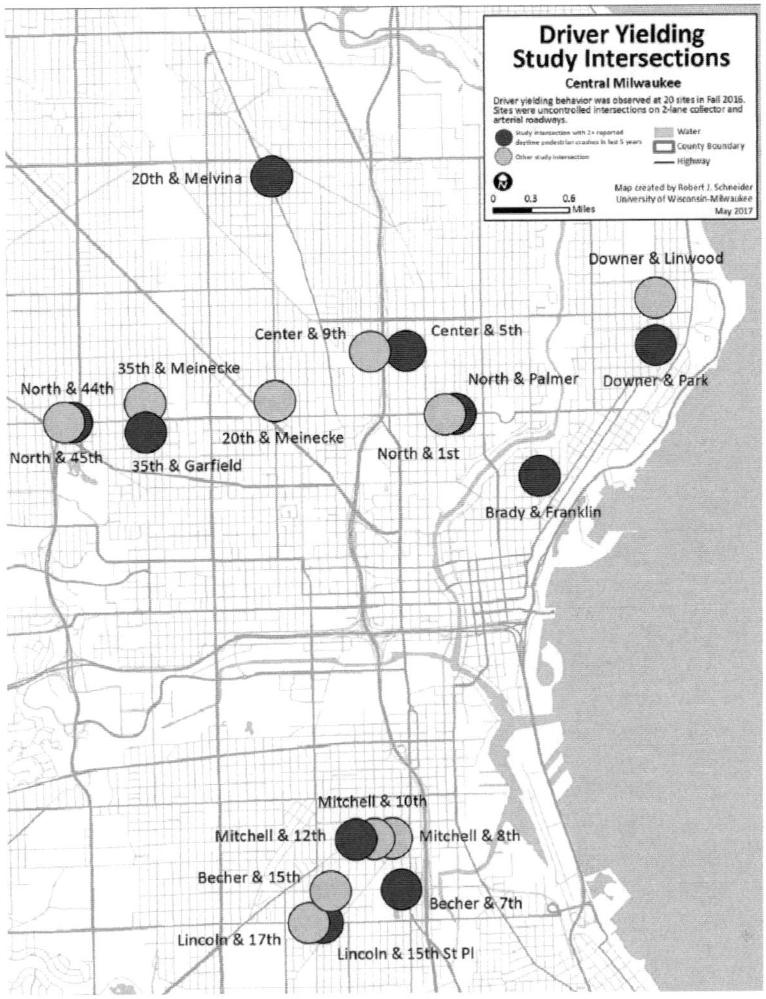

Figure 1. Milwaukee study intersection locations.

sufficient sample of pedestrian crossings during relatively short field data collection periods at each site.

Driver Yielding Behavior

Three data collectors made field observations for 2 hours during weekday evening travel periods (Monday through Thursday, typically 5 pm to 7 pm). Unstaged pedestrian crossings were observed—pedestrians and drivers of all types of motor vehicles interacting naturally in public. Pedestrian crossings were only observed for the mainline roadway crosswalks and were only considered when pedestrians started within the crosswalk lines. Pedestrians were observed when crossing either from the driver's left or right. After a pedestrian arrived at the crossing, data collectors observed the first driver from either direction with an opportunity to yield. Drivers were considered as having an opportunity to yield if they were beyond a minimum distance away from the crosswalk when the pedestrian arrived at the curb (which is slightly different than state law: drivers must yield the right-of-way when a pedestrian puts at least one foot in the crosswalk). The method described by Van Houten et al. (*34*) was used to calculate the minimum safe stopping distance for drivers. For example, based on a driver reaction time of 2.5 s, the posted speed limit in feet per second, and a conservative deceleration rate of 11.2 feet (3.41 m) per second, the safe stopping distance for vehicles traveling at 30 mph (48 km/h) on a flat grade is 196 feet (59.7 m). Drivers within this distance when the pedestrian arrived were not considered to have an opportunity to yield. Overall, 473 pedestrian crossings were observed across the 20 study sites, and drivers had an opportunity to yield for 364 of these crossings.

Explanatory Variables

Several categories of explanatory variables were collected, including site characteristics, pedestrian and driver demographics, and pedestrian group size and behavior characteristics.

Roadway and crosswalk characteristics were measured at each intersection (Table 1). All of the study intersections were in corridors with on-street parking. Annualized average daily traffic (AADT) volumes in these corridors ranged from approximately 5,000 to 16,000, and posted speed limits were either 25 or 30 mph. Some corridors had bicycle lanes, while others did not. This feature was correlated with crossing distance, which ranged from 35 to 52 feet. All study intersections had four legs. Some intersections had both crosswalks marked across the major roadway, while others only had one of the two crosswalks marked. Some intersections had standard crosswalk warning signs (MUTCD W11-2 or W11-2A). None of the study intersections had designated left-turn lanes, designated right-turn lanes, far-side bus stops, in-street pedestrian crossing signs (MUTCD R1-6), curb extensions, or median islands.

The field data collectors also assessed pedestrian assertiveness, observed pedestrian crossing group size, and estimated pedestrian and driver demographic characteristics. Table 2 defines and summarizes the variables corresponding with each of the 364 driver yielding opportunities. How these variables may be related to driver yielding based on previous research was hypothesized. To the authors' knowledge, few previous studies have evaluated as broad of a combination of demographic, behavioral, and site variables. More details about the data collection protocol and specific variables are provided in the study report (*35*).

Binary Logistic Regression Model Structure

A series of binary logistic regression models was developed to identify explanatory variables that may be associated with drivers choosing to yield (or not yield) to pedestrians. The binary logistic regression model is specified as:

$$g(X) = Ln\left[\frac{P(x_i)}{1-P(x_i)}\right] = \beta_0 + \beta_1 X_{1i} + \beta_2 X_{2i} + \cdots + \beta_k X_{ki} + \varepsilon_i \quad (1)$$

where $P(x_i)$ is the probability of a driver yielding to a pedestrian at any of the $i = 364$ pedestrian crossing opportunities, X_{ki} is a vector representing the kth explanatory variables (e.g., roadway features, pedestrian and driver demographic characteristics, pedestrian behaviors) for the ith observation, β_k is a vector of parameters that express the relationship between each explanatory variable in X_{ki} and the probability of a driver yielding to a pedestrian, and ε_i is the error term. The parameters β_k were estimated using maximum likelihood estimation. Using the equation specified above, the probability of driver yielding to pedestrian can be written as:

$$P(x_i) = \frac{1}{1+exp\left[-\left(\beta_0 + \beta_1 X_{1i} + \beta_2 X_{2i} + \ldots + \beta_k X_{ki}\right)\right]} \quad (2)$$

Since observations were collected at 20 distinct intersections, it is possible that driver yielding behavior is correlated between drivers at a particular intersection as a result of some unmeasured factors. To evaluate the unobserved correlation in driver yielding at particular intersections (or any other unmeasured intersection-related characteristics contributing to the driver yielding decision), a random-effect logistic regression model was also developed and tested. This was done by introducing 20 intersection-specific indicator variables to the model. However, these indicator variables were not significant, and calculations of null deviance, residual deviance, and Akaike Information Criterion (AIC) showed that this model did not have better performance, so the final results from our binary logistic regression model are presented.

Modeling Process

All 23 explanatory variables were tested in a series of models, starting with only demographic and behavior variables, assuming that driver yielding behavior may depend on driver and pedestrian age, gender, and race/ethnicity, as well as pedestrian assertiveness. Then site-specific variables were added. Variables were selected for more refined models using random forest and step-wise procedures. Explanatory variables were removed that did not have a significant relationship (90% confidence level) with driver yielding, starting with the least significant variables. Theoretical relationships and practical expertise were used to prevent statistically correlated variables from being included in the same model.

Results

In total, 473 pedestrian crossings at 20 uncontrolled intersections in Milwaukee were documented (Table 1). Of these crossings, 364 had opportunities for drivers to yield to a pedestrian. Sixty of these drivers were observed to yield, producing an overall yielding rate of 16%. Yielding rates differed between intersections, ranging from a high of 60% to a low of 0%. Note that these results apply to the specific urban context of Milwaukee and are limited to the range of values available in this dataset (see Table 2).

Overall Model Fit

The final binary logistic regression model of driver yielding included eight explanatory variables (Table 3). It had the lowest AIC of all models tested (261), suggesting the best fit. All explanatory variables in the final model were significant at the 95% confidence level. Table 4 provides an example of how the model-estimated percentage of drivers who yield is

Table 1. Milwaukee Study Intersection Characteristics and Driver Yielding Rates

Study intersection	Major road AADT	Major road posted speed limit (miles per hour)	Crossing distance[a] (feet)	Distance to on-street parking[b] (feet)	Bus stop (1 = Yes, 0 = No)	Both crosswalks marked (1 = Yes, 0 = No)	Crosswalk sign (MUTCD W11-2 or W11-2A) (1 = Yes, 0 = No)	Adjacent commercial land use (1 = Yes, 0 = No)	Intersection had 2+ reported crashes in 5-year period (1 = Yes, 0 = No)	Field data collection date	Field data collection time period (pm)	Pedestrian crossings	Driver yielding opportunities	Number of drivers who yielded	Percent of drivers who yielded
E North Ave & N Palmer St	16400	30	51	100	0	1	1	0	1	8/15/16	5:00–7:00	17	17	0	0%
W North Ave & N 1st St	16400	30	52	200	1	1	0	0	0	11/1/16	4:30–6:30	14	10	2	20%
N 35th St & W Garfield Ave	14500	30	56	500	1	1	0	1	1	8/24/16	5:15–7:15	31	28	1	4%
W Lincoln Ave & S 15th St Pl	14200	30	55	20	0	0	0	1	1	8/25/16	5:00–7:00	38	34	0	0%
N 35th St & W Meinecke Ave	13800	30	51	400	0	1	1	1	0	9/13/16	4:45–6:45	26	22	1	5%
W North Ave & N 44th St	13200	30	56	25	0	0	0	1	1	8/18/16	5:15–7:15	17	16	0	0%
W Lincoln Ave & S 17th St	13000	30	51	25	0	0	0	0	0	10/18/16	5:00–7:00	19	13	5	38%
W North Ave & N 45th St	11000	30	56	200	1	1	0	1	0	9/6/16	4:30–6:30	30	27	8	30%
E Brady St & N Franklin Pl	10100	25	40	18	0	1	1	1	1	8/10/16	5:00–7:00	38	27	14	52%
W Center St & N 5th St	10000	30	54	80	1	0	0	1	1	8/16/16	5:00–7:00	32	29	1	3%
W Becher St & S 7th St	9600	30	49	300	0	0	1	0	1	8/29/16	5:00–7:00	24	11	1	9%
W Center St & N 9th St	9100	30	59	400	0	0	1	0	0	8/31/16	4:30–6:30	16	16	0	0%
W Mitchell St & S 8th St	9000	25	56	25	0	1	0	1	0	10/19/16	4:30–6:30	16	10	1	10%
W Mitchell St & S 10th St	8400	25	54	30	0	1	0	1	0	9/1/16	5:00–7:00	20	18	6	33%
N Downer Ave & E Park Pl	7200	30	50	30	1	1	1	1	1	8/11/16	5:00–7:00	58	29	6	21%
N 20th St & W Meinecke Ave	6800	30	50	120	0	1	0	0	0	10/25/16	4:30–6:30	23	18	4	22%
W Becher St & S 15th St	6600	30	46	200	0	0	0	0	0	10/11/16	4:30–6:30	15	7	0	0%
W Mitchell St & S 12th St	6300	25	5w1	30	0	1	0	1	1	8/24/16	5:00–7:00	18	16	4	25%
N Downer Ave & E Linnwood Ave	5100	30	58	200	1	0	0	0	0	8/30/16	5:00–7:00	15	10	6	60%
N 20th St & W Melvina St	4900	30	49	90	0	0	0	0	1	8/17/16	5:00–7:00	6	6	0	0%
Average of 20 Study Sites	10280	29	52	150	0.30	0.55	0.30	0.55	0.50			473	364	60	16%

Note: 10 miles per hour = 16.1 kilometers per hour; 1 foot = 0.305 meters.

[a]The crosswalk crossing distance was the shortest curb-to-curb distance within the crosswalk. The distance shown is the average of the two major road crosswalks.

[b]Distance from the crosswalk to street parking was the distance to the closest car parked on the street in advance of the crossing. The distance shown is the average of the two mainline crosswalks. All of the study intersections were two lanes with on-street parking during most times of day. However, two intersections (N. 35th Street and W. Meinecke Avenue and N. 35th Street and W. Garfield Avenue) had peak-hour parking restrictions on between 3:30 pm and 5:30 pm, so they had four operating traffic lanes and no on-street parking for part of the time when data were collected (approximately 45 minutes at N. 35th Street and W. Meinecke Avenue and approximately 15 minutes at N. 35th Street and W. Garfield Avenue).

Table 2. Explanatory Variables Tested in Binary Logistic Regression Models

	Mean	Std. dev.	Min	Max	Expected relationship with yielding
Demographic variables[a]					
Pedestrian race/ethnicity = White (1 = yes; 0 = no)	0.291	0.455	0	1	+
Pedestrian gender = Male (1 = yes; 0 = no)	0.632	0.483	0	1	+
Pedestrian age <25 years (1 = yes; 0 = no)	0.179	0.384	0	1	+
Driver race/ethnicity = White (1 = yes; 0 = no)	0.368	0.483	0	1	+
Driver gender = Male (1 = yes; 0 = no)	0.640	0.481	0	1	–
Driver age <25 years (1 = yes; 0 = no)	0.440	0.497	0	1	–
Behavior variables					
Pedestrian standing in the street (1 = yes; 0 = no)[b]	0.819	0.386	0	1	+
Pedestrian acting assertively (1 = yes; 0 = no)[c]	0.431	0.496	0	1	+
Pedestrian group size = 1 (1 = yes; 0 = no)[d]	0.731	0.444	0	1	–
Site variables					
Traffic volume (AADT)	10800	3260	4900	16400	–
Posted speed limit (miles per hour)	29.0	1.98	25	30	–
Average crossing distance (feet)[e]	52.3	4.53	40	59	–
Bus stop present (1 = yes; 0 = no)	0.365	0.482	0	1	+
Distance from upstream signal (feet)[f]	720	421	253	2000	–
Distance to downstream signal (feet)[f]	779	452	253	2000	+
Distance from crosswalk to street parking (feet)[g]	148	157	18	500	+
Both crosswalks marked (1 = yes; 0 = no)	0.610	0.488	0	1	+
Crosswalk sign present (1 = yes; 0 = no)[h]	0.335	0.473	0	1	+
Right-turn area present (1 = yes; 0 = no)[i]	0.508	0.501	0	1	–
Adjacent commercial land use (1 = yes; 0 = no)[j]	0.703	0.457	0	1	+
Majority of pedestrians are White (1 = yes; 0 = no)	0.228	0.420	0	1	+
Majority of drivers are White (1 = yes; 0 = no)	0.302	0.460	0	1	+
Intersection had 2+ pedestrian crashes (1 = yes; 0 = no)[k]	0.585	0.493	0	1	–

Note: 10 miles per hour = 16.1 kilometers per hour; 1 foot = 0.305 meters.
[a]The demographic characteristics of pedestrians and drivers were estimated by field data collectors.
[b]"Standing in the street" indicates that pedestrians waited in the crosswalk with at least one foot in the street.
[c]"Acting assertively" indicates that pedestrians exhibited any one of the following characteristics: (1) the pedestrian actively leaned toward the opposite side of the roadway when in the crosswalk, (2) the pedestrian directed his or her eyes toward approaching drivers for more than 3 seconds, or (3) the pedestrian pointed his or her arms or fingers toward the crosswalk.
[d]Group size was the total number of pedestrians waiting to cross at one time. The group was defined by waiting together; they did not all need to end up crossing at the same time.
[e]The crosswalk crossing distance was the shortest distance from the curb on one side of the street to the curb on the other side of the street within the crosswalk. Average crossing distance for each intersection was the average of the two crosswalks.
[f]Distance from upstream signal is the distance from the upstream signal to the center of the intersection, and distance to downstream signal is the distance from the downstream signal to the center of the intersection. These variables were calculated for the direction of travel of each individual approaching vehicle.
[g]Distance from the crosswalk to street parking was the distance from the crosswalk at the edge of the intersection to the closest car parked on the street in advance of the crossing (average distance from both mainline crosswalks).
[h]Intersection had a crosswalk sign (MUTCD W11-2 or W11-2A). No site had warning signs in advance of the crosswalks.
[i]A right-turn area was noted if there was a separate area to the right of the travel lane that was commonly used by right-turning cars to move out of the traffic stream prior to turning right. This operated like a right-turn lane but was not designated.
[j]Adjacent commercial land use indicates that the intersection was in a downtown or neighborhood commercial district.
[k]Intersection experienced at least two reported daytime pedestrian crashes during the 5-year period, 2010–2014.

associated with changes in each individual variable. These associations between each variable and driver yielding behavior are discussed below.

Demographic Variables Associated with Driver Yielding

The final model showed that drivers were more likely to yield to pedestrians who were White than pedestrians of other racial and ethnic backgrounds. This supports the findings of Goddard et al. (*29*). However, this result should be interpreted with caution. There were notable differences in the percentage of White pedestrians among the study intersections (12 intersections had 15% or fewer White pedestrians; three intersections had 85% or more White pedestrians). Therefore, the variable indicating that the pedestrian was White could be capturing other unmeasured characteristics that were different among the intersections (e.g., neighborhood income

Table 3. Final Driver Yielding Model

Variables	Coefficient estimate	Std. error	*Z*-value	Pr(>|z|)
Intercept	15.580	3.751	4.155	0.000
Demographic variables				
Pedestrian race/ethnicity = White	1.005	0.423	2.376	0.017
Behavior variables				
Pedestrian standing in the street	1.284	0.548	2.341	0.019
Pedestrian acting assertively	0.946	0.405	2.330	0.019
Site variables				
Traffic volume (AADT)	−0.000125	0.000062	−2.071	0.038
Posted speed limit (miles per hour)	−0.384	0.099	−3.869	0.000
Average crossing distance (feet)	−0.129	0.052	−2.482	0.013
Bus stop present	1.669	0.520	3.208	0.001
Intersection had 2+ pedestrian crashes	−1.928	0.462	−4.168	0.000
Sample size (*n*)			364	
Null deviance			325.85 on 363 degrees of freedom	
Residual deviance			242.52 on 355 degrees of freedom	
AIC			260.52	

Note: 10 miles per hour = 16.1 kilometers per hour; 1 foot = 0.305 meters.

levels, which may relate to pedestrian clothing type and appearance; neighborhood driver behavioral norms, including acceptance of speeding; neighborhood enforcement of traffic laws). The data cannot determine whether specific drivers are less likely to yield because a pedestrian is a person of color or whether drivers in neighborhoods where there are more people of color crossing the street are less likely to yield in general. Nonetheless, this finding points to the importance of equity in pedestrian safety efforts.

Pedestrian age and gender were not significant. No driver demographic variables (age, gender, or race/ethnicity) were significant in the final model.

Behavior Variables Associated with Driver Yielding

Two pedestrian behaviors were significant in the final model. Pedestrians who waited in the street rather than on the curb were more likely to have drivers yield to them. In addition, pedestrians who indicated their intent to cross the street assertively were more likely to have drivers yield. Waiting in the street and taking an assertive stance in the crosswalk may make drivers more aware of pedestrians and may also more clearly indicate an intent to cross. Pedestrian group size was not significant in the model.

Site Variables Associated with Driver Yielding

Within this study context of two-lane roadways with speed limits of 25 to 30 mph, all else equal, drivers were less likely to yield at intersections with higher traffic volumes and higher posted speed limits. Streets with these design characteristics may be perceived by drivers and pedestrians as thoroughfares for automobiles, so drivers may be less aware of pedestrians and be less concerned about yielding to pedestrians at a crosswalk. In addition, it is more difficult to yield to a pedestrian when traveling at a higher speed, since it requires seeing the pedestrian sooner and decelerating more quickly than when traveling more slowly. Plus, on streets with more traffic and higher speeds, drivers may be more concerned about being rear-ended by other cars or about being passed recklessly on the right or left when the pedestrian is in the crosswalk.

Drivers were less likely to yield when the pedestrian crossing distance was longer. While this study focused on two-lane roadways, there was still some variation in road width. Wider roads generally make it easier for other drivers to pass on the right, especially when they think that a car stopped in front of them for a pedestrian is turning left. In addition, wider roads may be associated with higher travel speeds, regardless of speed limit, making it more difficult for drivers to yield.

There was evidence that intersections with bus stops tend to have a higher likelihood of driver yielding. This could be as a result of higher pedestrian volumes (pedestrians crossing to catch the bus), buses slowing traffic near the intersection as they exit and enter the traffic stream, or other factors. All of the bus stops were on the near side of the study intersections (far-side stops are rare in Milwaukee). This does not suggest that near-side bus stops are safer for pedestrians than far-side bus stops. To the contrary, the Federal Highway Administration recommends placing bus stops on the far side of intersections when possible to reduce the chances of multiple-threat pedestrian crashes (*36*). Future studies should examine the relationship between bus stop placement and driver yielding.

Drivers were less likely to yield to pedestrians who were crossing at intersections that had experienced two or more

Table 4. Model Sensitivity Example Using Hypothetical Site Values

Model variables	Hypothetical site variable initial value[a]	Hypothetical site variable changed value	Percent change in driver yielding associated with hypothetical variable change[b]
Pedestrian race/ethnicity = White (1 = yes; 0 = no)	0	1	+2.0%
Pedestrian standing in the street (1 = yes; 0 = no)	0	1	+3.0%
Pedestrian acting assertively (1 = yes; 0 = no)	0	1	+1.8%
Traffic volume (AADT)	6,000	5,000	+0.2%
Posted speed limit (miles per hour)	30	25	+6.4%
Average crossing distance (feet)	45	40	+1.1%
Bus stop present (1 = yes; 0 = no)	0	1	+4.8%
Intersection had 2+ pedestrian crashes (1 = yes; 0 = no)	1	0	+6.4%

[a]Hypothetical site variable initial values were chosen to produce a low model-estimated probability of driver yielding. This made it easier to change each individual variable in the same direction (i.e., to increase the model-estimated probability).
[b]The model-estimated probability of a driver yielding at a site with the hypothetical initial values for all variables is 1.2%. For example, changing the base value of White = no to White = yes is associated with a 2.0% increase in the model-estimated driver yielding percentage (from 1.2% to 3.2%).

daytime pedestrian crashes in the last 5 years. Given that factors such as traffic volume, speed limit, and crossing distance are already captured in the model, this variable may be capturing other unmeasured roadway or behavior characteristics that are associated with higher pedestrian risk at these particular crossings (i.e., driving while intoxicated or distracted; speeding; walking while intoxicated). This result also suggests an important relationship between higher driver yielding rates and enhanced pedestrian safety.

Several other site characteristics were tested but not significant in the final model, including marked crosswalks, crosswalk signs, right-turn areas, the majority of drivers and pedestrians at the intersection being White, and adjacent commercial land use. While these variables were not significant in this particular model of behavior at a limited set of crosswalks in Milwaukee, they may still have a relationship with driver yielding and pedestrian safety.

Implications for Practice

This model identifies associations between driver yielding and specific roadway and behavior characteristics. While these results do not imply direct causation, they provide support for several strategies to increase driver yielding to pedestrians in uncontrolled crosswalks.

Reduce Roadway Design Speeds and Reduce Posted Speed Limits

Lower speed limits were associated with a higher likelihood of drivers yielding to pedestrians. This finding complements other research that finds actual approach speeds are inversely related to driver yielding rates (*17*). As higher speeds are also associated with higher pedestrian injury severity (*37*), reducing motor vehicle speeds is an important pedestrian safety strategy. Design strategies to discourage speeding along roadway corridors may include:

- Minimize the total number of automobile lanes.
- Minimize automobile lane widths.
- Introduce on-street parking or other features that provide visual cues to travel slowly.

Reduce Pedestrian Crossing Distances

Shorter pedestrian crossing distances were associated with a higher likelihood of drivers yielding to pedestrians. Shorter crossings also reduce the amount of time that pedestrians are in the roadway and exposed to traffic. Several design treatments can be used to shorten pedestrian crossing distances:

- Install curb extensions on roadways with on-street parking. In addition to reducing the crossing distance, this treatment helps keep sight lines between pedestrians and drivers from being blocked by parked cars. Curb extensions can also reduce vehicle turning speeds.
- Install raised median islands. Median islands divide the crosswalk into two distinct crossings, each shorter than the original crossing. This also provides a refuge for pedestrians, allowing them to cross one direction of traffic at a time.
- Reduce the number of automobile lanes and reduce lane widths. Reducing the roadway width allocated to automobile travel will reduce the distance that pedestrians will be exposed to potential conflicts with motor vehicles. Reducing the number of automobile lanes has been shown to benefit pedestrian, bicyclist, and motorist safety (*38*). This also provides additional space for bicycle lanes, shoulders, landscaped buffers, sidewalks, or other street features.

Help Pedestrians Communicate Intent to Cross

Drivers were more likely to yield to pedestrians who acted assertively. One simple strategy to increase assertiveness is to encourage pedestrians to point across the roadway after arriving at a crosswalk and make eye contact with approaching drivers. Pointing has been shown to increase driver yielding rates (*33*). However, increasing pedestrian assertiveness in a community is challenging because it involves modifying social norms that may have been established over decades and because pedestrian assertiveness is likely tied to expectations of driver yielding behavior (*7*). Helping pedestrians feel safe enough (and trust drivers enough) to act assertively is likely to require a comprehensive approach that includes:

- Education (e.g., emphasize that drivers should anticipate pedestrians and must yield the right-of-way to pedestrians who enter uncontrolled crosswalks; emphasize that pedestrians should make eye contact with drivers and confidently assert their right-of-way in crosswalks).
- Enforcement (e.g., implement high-visibility enforcement [HVE] programs that combine enforcement of pedestrian right-of-way laws with public messaging and media outreach [*10*, *34*]).
- Engineering (e.g., use roadway design principles supported by this study and other promising treatments described in the literature review).

These types of comprehensive pedestrian safety programs have increased driver yielding (*9*, *39*). Further, given that these findings suggest that drivers may be more likely to yield to White pedestrians, local leaders and agency staff should ensure that comprehensive education, enforcement, and engineering strategies are applied throughout a jurisdiction, including neighborhoods with many people of color. Additional focus may be needed to ensure that pedestrians of color experience crossing the street as comfortably and safely as all other community members.

Considerations and Future Research

This research studied driver yielding behavior at uncontrolled crossings of two-lane arterial and collector roadways. Although several significant demographic, behavior, and site variables were identified, there are other variables that should be included in future studies, such as actual vehicle speeds, pedestrian clothing type and brightness, and a wider range roadway design characteristics (e.g., number of travel lanes, designated turn lanes, median islands, curb extensions, and different combinations of signs, markings, and beacons). In addition, there are other contexts where driver yielding behavior at uncontrolled crossings should be documented, such as on urban residential streets and on suburban and rural roadways. A related line of research should also be undertaken to study driver yielding behavior when turning at controlled and uncontrolled intersections.

Data were collected at 20 study sites in Milwaukee. Although the multiple data collection sites allowed analysis of several different roadway features, the sample of sites is small. Therefore, data should be collected at more locations in Milwaukee as well as sites in other communities to examine the consistency of these results and to explore additional variables that may be related to driver yielding behavior. In particular, more study sites would help provide more data to explore the association between pedestrian and driver race/ethnicity and driver yielding and better understand the underlying reasons for any inequities.

Like most other driver yielding studies, this study collected data during daylight hours. However, 31% of the pedestrian crashes in Wisconsin between 2011 and 2015 where driver failure to yield was a contributing circumstance occurred during darkness (*35*). Lighting and visibility are likely to influence pedestrian and driver behaviors at uncontrolled crosswalks, so more research is needed to identify the potentially different factors associated with driver yielding at night.

Demographic and behavior data were collected in the field, so observation, judgment, and recording errors made by field data collectors were accepted. Other researchers have addressed this challenge by conducting training to improve inter-observer agreement (*10*) or using video in driver yielding studies (*13–15*, *31*). However, given available resources, field observations were the only method available that allowed observation of pedestrian and driver demographic characteristics as well as behaviors at 20 different sites. By comparison, a previous study of unstaged pedestrian crossings in Raleigh used video to observe driver and pedestrian demographic characteristics and behaviors but only included two sites (*14*, *15*). Collecting data at multiple sites allowed testing of the relationship between several roadway design variables and driver yielding.

The definition of driver yielding opportunities did not match state law precisely. If the study had specified that driver yielding opportunities only existed when pedestrians placed at least one foot in the crosswalk, as defined by state law, it would have found that 54 of 298 drivers yielded (18% of the total rather than 16%). Yet, narrowing the definition of driver yielding opportunity to the state law would have excluded 66 pedestrians (18% of all observed yielding opportunities) who were hesitant to step in the roadway but still intended to cross the street. Of these 66 people, 17 (26%) assertively indicated their intention to cross from the curb. The broader definition of driver yielding opportunity allowed the authors to account for all people who wanted to cross and all people exhibiting assertive and non-assertive behavior so that pedestrian crossing behavior could be analyzed more completely.

This study points to several lines of additional research. In particular, future studies should:

- Develop a consistent definition of pedestrian assertiveness. Previous studies have used different definitions, but the important relationship between this behavior and driver yielding underscores the need for a similar measure across multiple studies.

- Collect additional data on driver yielding to pedestrians with disabilities. There were only four driver yielding opportunities to people who used wheelchairs or walkers or had other visible physical disabilities, so a separate analysis for this group was not conducted. However, people with disabilities are particularly important because they may cross at slower speeds (they may only find a sufficient gap in traffic to cross at an uncontrolled location if drivers yield) and may be less able to take evasive maneuvers if drivers do not yield.
- Count total pedestrian crossing volumes. While it was noted whether or not pedestrians crossed in groups, the overall intersection pedestrian volume could also be associated with the general level of driver awareness of pedestrians at a crossing location. Higher pedestrian volumes may help drivers anticipate yielding opportunities, even for a single pedestrian.
- Explore the association between bus stops and driver yielding behavior. Pedestrians may be more assertive when hurrying to board approaching buses, and drivers may be more cautious when approaching buses that are dropping passengers off at the intersection, but these behaviors were not observed in this study.
- Record vehicles traveling in platoons. Cars traveling in groups may influence driver yielding (*15*). The study attempted to observe the total number of cars that did not yield after the pedestrian arrived as proxy for the effect of platooning. Yet, the data collectors did not have sufficient capacity to observe this characteristic while documenting other important behaviors.
- Use staged pedestrian crossings to complement naturalistic observations. Staged pedestrian crossings could help address the issues in the bullets above and control for characteristics such as pedestrian attire and pedestrian skin color.
- Incorporate video-based observations at more sites. This would help produce more reliable measurements of driver yielding and pedestrian assertiveness while still capturing data at multiple sites with a variety of roadway characteristics.
- Collect complementary survey data on local social norms related to driver yielding behavior. Some drivers may follow how the majority of other drivers behave in their local community, despite pedestrian behavior, roadway design, or other site characteristics. Therefore, as future studies begin to collect and analyze driver yielding data from multiple communities, it will be important to control for social norms as a model variable.

Importantly, although only ten sites with reported crashes and ten sites without were analyzed, it was found that intersections with higher yielding rates were associated with fewer pedestrian crashes. This contributes to a growing body of research showing similar results: a pedestrian enforcement campaign led to higher yielding and fewer pedestrian crashes in St. Johns, Newfoundland and Fredericton, New Brunswick (*9*), and the Gainesville HVE study found that higher driver yielding rates were associated with significant decreases in pedestrian crashes across the city (*10*). However, there is a need for additional studies to explore the relationship between driver yielding rates and pedestrian crash rates at specific locations in a variety of communities and roadway environments.

Conclusion

This exploratory study adds to the knowledge about factors affecting driver yielding behavior. Data were collected at 20 uncontrolled intersections along two-lane arterial and collector roadways in Milwaukee to examine how often drivers yield to pedestrians, and to identify characteristics of drivers, pedestrians, and crosswalk locations that are associated with this behavior. The binary logistic model suggested that drivers were more likely to yield when the major roadway had a lower speed limit or less traffic; when the intersection had a shorter crossing distance or a bus stop; and when the pedestrian was White, standing in the street, or acting assertively. These results support roadway design strategies such as reducing roadway design speeds and narrowing roadway crossing distances to increase the probability of drivers yielding to pedestrians. They also suggest the importance of educating pedestrians and drivers so that pedestrians can carefully and confidently assert their right-of-way in uncontrolled crosswalks. Additional driver yielding research at a wider range of sites is recommended, in more communities, and at different times of day, especially at night. Ultimately, it is hoped that this study will help improve pedestrian safety programs and reduce pedestrian injuries and fatalities.

Acknowledgments

Thanks to the Wisconsin Department of Transportation for providing grant funding and guidance for this study.

Author Contributions

The authors confirm contribution to the paper as follows: study conception and design: Schneider, Qin; data collection: Sanatizadeh, Shaon, He; analysis and interpretation of results: Schneider, Sanatizadeh, Shaon; draft manuscript preparation: Schneider, Sanatizadeh. All authors reviewed the results and approved the final version of the manuscript.

References

1. National Highway Traffic Safety Administration (NHTSA). *Fatality Analysis Reporting System (FARS) Encyclopedia, FARS Data Tables*, https://www-fars.nhtsa.dot.gov/Main/index.aspx. Accessed July 17, 2017.

2. Retting, R. *Pedestrian Traffic Fatalities by State: 2016 Preliminary Data*. Governors Highway Safety Association, 2017.
3. U.S. Department of Transportation, National Highway Traffic Safety Administration. *Traffic Safety Facts 1993: Pedestrians*. U.S. Department of Transportation, National Highway Traffic Safety Administration, Washington, D.C., 1993.
4. Federal Highway Administration. *Zero Deaths*, 2017. https://safety.fhwa.dot.gov/tzd/. Accessed July 17, 2017.
5. Toward Zero Deaths, National Strategy on Highway Safety. *American Association of State Highway and Transportation Officials Toward Zero Deaths Resolution*, 2011. http://www.towardzerodeaths.org/resource/aashto-toward-zero-deaths-resolution/, Accessed July 17, 2017.
6. Vision Zero Network. *Vision Zero Cities*. http://visionzeronetwork.org/about/elevating-efforts-in-vision-zero-cities-across-the-u-s/. Accessed July 17, 2017.
7. Schneider, R. J., and R. L. Sanders. Pedestrian Safety Practitioners' Perspectives of Driver Yielding Behavior across North America. *Transportation Research Record: Journal of the Transportation Research Board*, 2015. 2519: 39–50.
8. Stutts, J. C., W. W. Hunter, and W. E. Pein. Pedestrian Crash Types: 1990s Update. *Transportation Research Record: Journal of the Transportation Research Board*, 1997. 1538: 68–74.
9. Malenfant, L., and R. Van Houten. Increasing the Percentage of Drivers Yielding to Pedestrians in Three Canadian Cities with a Multifaceted Safety Program. *Health Education Research*, Vol. 5, 1990, pp. 275–279.
10. Van Houten, R., L. Malenfant, R. D. Blomberg, and B. E. Huitema. *The Effect of High-Visibility Enforcement on Driver Compliance with Pedestrian Right-of-Way Laws: 4-year Follow-Up*. National Highway Traffic Safety Administration, 2017.
11. Schneider, R. J., and J. Stefanich. *Wisconsin Pedestrian and Bicycle Crash Analysis: 2011-2013*. Wisconsin Department of Transportation, Bureau of Transportation Safety, 2015.
12. Salamati, K., B. J. Schroeder, D. R. Geruschat, and N. M. Rouphail. Event-Based Modeling of Driver Yielding Behavior to Pedestrians at Two-Lane Roundabout Approaches. *Transportation Research Record: Journal of the Transportation Research Board*, 2013. 2389: 1–11.
13. Stapleton, S., T. Kirsch, T.J. Gates, and P.T. Savolainen. "Factors Affecting Driver Yielding Compliance at Uncontrolled Midblock Crosswalks on Low-Speed Roadways," *Transportation Research Record: Journal of the Transportation Research Board*, 2017. 2661: 95–102.
14. Schroeder, B. A Behavior-Based Methodology for Evaluating Pedestrian-Vehicle Interaction at Crosswalks. PhD thesis. North Carolina State University, 2008.
15. Schroeder, B. J., and N. M. Rouphail. Event-Based Modeling of Driver Yielding Behavior at Unsignalized Crosswalks. *Journal of Transportation Engineering*, Vol. 137, No 7, pp. 455–465, 2011.
16. Fitzpatrick, K., S. Turner, M. Brewer, P. Carlson, B. Ullman, N. Trout, E. S. Park, J. Whitacre, N. Lalani, and D. Lord. *NCHRP Report 562: Improving Pedestrian Safety at Unsignalized Crossings*. Transportation Research Board of the National Academies, Washington, D.C., 2006.
17. Bertulis, T., and D. M. Dulaski. Driver Approach Speed and Its Impact on Driver Yielding to Pedestrian Behavior at Un-signalized Crosswalks. *Transportation Research Record: Journal of the Transportation Research Board*, 2014. 2464: 46–51.
18. Fitzpatrick, K., M. A. Brewer, R. Avelar, and T. Lindheimer. *Will You Stop for Me? Roadway Design and Traffic Control Device Influences on Drivers Yielding to Pedestrians in a Crosswalk with a Rectangular Rapid-Flashing Beacon*. Center for Transportation Safety, Texas A&M Transportation Institute, June 2016.
19. Huang, H., C. Zegeer, R. Nassi, and B. Fairfax. *The Effects of Innovative Pedestrian Signs at Unsignalized Locations: A Tale of Three Treatments*. Federal Highway Administration, August 2000.
20. Bennett, M. K., and R. Van Houten. Variables Influencing Efficacy of Gateway In-Street Sign Configuration on Yielding at Crosswalks. *Transportation Research Record: Journal of the Transportation Research Board*, 2016. 2586: 100–105.
21. Van Houten, R., J. E. L. Malenfant, and D. McCusker. Advance Yield Markings: Reducing Motor Vehicle-Pedestrian Conflicts at Multilane Crosswalks with Uncontrolled Approach. *Transportation Research Record: Journal of the Transportation Research Board*, 2001. 1773: 69–74.
22. Van Houten, R., R. Ellis, and E. Marmolejo. Stutter-Flash Light-Emitting-Diode Beacons to Increase Yielding to Pedestrians at Crosswalks. *Transportation Research Record: Journal of the Transportation Research Board*, 2008. 2073: 69–78.
23. Shurbutt, J., and R. Van Houten. *Effects of Yellow Rectangular Rapid-Flashing Beacons on Yielding at Multilane Uncontrolled Crosswalks*. Federal Highway Administration, September 2010.
24. Ross, J., D. Serpico, and R. Lewis. *Assessment of Driver Yielding Rates Pre- and Post-RRFB Installation*, Bend, Oregon, Oregon Department of Transportation, 2011.
25. Fitzpatrick, K., S. T. Chrysler, R. Van Houten, W.W. Hunter, and S. Turner. *Evaluation of Pedestrian and Bicycle Engineering Countermeasures: Rectangular Rapid-Flashing Beacons, HAWKs, Sharrows, Crosswalk Markings, and the Development of an Evaluation Methods Report*, Federal Highway Administration, April 2011.
26. Van Houten, R., R. A. Retting, C. M. Farmer, and J. Van Houten. Field Evaluation of a Leading Pedestrian Interval Signal Phase at Three Urban Intersections. *Transportation Research Record: Journal of the Transportation Research Board*, 2000. 1734: 86–92.
27. Harrell, W. A. Driver Response to a Disabled Pedestrian Using a Dangerous Crosswalk. *Journal of Environmental Psychology*, Vol. 12, No. 4, 1992, pp. 345–354.
28. Harrell, W. A. The Impact of Pedestrian Visibility and Assertiveness on Motorist Yielding. *Journal of Social Psychology*, Vol. 133, No 3, 1993, pp. 353–360.
29. Goddard, T., K. B. Khan, and A. Adkins. Racial Bias in Driver Yielding Behavior at Crosswalks. *Transportation Research Part F*, Vol. 33, 2015, pp. 1–6.
30. Coughenour, C., S. Clark, A. Singh, E. Claw, J. Abelar, and J. Huebner. Examining Racial Bias as a Potential Factor in Pedestrian Crashes. *Accident Analysis & Prevention*, Vol. 98, pp. 96–100, 2017.
31. Sun, D., S. V. S. K Ukkusuri, R. F. Benekohal, and S. T. Waller. Modeling of Motorist-Pedestrian Interaction at Uncontrolled Mid-Block Crosswalks. Presented at 82nd Annual Meeting of the Transportation Research Board, Washington, D.C., 2003.

32. Piff, P. K., D. M. Stancato, R. Mendoza-Denton, D. Keltner, and S. Coteb. Higher Social Class Predicts Increased Unethical Behavior. *Proceedings of the National Academy of Sciences of the United States of America*, Vol. 109, No. 11, 2012, pp. 4086–4091.
33. Crowley-Koch, B. J., R. Van Houten, and E. Lim. Effects of Pedestrian Prompts on Motorist Yielding at Crosswalks. *Journal of Applied Behavior Analysis*, Vol. 44, 2011, pp. 121–126.
34. Van Houten, R., L. Malenfant, R. D. Blomberg, B. E. Huitema, and S. Casella. *High-Visibility Enforcement on Driver Compliance with Pedestrian Right-of-Way Laws*. National Highway Traffic Safety Administration, August 2013.
35. Schneider, R. J., X. Qin, M. R. R. Shaon, A. Sanatizadeh, Z. He, P. Wkyhuis, B. Block, K. Santiago, and A. Bill. *Evaluation of Driver Yielding to Pedestrians at Uncontrolled Crosswalks*. Prepared by the University of Wisconsin-Milwaukee and University of Wisconsin-Madison Traffic Operations and Safety (TOPS) Laboratory for the Wisconsin Department of Transportation, December 2017.
36. Zegeer, C. V., D. Nabors, and P. Lagerwey. *PEDSAFE: Pedestrian Safety Guide and Countermeasure Selection System*. Federal Highway Administration, August 2013.
37. Tefft, B. Impact Speed and a Pedestrian's Risk of Severe Injury or Death. *Accident Analysis and Prevention*, Vol. 50, 2013, pp. 871–878.
38. Knapp, K., B. Chandler, J. Atkinson, T. Welch, H. Rigdon, R. Retting, S. Meekins, E. Widstrand, and R. J. Porter. *Road Diet Informational Guide*. Federal Highway Administration, November 2014.
39. Sandt, L. S., S. W. Marshall, D. A. Rodriguez, K. R. Evenson, S. T. Ennett, and W. R. Robinson. Effect of a Community-Based Pedestrian Injury Prevention Program on Driver Yielding Behavior at Marked Crosswalks. *Accident Analysis & Prevention*, Vol. 93, 2016, pp. 169–178.

The Pedestrians and Cycles Section (ANF00) peer-reviewed this paper (18-05812).

TRR
JOURNAL OF THE TRANSPORTATION RESEARCH BOARD

Article

Transportation Research Record
2018, Vol. 2672(35) 33–45

DOI: 10.1177/0361198118774182
journals.sagepub.com/home/trr
SAGE

Validation of an Agent-based Microscopic Pedestrian Simulation Model at the Pedestrian Walkway of Brooklyn Bridge

Mohamed Hussein[1] and Tarek Sayed[1]

Abstract

The objective of this study is to validate a recently developed agent-based pedestrian simulation model, using data collected at the pedestrian walkway of Brooklyn Bridge. Video data were collected at the walkway and the trajectories of 294 pedestrians were extracted using computer vision. A genetic algorithm was applied to identify the optimum model parameters that minimize the error between the simulated and the actual trajectories of the calibration dataset. The simulation model was then applied to reproduce the trajectories of 214 pedestrians, considered for validation. The validation results showed that the model was capable of producing pedestrian trajectories with high accuracy, as the average location error between actual and simulated trajectories was for 0.32 m, while the average speed error was 0.06 m/s. Macroscopic results of the model were assessed by comparing the density–speed relationship in both actual data and the simulation. Finally, the accuracy of the model in reproducing the actual behavior of pedestrians during different interactions was evaluated. Results showed that the model was capable of handling these interactions with high accuracy, ranged between 79% and 100%.

Encouraging road users to rely on active modes of transportation, such as walking, has become a fundamental objective for transportation agencies all over the world. Many researchers and transportation professionals have recently shown interest in developing a solid understanding of the behavior of active road users and introducing new planning and design concepts that meet their requirements. Satisfying these objectives requires the development of comprehensive methodologies and innovative tools that enable the analysis of active road users' data, particularly pedestrians, and an understanding of their behavior. Computer simulation of pedestrian dynamics is a promising tool that has shown recent success in analyzing pedestrian behavior in different situations, aiding the design of pedestrian facilities, and investigating scenarios that are difficult to study using actual pedestrian data. Despite its recent success, computer simulation of pedestrian behavior faces many challenges, mainly due to the complex nature of the pedestrian systems. Pedestrian systems encompass a large degree of randomness, and stochasticity. Pedestrians interact with other elements in their system using complicated and sometimes unpredictable rules. Modeling a pedestrian system involves capturing many attributes and decision-making parameters, which are not easy to obtain in most of the cases, to define the behavior of the pedestrians in the model and to calibrate its parameters (*1*).

Recently, an agent-based pedestrian simulation model was developed at the University of British Columbia to study the behavior of pedestrians during interactions with other pedestrians and objects in the walking environment (*2*, *3*). The model focuses on producing accurate pedestrian trajectories that reflect the actual behavior of pedestrians during interactions. Behavior rules that control pedestrian behavior during interactions were extracted from a detailed pedestrian behavior study, which was conducted in the city of Vancouver (*4*). The model was validated using two datasets of pedestrian video data, collected at a typical signalized intersection in the city of Vancouver (*3*), and a scramble phase signalized intersection in Oakland (*5*).

The objective of this study is to apply the microsimulation model to evaluate pedestrian interactions in a new walking environment and thus confirm the model robustness and ensure the fitness of the calibration and

[1]Department of Civil Engineering, The University of British Columbia, Vancouver, BC, Canada

Corresponding Author:
Address correspondence to Mohamed Hussein: m.hussein@civil.ubc.ca

validation methodology introduced previously (*2*, *3*). As such, the model was utilized to simulate pedestrian interactions at part of the walkway of Brooklyn Bridge in New York City. The walkway is part of the separated pedestrian and cyclist path that exists in the center of the bridge, providing access for pedestrians and cyclists to walk in a relaxed environment, exercise, and enjoy this spectacular landmark. The walkway experiences high pedestrian volumes, particularly in the afternoon, which create considerable interactions between pedestrians. In addition, many interactions are observed between pedestrians and cyclists, especially when the density of active road users is high, causing cyclists to ride close to or on the pedestrian walkway to overtake slower bikes on the bike path. The paper presents the details of collecting data required for the calibration and validation, the simulation process, and the validation of the simulated trajectories, the macroscopic results of the simulation model, and the accuracy of the model in reproducing pedestrian behavior during interactions with other road users.

Literature Review

Several microscopic pedestrian simulation models that were developed using different modeling approaches can be found in the literature. Among the existing models, the social force model (SFM) (*6*) and the cellular automata (CA) model (*7*) can be considered the most recognized simulation models that were successfully applied to analyze pedestrian dynamics in different applications. Several adaptations and modifications to the original SFM and CA were introduced to improve their accuracy in handling specific applications (e.g., *8–10*). Other simulation models were also developed based on other modeling approaches and can also be found in the literature (e.g., *11–13*).

Existing simulation models were applied to study various applications related to pedestrians, including evacuation studies (e.g., *14*) and crowd movement in large events (e.g., *15*, *16*). The multi-agent simulation model (legion) (*17*) was adopted by several studies to analyze the movement of large crowds at large events (e.g., *18*, *19*). Pretto et al. developed a simulation model based on a modeling concept that combines the social force approach and the behavior rules concept to analyze pedestrian gap acceptance behavior (*20*). Kim et al. utilized the social force theory to study pedestrian queuing at a cinema ticketing booth (*21*). Hussein and Sayed developed an agent-based pedestrian simulation model for modeling detailed pedestrian interactions (*2*, *3*). The model was applied successfully to address pedestrian interactions at an untypical signalized intersection with a scramble pedestrian phase in Oakland, California (*5*).

The key shortcomings of existing models were documented in several studies, available in the literature. Shiwakoti et al. addressed the shortcomings of the SFM as it does not consider the pedestrians' intelligence and the heterogeneity of the pedestrian system (*22*). Pan reported that the individual variation in pedestrian parameters, particularly the pedestrian walking speed, is difficult to consider for the CA model (*23*). Zhou et al. reviewed existing simulation models and identified the approaches used to calibrate a model's parameters as the major issue associated with current simulations (*24*). According to Duives et al., current simulation models are not sufficient to address pedestrian applications that require both precision in pedestrian trajectories and speed of calculations (*25*).

Methodology

Video data was collected at one section of Brooklyn bridge walk way in New York City. Pedestrian trajectories were automatically extracted from the video footage by means of computer vision. The extracted trajectories were considered the main source of data used to calibrate the model parameters and validate its results. Model parameters were calibrated either directly from the data for the measurable parameters or indirectly using a genetic algorithm (GA) for non-measurable parameters. Once optimum parameters were identified, the characteristics of the studied section and basic pedestrian attributes were entered into the simulation model. Each pedestrian was assigned to an origin and a goal extracted from the corresponding pedestrian in the actual data. Pedestrians were allowed to move toward their destination according to the rules of the simulation model. The simulated trajectories were collected and compared with actual trajectories to evaluate their accuracy. In addition, the accuracy of executing different interactions and the accuracy of the macroscopic results of simulation were also assessed. The following section summarizes the methodology applied in each of the following tasks: data collection, pedestrian trajectory extraction, simulation of pedestrian movement.

Data Collection

Two hours of video data were collected at one section of the pedestrian walkway of Brooklyn Bridge in New York City. Data were collected on September 20, 2016 from 8:00 a.m. to 9:00 a.m. and from 5:00 p.m. to 6:00 p.m. near the west entrance of the bridge, as shown in Figure 1*b*. The Brooklyn Bridge is one of the oldest bridges in the United States, which is considered one the important landmarks of New York City that has both cultural and architectural significance. As shown in

Figure 1. Study location and trajectory extraction process: (*a*) study location; (*b*) ortho image of the studied section; (*c*) camera view; (*d*) a typical pedestrian flow in the morning; (*e*) a typical pedestrian flow in the afternoon; (*f*) raw video footage of the trajectory extraction process; (*g*) camera calibration in the trajectory extraction process; (*h*) feature tracking in the trajectory extraction process; and (*i*) tracked trajectories in the trajectory extraction process.

Figure 1*a*, the bridge spans the East river to provide connection between Manhattan and Brooklyn. In addition to three traffic lanes in each direction, the Brooklyn Bridge has a wide pedestrian/cyclist path, with an average width of 5.5 m, in the center of the bridge (Figure 1*c*). The pedestrian/cyclist path carries more than

4,000 pedestrians and 3,100 bikes every day (*26*). The pedestrian flow varies significantly between the morning and the afternoon. Whereas the morning hours experiences light pedestrian flow crossing the bridge (Figure 1*d*), the pedestrian flow increases significantly in the afternoon, as shown in Figure 1*e*, when locals and visitors walk and bike over the bridge for leisure, sport, and to enjoy the spectacular view. The 2 hours of video data were reviewed to identify the pedestrians who are involved in interactions with each other or with other objects in the walking environment. The data mining resulted in identifying a total of 294 pedestrians for the analysis. Among the selected pedestrians, 80 were considered for the calibration process to match the number of pedestrians considered for the original calibration of the model parameter (*3*), whereas the remaining 214 pedestrians were reserved for the model validation.

Extracting Pedestrian Trajectories

The automatic extraction of pedestrian trajectories was conducted using an automated video processing system, developed at the University of British Columbia. As shown in Figure 1, *f* and *g*, the procedure starts with the camera calibration, in which transformation matrix is generated to create a mapping between the video and the real world coordinate system, as described in detail in (*27*). In the following step (the feature tracking), computer vision algorithms are used to detect any discrete points (features) on moving objects (e.g., pedestrians) in the video scene (Figure 1*h*). Features are identified and tracked using an implementation of the well-known Kanade–Lucas–Tomasi Feature Tracker algorithm, as described in detail in (*28*). Features that are close to each other and have similar movement patterns are then grouped to form objects in the feature-grouping step (Figure 1*i*). Grouped objects are tracked on each video frame, forming the trajectory of road users.

Pedestrian Simulation

The model utilized in this study is as agent-based microscopic simulation model, in which pedestrians are modeled as autonomous agents. Agents are characterized by a set of parameters, attributes, and behavior rules that control pedestrian behavior during different interactions with other pedestrians or other elements in the walking environment. The model handles pedestrian interaction with other pedestrians, either slower pedestrians moving in the same direction (uni-directional interactions) or opposing pedestrians (bi-directional interactions) (*2*, *3*). The model also addresses the behavior of pedestrians interacting with static objects in the walking environment and the behavior of pedestrians walking in groups (*3*). Pedestrian groups interacting with other elements in the environment were classified into strict groups and flexible groups. Strict groups prefer to maintain the unity of the group, so they do not split up during interactions. In contrast, members of flexible groups prefer to keep their relatively high walking speed; they can thus split up and move in different directions while avoiding other pedestrians or elements in the walking environment (*3*). The logic of pedestrian agents' behavior within the simulation is summarized in Figure 2.

Model Parameters' Calibration

Direct Calibration

Parameters that could be measured from the data, mainly, desired speed and preferred lateral distance at crossing, were calibrated directly from the actual trajectories. Desired speed (V_d) is the preferred walking speed for pedestrians when they are walking freely in the walking environment and can be expressed as the average walking speed during segments in which the pedestrians are not interacting with other elements in the walking environment (*3*). Segments at which pedestrians are walking freely were manually identified for each pedestrian, and the average speed of each pedestrian during these segments was considered as the desired speed. The preferred lateral distance at crossing (L) was extracted from 113 crossing situations observed in the data (55 situations in the morning and 58 in the afternoon). As the pedestrian density in the afternoon was significantly higher than the morning, it was observed that the value of L varies among the 2 hours. As such, two sets of values of L were extracted, one for the morning hour and another for the afternoon. The mean value of L was found to be 0.85 m with a standard deviation of 0.22 m in the morning, while in the afternoon the value was significantly lower (0.68 m with a standard deviation of 0.2). Both morning and afternoon values showed a good fit to a normal distribution, which was confirmed by the Chi square test. The model assigns a random value for L to each pedestrian, derived from these two distributions, depending on the time being analyzed.

Indirect Calibration

The calibration of the rest of model parameters that are not measurable was conducted indirectly using a GA (*29*). The goal of the indirect calibration was to select a combination of parameters that minimizes the location and speed error between the actual trajectories, extracted using computer vision, and the simulated trajectories, resulting from the simulation model. The cost function was used to assess the parameter set being examined:

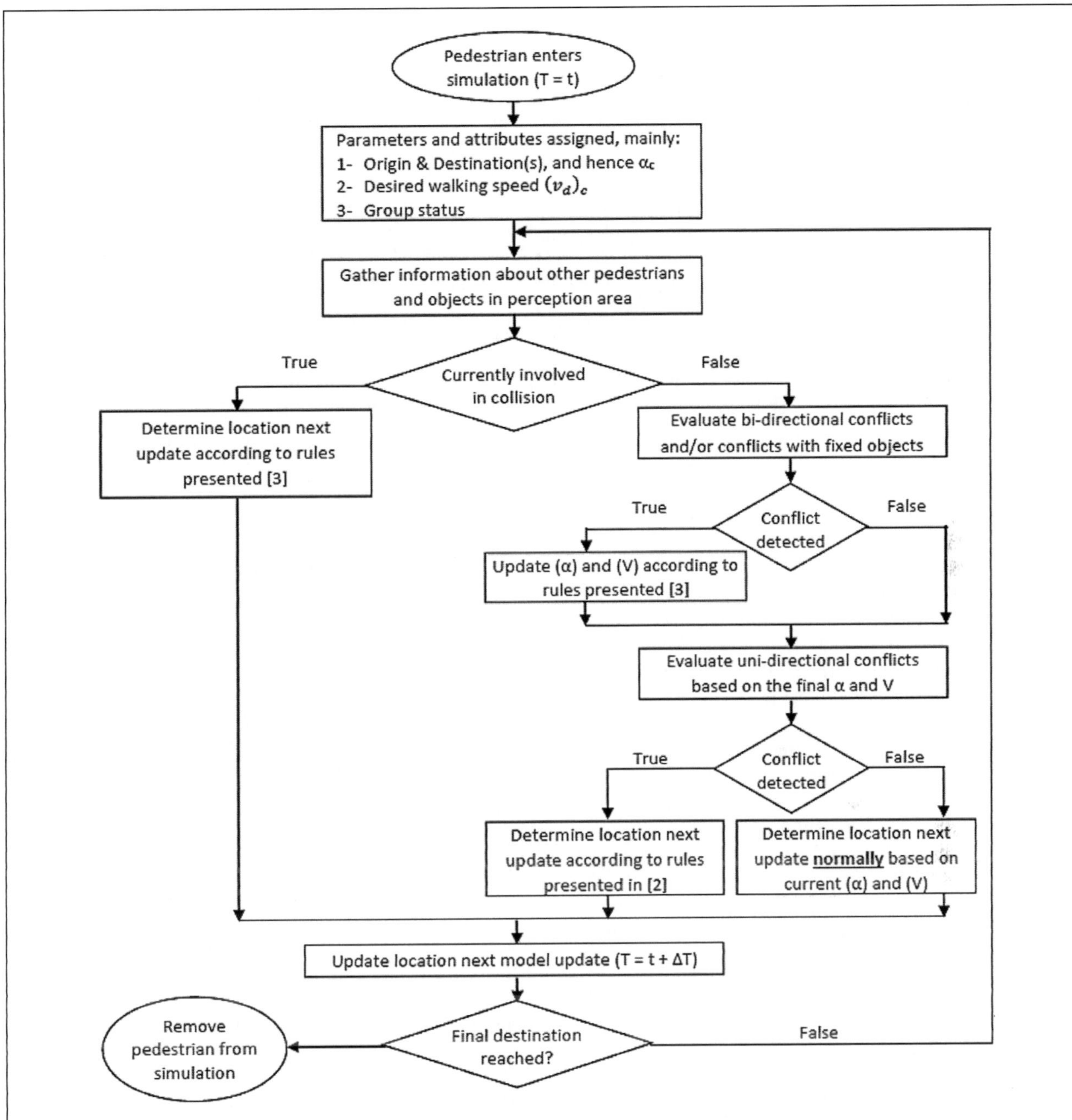

Figure 2. The life cycle of pedestrian agent in the simulation model.

$$C = \left(0.5 \times \frac{\sum_{i=1}^{n} \mathrm{LE}_i}{n}\right) + \left(0.5 \times \frac{\sum_{i=1}^{n} \mathrm{SE}_i}{n}\right) \quad (1)$$

In which, LE_i and SE_i are the average location and average speed error between the actual and simulated trajectory for the ith trajectory, and n is the number of trajectories considered in the calibration process. The whole process is described in much more detail in (*3*). Table 1 summarizes the key parameters of the model and the values resulting from the indirect calibration process. Compared with parameters obtained from pedestrian data on a typical cross walk in Vancouver (*3*), two main differences can be noted. The speed required to overtake is relatively high, particularly for lower density levels,

Table 1. Model Key Parameters, Indirectly Calibrated Using GA

Parameter	Description	Value
Perception area (*A*)	*A* is the area in front of each pedestrian defined by the angle of vision (θ) and a radius (*R*). Pedestrians only consider objects in this area when making decisions	(θ) was set to 170° R = 9.0 m
Speed required to overtake ($Vp_{\|PD}$):	This parameter defines the minimum walking speed that enables a pedestrian to overtake slower pedestrians, according to the pedestrian's perceived density (PD) of the environment.	($Vp_{\|PD}$) = 1.35 m/s ... for all densities
Swerving distance (*D*)	*D* is the distance at which a pedestrian starts to swerve to avoid an opposing pedestrian. According to (*3*), *D* is defined as follows: $D = (d_{speed}{}^{*}V_d) + (d_{group}{}^{*}\ \text{Group}) + (d_{opposing\_group}{}^{*}\ \text{Opposing\_Group})$ where: "Group" and "Opposing_Group" are dummy variables that equal 1 if the current pedestrian and the opposing pedestrian are walking in a group, respectively	D_{speed} = 1 D_{group} = −2.1 $D_{opposing\_group}$ = 1 D_{min} = 1.70 m
Strict Group Threshold (G_{th}):	If the average distance between opposing group members over time (t_g) is less than (G_{th}), the current pedestrian will treat the opposing group as a strict group	(t_g) was set to 1.0 seconds G_{th} = 1.00 m
Opposing pedestrians range (η):	Opposing pedestrians within this range will be considered during interactions with multiple pedestrians	η = 5.0 m

and it is not affected by the changes in walking environment density. In the crosswalk data, this parameter was sensitive to density levels so that different speed values were defined for different density levels. The current value of the parameter suggests that only pedestrians with higher walking speeds decide to overtake slower pedestrians, regardless of the density. In addition, the strict group threshold (G_{th}) is much higher for the current application (1.0 m compared with 0.6 m on the typical crosswalk), which means that pedestrians will not consider going through opposing group members unless the distance between the members of the opposing group is obviously high. Both findings can be explained by the different nature of walking on the bridge (mainly leisure, more relaxed pedestrians) compared with the crosswalk (commuters, more aggressive pedestrians).

Model Validation

Two hundred and fourteen pedestrians selected from the 2 h video data were considered in the validation process. The validation dataset includes 199 pedestrians with an average walking speed of (1.34 ± 0.23) m/s, and 15 joggers with an average running speed of (3.11 ± 0.59) m/s. The 214 pedestrians were simulated using the parameter set obtained during the calibration process. Figure 3, *a* and *b* show the actual trajectories and the corresponding simulated trajectories for the 214 pedestrians considered in the validation process. The following subsections provide a detailed evaluation of the accuracy of the resulting trajectories, as well as a detailed evaluation of both the macroscopic results of the model and its accuracy in reproducing the actual behavior during different interactions.

Overall Trajectory Evaluation

The overall assessment of simulated trajectories showed that the model is capable of reproducing pedestrian trajectories with very high accuracy. On average, the error between the actual and simulated trajectories was assessed to be 0.32 m. Figure 3*c* shows examples of selected pedestrian trajectories and the corresponding simulated trajectories. The figure confirms the high accuracy of the simulated trajectories in terms of the closeness to the actual trajectories and the ability to execute the same decision taken by the pedestrian in the actual data (e.g., when to swerve to resolve a conflict). The average location error for the shown examples ranges between 0.06 and 0.20 m, as shown in Figure 3*c*.

Speed Validation

The model showed high accuracy in simulating the pedestrian walking speed, with an average error in of just 0.06 m/s. The accuracy of the resulting pedestrian simulated speed was assessed by plotting the average crossing speed of each pedestrian against the corresponding average crossing speed of the actual trajectory, as shown in Figure 4. As shown in the figure, the relation between the actual and the predicted walking speed is very accurate, compared with an ideal 45° line (the root mean square error = 0.096).

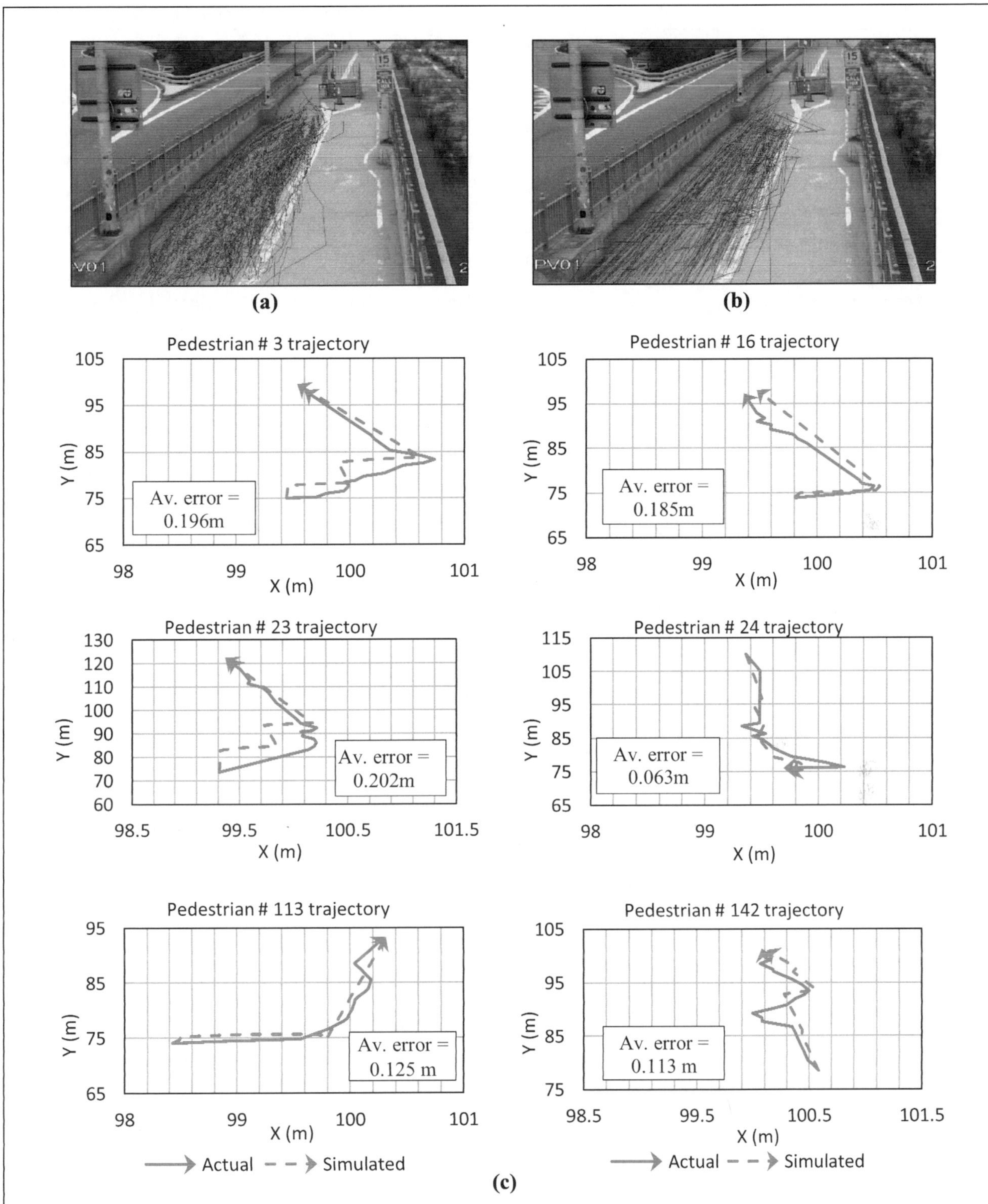

Figure 3. Overall accuracy of simulated trajectories: (*a*) actual trajectories of pedestrians considered in the validation process; (*b*) corresponding simulated trajectories; and (*c*) examples of pedestrian actual and simulated trajectories

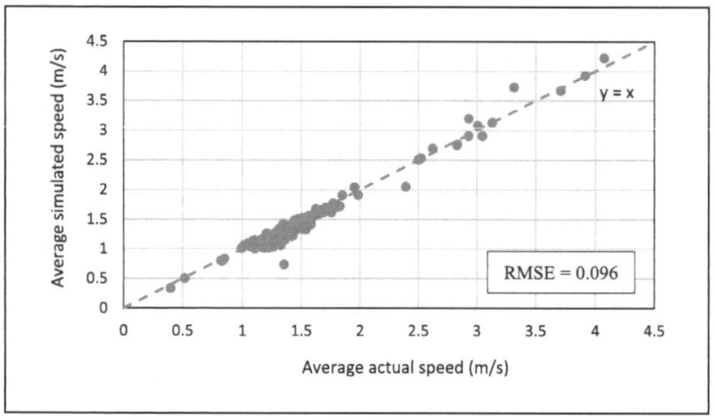

Figure 4. Pedestrian speed validation.

Macroscopic Validation

The macroscopic performance of the model was evaluated by comparing the density–speed relationship resulted from the simulation to the actual density-speed relationship observed in the data. The number of pedestrian existed in the data extraction area, shown in Figure 5*a*, was automatically estimated each 2 s in both the actual and simulation data. For each pedestrian in the density extracted area, the speed in the previous second was calculated to represent the current pedestrian speed at the current density level. The speed of all pedestrians in the calculation area at the current time was averaged and was considered to represent the speed corresponding to the current density. The resulting density–speed relations in both actual data and simulation are presented in Figure 5, *b* and *c*, respectively, and the both relations are compared in Figure 5*d*. As shown in the figure, the density–speed relation, resulting from the simulation model, is accurate compared with the actual relation (average error = 9%). The actual and simulated curves have a higher discrepancy at lower density levels, as pedestrians tend to walk with their free walking speed. The error in this part of the curve could be attributed more to the error associated with estimating the free walking speed. As density increases, more interactions begin to occur between pedestrians, and the choice of speed becomes more dependent on the interaction rules of the model. It can be observed that at higher density levels, the average speeds in both actual data and simulation are almost identical.

Validation of Interaction Rules

In addition to the overall accuracy of the simulated trajectories and the average values of walking speed, the accuracy of simulating the behavior of pedestrian during interactions with other pedestrians and objects in the walking environment was also investigated. The simulated interactions were considered accurate if they were executed in the model similar to the corresponding behaviors observed in the actual data. Considered interactions include pedestrian interactions with static objects (e.g., stopping pedestrians), moving objects (e.g., bikes), slower pedestrians moving in the same direction (uni-directional interactions), opposing pedestrians (bi-directional interactions), and the behavior of pedestrians walking in groups. Results showed that the model was capable of reproducing the actual pedestrian behavior during different interactions with a very good accuracy that ranges from 79% and 100%. The details of the evaluation of each interaction are provided in the following section.

Bi-directional Interactions

In the validation dataset, 445 interactions between opposing pedestrians (or groups of pedestrians) were observed. Pedestrians involved in this interaction have to swerve to avoid collision with the opposing pedestrian. The interaction was considered to be simulated correctly if the pedestrians involved in the conflict swerved to the same side observed in the actual interaction. The lateral distance between each pair of pedestrians at the moment of crossing was considered as a measure of accuracy for simulating this interaction. Results showed that the model correctly predicted the side of swerving in 411 incidents, (simulation accuracy = 90.3%). The average values of the lateral distance at crossing were estimated to be 1.13 and 1.25 m in both the simulation and actual data, respectively. Figure 6*a* presents the actual and simulated lateral distance for the correctly simulated interactions. A positive value of the lateral distance indicates that the pedestrians involved in the interaction swerve to the right side to resolve the conflict, while a negative value indicates a swerve to the left side. As shown in the figure, the simulated lateral distance showed a good fit with the actual values for the interactions investigated.

Uni-directional Interactions

Pedestrians involved in a uni-directional interaction may resolve the conflict using two strategies. Faster pedestrians can change the direction of movement to overtake the leading pedestrians or they may reduce speed and follow the slower leading pedestrians, especially if the walking environment's density is high and it is difficult for the faster pedestrian to change movement direction and walking speed (*4*). The uni-directional interactions were observed 55 times during the time analyzed. Pedestrians involved in uni-directional interactions chose to follow slower pedestrians in 41 cases, while they overtook the

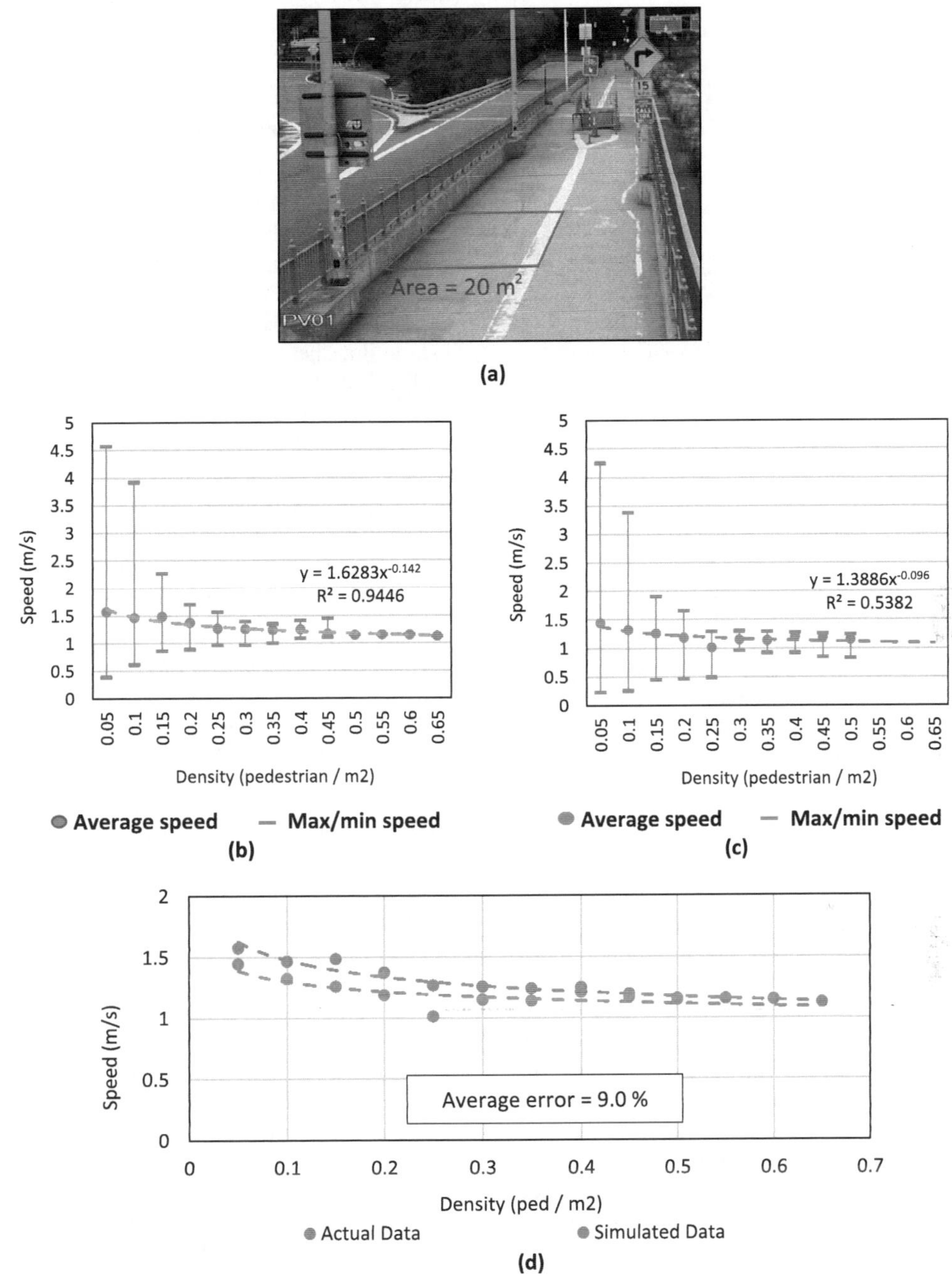

Figure 5. Macroscopic validation of the model: (*a*) density extraction area; (*b*) density–speed relationship in actual data; (*c*) density–speed relationship in simulation; and (*d*) comparison between density–speed relationship in both actual and simulated data.

leading pedestrian in 14 cases. The model correctly predicted the correct conflict resolution strategy in 52 cases, which represents an accuracy of 94.5 %. Details of the evaluation of the two conflict resolution strategies are presented below.

Following Behavior. Pedestrians in the simulation model chose to follow slower leading pedestrians in all situations where pedestrians in the actual data chose the same strategy (accuracy of executing this behavior = 100%). However, the average following distance and the average

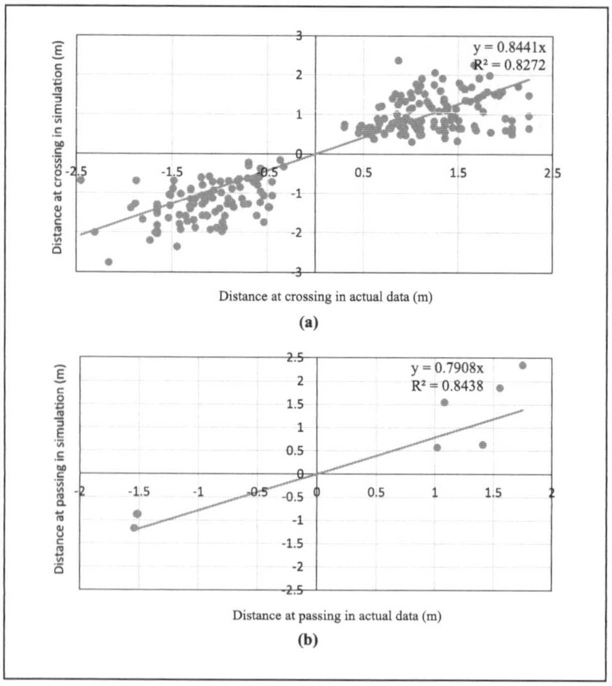

Figure 6. Evaluation of uni-directional and bi-directional interactions: (*a*) actual and simulated distance at crossing for bi-directional interactions and (*b*) actual and simulated distance at crossing for passing maneuvers.

time spent by a pedestrian following a slower pedestrian resulting from the simulation was lower than the actual values. The average following distance in both simulation and actual data was 1.00 and 2.15 m, respectively. The average following time was 3.3 s in the simulation while the average value in the actual data was 6 s. This difference is mainly attributed to an untypical behavior executed by six pedestrians (two groups of three pedestrians each). The faster group in the actual data reduced speed significantly and followed the leading group at a relatively far distance from the leading group. This behavior was different from all other following behaviors observed in the data, where pedestrians come close to the slower pedestrians before reducing speed and deciding to follow.

Overtaking Behavior. This conflict resolution strategy was only executed in 14 uni-directional conflicts. The passing maneuvers were executed in 11 cases in the model (accuracy = 78.6%), whereas in 3 cases, the pedestrian in the simulation chose to follow slower pedestrians instead of overtaking them. The lateral distances at the moment of passing for the 11 cases are presented in Figure 6*b*. As shown in the figure, the model was capable of predicting the passing distance with a good accuracy. It should be acknowledged that the number of interactions for this behavior was quite low to draw a statistically significant evaluation.

Interaction with Static Objects

Interactions with static objects involved interactions between pedestrians and stopping pedestrians who stopped to take photos or enjoy the view of the bridge, as shown in Figure 7*a*. In addition, pedestrians were also involved in interactions with various objects like a food cart serving food in the afternoon period (Figure 7*a*). Throughout the analysis, this interaction was observed in 79 incidents. The simulation model successfully predicted the side of the maneuver in 69 cases, which represent an accuracy of 87.3%. Figure 7*b* shows the actual and simulated values of the lateral distance between the pedestrian involved in the conflict and the static object at the moment of crossing. As shown in the figure, the accuracy of predicting the crossing distance is quite high and the data regression line is close to an ideal 45° line.

Interaction with Moving Objects

Throughout the analysis, pedestrians periodically interacted with moving objects, especially bikes that moved close to pedestrian paths and forced pedestrians to take action to avoid collision (Figure 7*c*). Furthermore, an interaction between four pedestrians and a cleaning vehicle that crossed the study area was also observed (Figure 7*c*). A total of 13 pedestrians were involved in this interaction. Pedestrians in the actual data changed movement direction to keep a safe distance with the moving object. Since the current model focuses on pedestrian behavior, the speed and acceleration profile of all moving objects were extracted from the actual trajectories and the moving objects were then introduced to the simulation and followed the exact path of the actual object. Pedestrians involved in the interaction were allowed to respond to the moving vehicle, according to the behavior rules. The model successfully predicted the correct side of swerving to avoid collision with the moving objects in all 13 cases. The actual and simulated values of the distance between pedestrian and moving objects at crossings are shown in Figure 7*d*.

Group Behavior Validation

The validation dataset includes 48 pedestrian groups that were involved in 71 interactions. As described earlier, pedestrian groups interacting with other elements in the environment may act as a strict group or a flexible group. The simulation model was successful in predicting the correct behavior of the group in 65 interactions of the 71

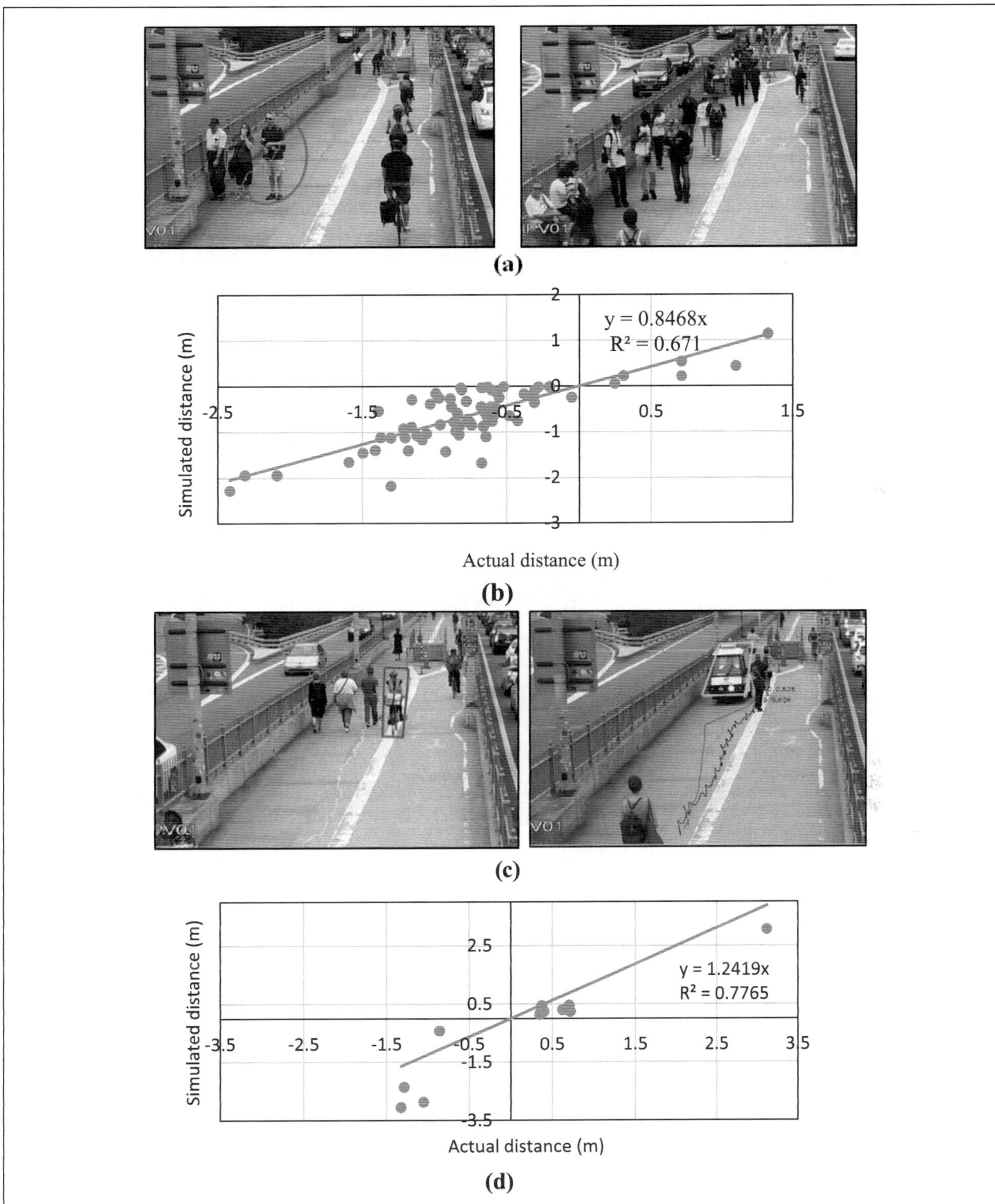

Figure 7. Evaluation of pedestrian interaction with static and moving objects: (*a*) examples of interactions with static objects; (*b*) actual and simulated lateral distance between pedestrians and static object at crossing, (*c*) examples of interactions with moving objects; and (*d*) actual and simulated lateral distance between pedestrians and moving object at crossing.

interactions observed (overall accuracy = 91.5%). In addition, the average distance between group members was estimated for the 48 groups in both actual and simulated trajectories. Results showed that the average distance resulting from the simulation model is accurate compared with actual data, with an average distance of 1.0 and 1.02 m for the actual data and the simulation, respectively.

Discussion and Future Research

The paper provided the details of the application of an agent-based pedestrian simulation model (*2*, *3*) to address pedestrian interactions on a part of the pedestrian walkway of Brooklyn Bridge in New York City. The walkway represents a walking environment where pedestrian behavior is expected to be significantly different from the typical behavior at signalized intersections addressed in previous validations of the simulation model (*3*, *5*). The data collected at the study location provided an opportunity to consider pedestrian interactions that were not addressed before, including pedestrian interactions with cyclists who cycle on the pedestrian path and interactions with slower pedestrians who walk slowly or stop to chat and take pictures in a touristy fashion.

Using computer vision techniques, 294 pedestrian trajectories were automatically extracted from video data to be used to calibrate the model parameter and validate its results. Measurable model parameters were directly calibrated by extracting their values from actual trajectories. Non-measurable parameters were indirectly calibrated using GA that aims at minimizing the error between actual and simulated trajectories. Once optimum parameters were identified, the model was applied to reproduce the trajectories of pedestrians in the validation dataset. Pedestrians interacted with each other and with other elements in the walking environment using predefined behavior rules. The simulated trajectories were collected and compared with actual trajectories to evaluate their accuracy. The average location error was found to be 0.32 m whereas the average speed error was 0.06 m/s. In addition, the accuracy of the macroscopic characteristics of the simulated pedestrians (density–speed relationship) was assessed to be very high compared with the corresponding characteristics of the actual data (accuracy = 91%). Furthermore, the model was capable of reproducing the actual behavior adopted by pedestrians during different interactions with an accuracy that ranges from 79% to 100%. The validation results showed that the model is accurate enough to be used to address many pedestrian applications that require solid understanding of pedestrian behavior and accurate representation of pedestrian trajectories.

Future research includes considering larger data sets that cover a variety of walking environments and pedestrian densities to further confirm the validity of model results in different conditions. As well, it is important to address pedestrian interactions with other road users in more detail. Developing the interaction rules that define pedestrian interactions with cyclists and vehicles will expand the applicability of the model to address several applications including pedestrian safety (e.g., traffic conflict studies) and pedestrian behavior at shared space facilities (e.g., complete streets). Despite the relative success of existing models in addressing some of these applications, the current model should provide more accurate results as it showed very high accuracy in producing individual trajectories and interaction behavior. In addition, it was developed based on a modelling approach that is appropriate to consider the complexity of the pedestrian system and rely on realistic interaction rules and calibration and validation methodologies.

Author Contributions

The authors confirm contribution to the paper as follows: study conception and design: Hussein and Sayed; data collection: Hussein and Sayed; analysis and interpretation of results: Hussein and Sayed; draft manuscript preparation: Hussein and Sayed. All authors reviewed the results and approved the final version of the manuscript.

References

1. Helbing, D. *Social Self-Organization: Agent-Based Simulations and Experiments to Study Emergent Social Behavior*. Springer, Berlin-Heidelberg, 2012.
2. Hussein, M., and T. Sayed. A Uni-Directional Agent Based Pedestrian Microscopic Model. *Canadian Journal of Civil Engineering*, Vol. 42, No. 12, 2015, pp. 1114–1124.
3. Hussein, M., and T. Sayed. A Bi-Directional Agent Based Pedestrian Microscopic Model. *Transportmetrica A: Transport Science*, Vol. 13, No. 4, 2017, pp. 326–355.
4. Hussein, M., and T. Sayed. Microscopic Pedestrian Interaction Behavior Analysis Using Gait Parameters. *Transportation Research Record: Journal of the Transportation Research Board*, 2015. 2519: 28–38.
5. Hussein, M., and T. Sayed. Validation of an Agent-Based Microscopic Pedestrian Simulation Model at a Scramble Phase Signalized Intersection. *Transportation Research Record: Journal of the Transportation Research Board*, 2017. 2661: 30–42.
6. Helbing, D., and P. Molnar. Social Force Model for Pedestrian Dynamics. *Physical Review E*, Vol. 51, No. 5, 1995, pp. 4282–4286.
7. Blue, V. J., and J. L. Adler. Cellular Automata Microsimulation of Bidirectional Pedestrian Flows. *Transportation Research Record: Journal of the Transportation Research Board*, 1999. 1678: 135–141.
8. Xi, H., S. Lee, and Y. J. Son. An Integrated Pedestrian Behavior Model Based on Extended Decision Field Theory

and Social Force Model. In *Human-in-the-Loop Simulations*, Springer, London, 2011, pp. 69–95.
9. Schadschneider, A. Cellular Automaton Approach to Pedestrian Dynamics-Theory. *arXiv preprint cond-mat/0112117*, 2002.
10. Sarmady, S., F. Haron, and A. Z. Talib. Simulating Crowd Movements Using Fine Grid Cellular Automata. *Proc., 12th International Conference on Computer Modelling and Simulation (UKSim)*, Cambridge, IEEE, New York, 2010, pp. 428–433.
11. Hoogendoorn, S. P., and P. H. L. Bovy. Pedestrian Route-Choice and Activity Scheduling Theory and Models. *Transportation Research Part B: Methodological*, Vol. 38, No. 2, 2004, pp. 169–190.
12. Wong, S. C., W. L. Leung, S. H. Chan, W. H. Lam, N. H. Yung, C. Y. Liu, and P. Zhang. Bidirectional Pedestrian Stream Model with Oblique Intersecting Angle. *Journal of Transportation Engineering*, Vol. 136, No. 3, 2010, pp. 234–242.
13. Robin, Th., G. Antonini, M. Bierlaire, and J. Cruz. Specification, Estimation and Validation of a Pedestrian Walking Behavior Model. *Transportation Research Part B: Methodological*, Vol. 43, No. 1, 2009, pp. 36–56.
14. Helbing, D., I. J. Farkas, P. Molnar, and T. Vicsek. Simulation of Pedestrian Crowds in Normal and Evacuation Situations. *Pedestrian and Evacuation Dynamics*, Vol. 21, No. 2, 2002, pp. 21–58.
15. Batty, M., J. Desyllas, and E. Duxbury. Safety in Numbers? Modelling Crowds and Designing Control for the Notting Hill Carnival. *Urban Studies*, Vol. 40, No. 8, 2003, pp. 1573–1590.
16. Klüpfel, H. The Simulation of Crowds at Very Large Events. In *Traffic and Granular Flow'05* (Schadschneider, A., T. Pöschel, R. Kühne, M. Schreckenberg, and W. Wolf, eds.), Springer, Berlin Heidelberg, 2007, pp. 341–346.
17. Berrou, J. L., J. Beecham, P. Quaglia, M. A. Kagarlis, and A. Gerodimos. Calibration and Validation of the Legion Simulation Model Using Empirical Data. In *Pedestrian and Evacuation Dynamics*, Springer, Berlin, Heidelberg, 2005, pp. 167–181.
18. Yue, Z., W. Jiangyan, D. Huang, and S. U. N. Shengyang. Pedestrian Simulation Modeling for World Expo 2010 Shanghai. *Journal of Transportation Systems Engineering and Information Technology*, Vol. 9, No. 2, 2009, pp. 141–146.
19. Monteleone, M., N. Veraart, S. Chough, and W. Aviles. Pedestrian Simulation Modeling Study for World Trade Center Memorial. *Transportation Research Record: Journal of the Transportation Research Board*, 2008. 2073: 49–57.
20. Pretto, C. O., H. B. B. Cybis, and A. Jacobsen. A Multi-layer Simulation Model for Vehicle and Pedestrian Interaction. Presented at 90th Annual Meeting of the Transportation Research Board, Washington, D.C., 2011.
21. Kim, I., R. Galiza, and L. Ferreira. Modeling Pedestrian Queuing Using Micro-Simulation. *Transportation Research Part A: Policy and Practice*, Vol. 49, 2013, pp. 232–240.
22. Shiwakoti, N., M. Sarvi, and G. Rose. Modelling Pedestrian Behaviour Under Emergency Conditions–State-of-the-art and Future Directions. In *31st Australasian Transport Research Forum*, Australia, 2008, pp. 457–473.
23. Pan, X. *Computational Modeling of Human and Social Behaviors for Emergency Egress Analysis*. Doctoral dissertation, Stanford University, 2006.
24. Zhou, S., D. Chen, W. Cai, L. Luo, M. Y. H. Low, F. Tian, V. S. H. Tay, D. W. S. Ong, and B. D. Hamilton. Crowd Modeling and Simulation Technologies. *ACM Transactions on Modeling and Computer Simulation (TOMACS)*, Vol. 20, No. 4, 2010, p. 20.
25. Duives, D. C., W. Daamen, and S. P. Hoogendoorn. State-of-the-art Crowd Motion Simulation Models. *Transportation Research Part C: Emerging Technologies*, Vol. 37, 2013, pp. 193–209.
26. NYCdot [Online]. http://www.nyc.gov. Accessed July 17, 2017.
27. Ismail, K., T. Sayed, and N. Saunier. A Methodology for Precise Camera Calibration for Data Collection Applications in Urban Traffic Scenes. *Canadian Journal of Civil Engineering*, Vol. 40, No. 1, 2013, pp. 57–67.
28. Saunier, N., and T. Sayed. A Feature-Based Tracking Algorithm for Vehicles in Intersections. *Proc., 3rd Canadian Conference on Computer and Robot Vision*, Québec, IEEE, New York, 2006, p. 59.
29. Golberg, D. E. *Genetic Algorithms in Search, Optimization, and Machine Learning*. Addion Wesley, Reading, Mass., 1989.

The Pedestrians and Cycles Section (ANF00) peer-reviewed this paper (18-01820).

Article

TRR
JOURNAL OF THE TRANSPORTATION RESEARCH BOARD

Transportation Research Record
2018, Vol. 2672(35) 46–57

DOI: 10.1177/0361198118780708
journals.sagepub.com/home/trr
SAGE

Assessing the Effect of Pedestrians' Use of Cell Phones on Their Walking Behavior: A Study Based on Automated Video Analysis

Rushdi Alsaleh[1], Tarek Sayed[1], and Mohamed H. Zaki[1]

Abstract

The objective of the study is to assess the effect of the use of cell phones while walking at urban crosswalks. The methodology uses recent findings in health science concerning the relationship between tempo-spatial characteristics of gait and the cognitive abilities of pedestrians. Gait measures are shown to be affected by the complexity of the task (e.g., talking and texting) performed during walking. This study focuses on the effect of distraction states, distraction types (visual such as texting/reading and auditory such as talking/listening), and pedestrian-vehicle interactions on the gait parameters of pedestrians at crosswalks. Experiments are performed on a video data set near a college campus in the city of Kamloops, British Columbia. The analysis relies on automated video-based data collection using a computer vision technique. The benefits of such an automated system include the ability to capture the natural movement of pedestrians and minimizing the risk of disturbing their behavior. Results show that pedestrians distracted by texting/reading (visually) or talking/listening (auditory) while walking tend to reduce and control their walking speed by adjusting their step length or step frequency, respectively. Pedestrians distracted by texting/reading (visually) have significantly lower step length and are less stable in walking. Distracted pedestrians involved in interactions with approaching vehicles tend to reduce and control their walking speeds by adjusting their step frequencies. This research can find applications in pedestrian facility design, modeling and calibrating pedestrian simulations, and pedestrian safety intervention programs and legislative actions.

The growing use of personal electronic devices during commuting has led, in recent decades, to a surge in distracted road user behavior. It is estimated that around 25% of road collisions involved distracted road users (*1*). While the danger of distracted drivers dominates public discussion, concerns have recently been raised about the risk of talking and texting while crossing vehicular traffic. Safety concerns attributed to pedestrian distraction are more evident in the younger population and college students than in other segments of the population. Young adults are frequently involved in using their smartphones while undertaking other activities such as commuting. It is estimated that between 2000 and 2011, more than 5,000 people under the age of 40 were treated for injuries related to distraction involving cell phone usage, making it a significant safety problem (*2*).

The limiting effect that the phone usage has on the cognitive functions of pedestrians is attributed to the inability of a pedestrian to manage more than one involved task at any given time. Dual-tasking competes for cognitive resources of pedestrians, which might lead to the prioritization of one task over another. When phone usage is prioritized, the pedestrian walking mechanism as well as the navigating decision procedure are affected. This dual-task state (e.g., texting and walking) therefore reduces situation awareness, limits the attention to possible surrounding hazards, and may increase unsafe behavior when crossing roads. Using cell phones while walking also limits the visual sensory of pedestrians required for controlling gait balance, resulting in abnormal walking function. This may lead to problems such as falling and musculoskeletal injuries (*3–5*). This effect was more evident for texting/reading as compared with phone conversing (*6,7*).

For distracted walking, developing safety intervention programs are hindered by limitations in data availability

[1]Department of Civil Engineering, University of British Columbia, Vancouver, BC, Canada

Corresponding Author:
Address correspondence to Mohamed H. Zaki: mzaki@civil.ubc.ca

and gaps in research related to the use of cell phones while crossing vehicular traffic. Such limitations are often attributed to the lack of means to collect the necessary data about the walking behavior and to develop necessary tools to analyze such data and derive the necessary knowledge about the extent and effect of cell phone-based distraction during walking. More precisely, collecting walking behavior is a challenging task when performed manually. Identifying the gait attribute along the pedestrian travel can be cost prohibitive if performed manually. Video-based analysis enables such processes to be performed automatically and accurately. Significant benefits of the automated system include the ability to capture the natural movement of pedestrians and minimizing the risk of disturbing their behavior. The richness of the data that can be extracted and the relatively high accuracy, consistency, and low costs make video analysis one of the prominent tools in modern transportation systems (*8*).

Limited studies have investigated the walking behavior of distracted pedestrians during different types of distraction (auditory distraction such as talking/listening by cell phone or visual distraction such as texting/reading by cell phone) and how their walking behavior and gait characteristics are affected during their interaction with vehicle traffic, especially young adult or college student pedestrians. The use of video technology for pedestrian distraction data collection has still not been explored in detail. This paper expands on earlier work on the automated analysis of walking behavior and gait parameters of distracted pedestrians to include analysis of the walking behavior, gait parameters, and gait variability and stability measures for different types of distracted pedestrians (*9*). This study focuses on the distraction caused by using cell phones while walking, for example, texting/reading or talking/listening while walking as examples of the manual-visual (phone held out in front) or manual-auditory (phone to ear) distraction types, respectively. The visual distraction involves any distraction that limits the visual ability of pedestrians, such as texting or reading using a cell phone. Similarly, the auditory distraction involves any types of distraction that requires cognitive processing of the sound such as talking or listening using a cell phone. Therefore, this paper has the following objectives:

- Expand the application of computer vision to the analysis of walking behavior, gait adaptation measures, and gait variability of distracted pedestrians.
- Demonstrate the effect of factors such as distraction state, distraction type (talking/listening or texting/reading), and pedestrian-vehicle interaction on the walking behavior, gait adaptation measures, and gait variability using a real case study near a college campus in Kamloops, British Columbia.

Literature Review

According to the British Columbia Injury Research and Prevention Unit, pedestrian error/confusion contributes to about 31% of pedestrian fatalities in motor vehicle crashes involving pedestrians (*10*). Distraction is considered as the leading contributing factor in motor vehicle crashes involving pedestrians. In the city of Seattle, Washington, observations at 20 intersections with the highest number of pedestrian injuries show that around 30% of all pedestrians crossing roads performed distracted activity including the use of headphones (11%), text-messaging (7%), and using a phone for conversation (6%). Young pedestrians constituted most of the distracted pedestrians (*11*). The effect of distraction on pedestrian safety was simulated in a semi-immersive virtual pedestrian environment (*12*). In this experiment, it was observed that distracted pedestrians have a higher risk of being involved in crashes with vehicle traffic while crossing roads compared with non-distracted pedestrians. Distracted pedestrians were also more likely to pay less attention to vehicle traffic and surrounding environment (*12*). Distracted pedestrians tend to miss safer opportunities to cross roads, take longer to initiate crossing when a safe gap is available, and pay less attention to vehicle traffic (*13*). In the U.S., the estimated injuries of pedestrian related to mobile phones in 2010 exceeded the estimate for drivers (1,162 injuries). However, the limited published research on the impact of using cell phones while walking constrained the development of pedestrian safety programs and associated legislation (*14*).

The prevalence of mobile device usage while walking has been raised, particularly among teens and young adults. A study that observed 34,325 school students crossing roads in a school zone found that one in five high school students and one in eight middle school students were distracted. Text-messaging was the most common distraction (39% of pedestrians), followed by talking on the phone (20%). Also, it was observed that females are more likely to walk while distracted (17% of all pedestrian) than males (14%) (*15*). A higher rate of distraction was observed among students at urban college campuses. Well et al. showed that over 40% of the students were distracted while crossing roadways (*16*). Headphones were the most common distraction (46% of distracted pedestrians), followed by text-messaging (19%) and talking on cell phones (12%).

Gait parameters are considered a useful biometric for human recognition. Pedestrians have a unique gait rhythm during walking (*17,18*). Given this unique characteristic, it was possible to use video analysis to identify

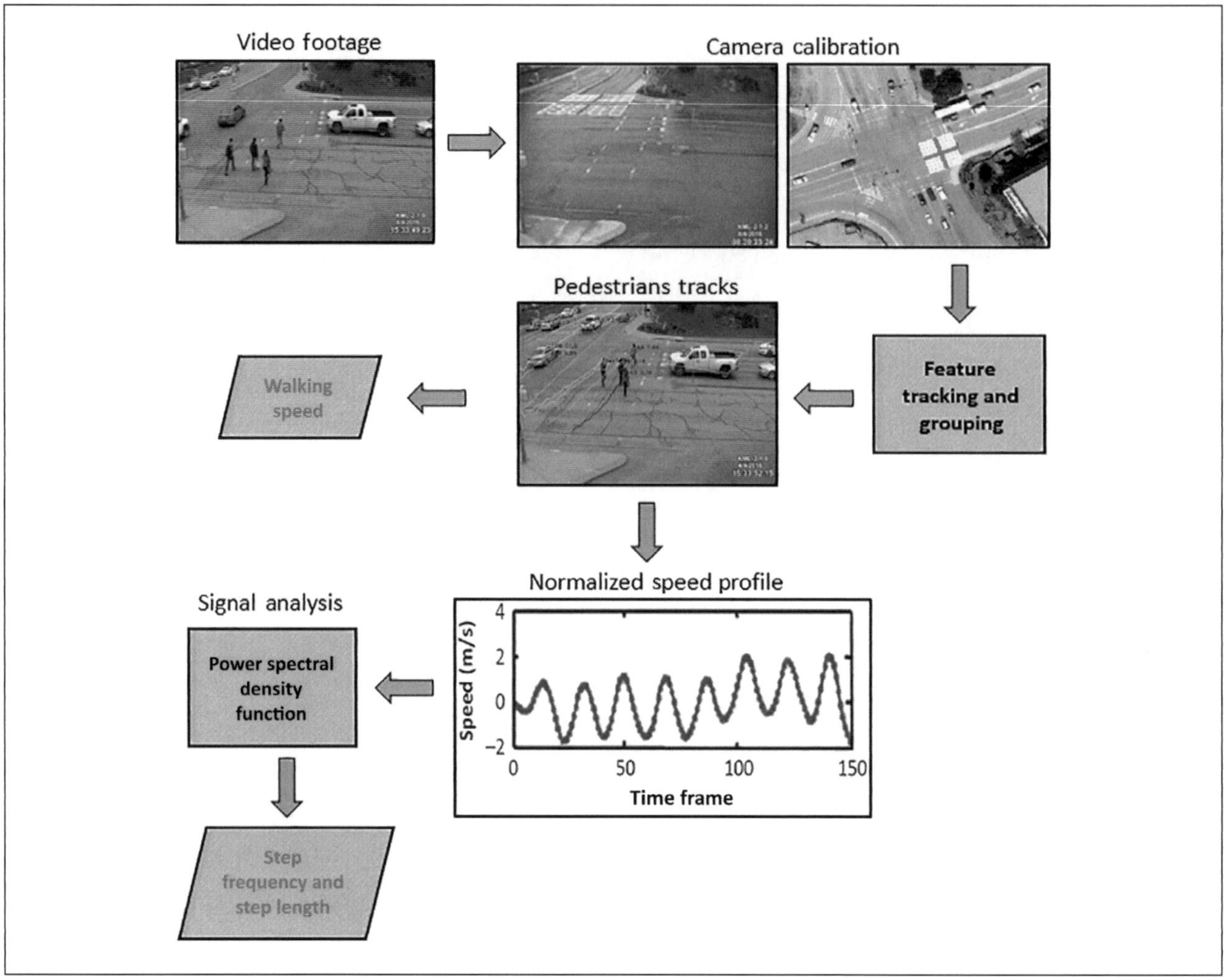

Figure 1. Process of tracking pedestrian and gait parameters extraction.

pedestrians and predict attributes such as their age and gender (*19–22*). Variations in the gait parameters (e.g., variations in the step length) can also reveal information about the health and mental state of pedestrians (*23–25*). For example, inconsistent step length can indicate a problem in the cognitive function of the individual. Further, the amount of variability in the walking pattern differs across age and gender. Gait variability measures and gait parameters were exploited by a computer vision application to detect distracted pedestrians in crosswalks (*9*).

Distraction affects the walking behavior and alters postural control (*26*). Such an effect was studied in different control experiments (*5, 27–29*). It was observed that pedestrians tend to change their gait pattern when walking while being distracted. Distracted pedestrians tend to have a slower gait speed (*30, 31*), a shorter step length, a lower stride frequency, and a higher deviation from gait path compared with non-distracted pedestrians.

Methodology

Pedestrian Tracking

Pedestrian trajectories and speed profiles are obtained from video footage using a video analysis system that automatically detects, tracks, and classifies road users in traffic scenes (Figure 1). Knowledge of the pedestrian spatial and temporal information relies on a mapping of video image plane coordinates and real-world coordinates. Camera calibration achieves this task. Camera calibration ensures the precise construction of the road user movement trajectories and then speed profiles in real-world coordinates. The effect of different viewing angles or parallax is minimal, and the accuracy of the system has been validated in several previous studies (*32*). A detailed description of the adopted camera calibration algorithm is presented in Ismail et al. (*32*). The detection and tracking of pedestrians are conducted using a computer vision algorithm [Kanade–Lucas–Tomasi Tracker (KLT)]. Details of the feature tracking algorithm can be found in Saunier and Sayed (*33*). The algorithm can differentiate between features that belong to road users and features that are part of the environment. The KLT algorithm can address the partial occlusion problem in which only some of the features of the moving object remain visible. The algorithm enables the continuous tracking of road users while continuing to track partially hidden objects.

In the tracking process, a road user can have multiple features. Clustering of features is necessary to determine which features belong to the same road user. The grouping (clustering) algorithm adopted in this study is implemented in Saunier and Sayed (*33*). The main cues for clustering tracked features are the spatial proximity and dynamical similarities. Differences in features' motion vectors are computed, and features with motion vectors differing by a specific threshold are not assumed to belong to the same object and are not clustered, regardless of their spatial proximity. The subsequent step in the analysis is the classification of road users. To discriminate between the trajectories of pedestrians and vehicles, a similarity-based approach is adopted. The main cues for classifying road users include the maximum speed and the frequency harmonics extracted from the speed profile (*20*). The accuracy of the system has been validated in several previous studies, and the tracking accuracy was satisfactory high (*8,20,32*).

Extraction of Gait Parameters

Pedestrian trajectories capture the movement of each pedestrian in the form of a sequence of spatial coordinates at each frame (1/30 s). A pedestrian trajectory is defined along the trajectory lifetime (*n* video frame) as a finite set of tuples (*32*):

$$T = \{(X_1, Y_1, V_{X_1}, V_{Y_1}, \ldots, X_i, Y_i, V_{X_i}, V_{Y_i}, \ldots, X_n, Y_n, V_{X_n}, V_{Y_n})\} \quad (1)$$

where $i = \{1, \ldots, n\}$, a discrete temporal index, X_i, Y_i are the spatial coordinates of the pedestrian at any frame (i), and V_{X_i}, V_{Y_i} are the corresponding velocities. A speed profile for each pedestrian is defined along the trajectory lifetime as $S = norm(Vx, Vy)$, with Vx and Vy as the velocity vectors of length n, for the X and Y coordinates, respectively. Each pedestrian step is observed to introduce a cyclic fluctuation in the speed profile, as shown in Figure 1 (*34*). This means that each cycle in the speed profile corresponds to a new step taken by the pedestrian. Therefore, the step frequency can be estimated from the reciprocal cycle (*19*).

The speed signal is first smoothed and then normalized by subtracting the mean speed from the instantaneous speed. Identifying step frequency includes detecting dominant periodicity in the speed signal. The power spectral density (PSD) is used to determine the step frequency of the speed profile. PSD is a quantitative measure that shows the strength of variation in frequencies of the speed signal; that is, the mean square of the Discrete Fourier Transform of the signal. The PSD estimate along n number of video frames for each frequency f is given by:

$$PSD = \frac{1}{F_s n} \left| \sum_{i=1}^{n} S_i . e^{-\frac{j(2\pi f)i}{F_s}} \right|^2 \quad (2)$$

where
F_S = sampling frequency,
f = frequency of power spectral density, and
j = imaginary unit.
The process of computing the gait parameters is shown in Figure 1. The speed profile of each pedestrian is used to compute the average step frequency for that pedestrian. The spatial coordinate of each pedestrian across time is used to compute the average walking speed. The average step length is then calculated from the following fundamental linear relationship (*35*):

$$Walking\ Speed = Step\ Frequency * Step\ Length \quad (3)$$

The acceleration profile can be computed by deriving the speed profile:

$$A = \frac{\Delta(S)}{\Delta t} = \frac{S_i - S_{i-\delta}}{\delta} \quad (4)$$

where A is the acceleration profile, and it is a time series of length $n - \delta$, and $i = \{\delta + 1, \ldots, n\}$ a discrete temporal index.

Gait Variations and Stability

The movement dynamics of pedestrian gait can be analyzed using gait variations and stability. Gait variability is a quantifiable measure of walking that is altered in regard to magnitude and dynamics. Gait stability is a measure of the capability of the system to return to a given state after a perturbation has driven the system away from that state. Local stability refers to the sensitivity of the human to internal perturbations associated with some physiological, muscle, or nervous control issues (*36*). Local stability is more relevant in this study. The following is a set of measures extracted from speed profile to quantify the gait variability and stability:

1. Walk Ratio (WR). The WR is a speed-independent indicator of walking pattern that describes the temporal and spatial coordination. WR is derived from the amplitude and frequency of the rhythmic leg movements when walking. The deviation from the normal WR reveals some degree of abnormal walking patterns. The WR is calculated as the average step length (L_{step}) divided by step frequency (F_{step}) (*25*):

$$WR = \frac{L_{step}}{F_{step}} \quad (5)$$

2. Acceleration Auto Correlation (AAC). The AAC is a measure of the pedestrian step similarity and regularity derived by examining the similarity in the acceleration profile shape or waveform. AAC is measured using Pearson's correlation coefficient with lag time defined as a function of an arbitrary number of steps. A greater degree of gait stability is associated with a higher ACC value (*24*).

Case Study

Site Characteristic

The study site is a busy intersection located near Thompson Rivers University at McGill and Summit Streets in Kamloops, British Columbia. McGill and Summit Streets are two-way commercial corridors with an active walking environment. The intersection is a signalized four-leg intersection. Each leg has two traffic lanes and one auxiliary left-turn lane. Only the eastbound and southbound approaches have channelized right turn lanes. Conventional marked pedestrian crosswalks are located on all four legs of the intersection, with two additional crosswalks in the channelized right turn lanes. Pedestrian activity at this intersection is high. Pedestrian-vehicle interaction is frequently observed at this intersection for vehicles turning right at a traffic light showing a red signal and at the channelized right turn lanes. Right turn on red permits vehicles at a traffic light showing a red signal to turn right (the direction of traffic nearer to the vehicle) after a complete stop when the way is clear, without having to wait for a green signal. Video footages were collected in daylight under mild weather conditions during the period April 4 and April 5, 2016. Three cameras covered vehicle and pedestrian movements at the intersection. Video data were collected from five crosswalks (Crosswalks 1–5), as shown in Figure 2.

Data Description

In this study, trajectories of 357 pedestrians are used to analyze the walking behavior and gait characteristics of pedestrians crossing roads at conventional crosswalks. Manual observation of tracked pedestrians is used to specify pedestrian distraction state, distraction type (visually or auditory), and pedestrian-vehicle interaction state. The trajectories consist of 136 trajectories of distracted pedestrians (Texting/reading or talking/listening on phones) and 221 trajectories of non-distracted pedestrians. About 58% of the pedestrians are males (207 pedestrians), whereas females form 42% (150 pedestrians). Most (86%) of the pedestrians are young adults (307 pedestrians). Around 12% of the pedestrians (43 pedestrians) are involved in interactions with approaching vehicles. Nearly 38% of all pedestrians observed performed a distraction activity while crossing. Text-messaging/reading was the most common distraction (95 pedestrians), followed by conversing on the phone (41 pedestrians). Most (80%) of the distracted pedestrians are young adults and 54% of them are males. The walking speeds for non-distracted and distracted pedestrians are found to follow normal distributions as shown in Figure 3. The mean values and the standard deviations of walking speed for 221 non-distracted pedestrians and 136 distracted pedestrians walking through the conventional crosswalks are found to be 1.66 ± 0.19 m/s, and 1.49 ± 0.24 m/s, respectively.

Results Summary

Comparison between Gait Parameters, Variability and Stability Values across Pedestrian Distraction State and Type

The walking behavior of pedestrians crossing within conventional marked crosswalks is analyzed. Crossing pedestrians are classified into two groups: distracted pedestrians and non-distracted pedestrians. Distracted pedestrians are then categorized into two sub-groups based on the distraction type: pedestrians distracted while text-messaging/reading and pedestrians distracted while talking/listening on cell phones. Table 1 summarizes the gait parameters (walking speed, step length, and step frequency) and the gait variability and stability measures (acceleration autocorrelation and walk ratio) values across the pedestrian categories. Tests of significance (two-sample *t*-tests) are performed to verify hypotheses of a difference in means between the categories of each attribute. The following observation can be made from Table 1:

- Compared with non-distracted pedestrians, pedestrians distracted by texting/reading tend to have a statistically significant slower average walking speed as well as a shorter average step length. However, no significant difference is found in the step frequency. The results may suggest that

Figure 2. Data collection site: (*a*) world image, (*b*) camera 1 image, (*c*) camera 2 image, and (*d*) camera 3 image.

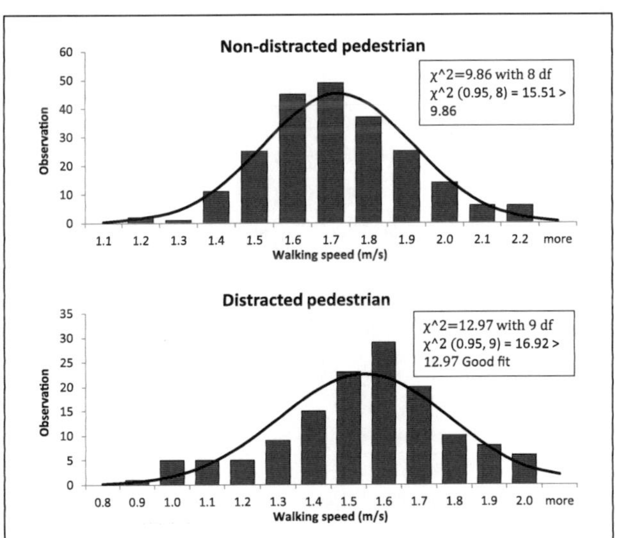

Figure 3. Walking speed distribution for non-distracted and distracted pedestrians.

pedestrians distracted by text-messaging/reading tend to reduce their walking speeds by adjusting their step lengths rather than step frequencies. Pedestrians distracted by texting/reading tend to have less stability and more variability in walking compared with non-distracted pedestrians as demonstrated by the statistically significant differences in average WR as well as average acceleration autocorrelation.

- Similar to pedestrians distracted by texting/reading, pedestrians distracted by talking/listening tend to have statistically significant slower average walking speed compared with non-distracted pedestrians. However, distracted pedestrians while talking/listening tend to reduce their walking speeds by adjusting their step frequencies rather than step lengths. The average step frequency is significantly lower for pedestrians distracted by talking/listening compared with non-distracted pedestrians. However, no significant difference is found in the step length. A high variability in walking behavior tends to be observed in pedestrians distracted by talking/listening compared with non-distracted pedestrians as demonstrated by the statistically significant lower average acceleration autocorrelation for the pedestrians distracted by talking/listening.

Table 1. Features' Significance across Distraction State and Type

Parameter	Mean [SD], distracted by texting/reading versus non-distracted		Mean [SD], distracted by talking/listening versus non-distracted		Mean [SD], total by distraction state		Mean [SD], distracted by talking/listening versus distracted by texting/reading		Total
	Non-distracted	Distracted by texting/reading	Non-distracted	Distracted by talking/listening	Non-distracted	Distracted (all)	Distracted by talking/listening	Distracted by texting/reading	
Count	221	95	221	41	221	136	41	95	357
Step frequency (step/s)	1.89 [0.33]	1.85 [0.32]	1.89 [0.33]	1.79[b] [0.33] (0.08)	1.89 [0.33]	1.83 [0.32]	1.79 [0.33]	1.85 [0.32]	1.87 [0.33]
Walking speed (m/s)	1.66 [0.19]	1.49[a] [0.24] (<0.01)	1.66 [0.19]	1.51[a] [0.22] (<0.01)	1.66 [0.19]	1.49[a] [0.24] (<0.01)	1.51 [0.22]	1.49 [0.24]	1.60 [0.23]
Step length (m/step)	0.92 [0.26]	0.82[a] [0.14] (<0.01)	0.92 [0.26]	0.88 [0.25]	0.92 [0.26]	0.84[a] [0.18] (<0.01)	0.88 [0.25]	0.82[a] [0.14] (0.05)	0.89 [0.24]
WR	1.08 [0.69]	0.94[a] [0.38] (0.05)	1.08 [0.69]	1.09 [0.60]	1.08 [0.69]	0.98 [0.46]	1.09 [0.60]	0.94[b] [0.38] (0.08)	1.04 [0.61]
Acceleration autocorrelation	0.35 [0.21]	0.26[a] [0.20] (<0.01)	0.35 [0.21]	0.24[a] [0.18] (<0.01)	0.35 [0.21]	0.25[a] [0.19] (<0.01)	0.24 [0.18]	0.26 [0.20]	0.31 [0.21]

Notes: Values in parentheses = *p*-value of the *t*-test; SD = standard deviation; 1 m = 3.28 ft.
[a]Statistically significant difference (at 95% confidence level) compared with the cell directly to the left.
[b]Statistically significant difference (at 90% confidence level) compared with the cell directly to the left.

- Compared with non-distracted pedestrians, distracted pedestrians (including both types of distraction states: distraction while texting/reading and distraction while talking/listening) tend to have a statistically significant slower average walking speed as well as a shorter average step length. Distracted pedestrians tend to have more variability in walking compared with the non-distracted pedestrian as demonstrated by the statistically significant lower average acceleration autocorrelation.
- For distracted pedestrians, the average walking speed and the average step frequency are not statistically different between pedestrians distracted by talking/listening on cell phones and those distracted by text-messaging/reading, even though pedestrians distracted by text-messaging/reading tend to have a significantly shorter average step length. Pedestrians distracted by text-messaging/reading tend to have less stability in walking behavior compared with pedestrians distracted by talking/listening as demonstrated by the statistically significant difference in the average walk ratio.

Analysis of Walking Behavior for Pedestrians Interacting with Approaching Vehicles

The walking behavior of pedestrians crossing within conventional marked crosswalks and involved in interactions with approaching vehicles is analyzed. Most of the pedestrian-vehicle interactions are observed at the crosswalk on the channelized right lane at Summit Street (Crosswalk 5 as shown in Figure 2). Pedestrians are categorized according to the interaction state (involved in an interaction with an approaching vehicle, and not involved in an interaction) and distraction state. Table 2 provides a summary of the gait parameter, gait variability and stability measure values across the interaction and distraction states. By the analysis of gait parameters and walking speed profile, the following observations can be made from Table 2:

- Distracted pedestrians involved in an interaction with a turning vehicle tend to have a significantly slower average walking speed and a lower average step frequency compared with distracted pedestrians not participating in an interaction. However, no significant difference is found in step length. This may suggest that distracted pedestrians control their walking speeds by adjusting their step frequencies instead of step lengths when involved in an interaction with a turning vehicle. No significant difference is found in gait variability and stability measures.
- For non-distracted pedestrians, the average walking speed is not statistically different between pedestrians involved in interactions with approaching vehicles, and those not involved in interactions, even though non-distracted pedestrians participating in an interaction with approaching vehicles tend to have a significantly lower average step frequency and a larger average step length. Significant differences in gait variability values are observed in non-distracted pedestrians across the interaction state. Non-distracted pedestrians involved in interactions with approaching vehicles have more variability in walking behavior compared with those not involved in an interaction as demonstrated by the significant difference in average acceleration autocorrelation. The change in these parameters can be related to pedestrians taking an evasive action. Similar results were found in (*37*, *38*).
- Regardless of the distraction state, pedestrians involved in interactions with approaching vehicles tend to have a different walking behavior than pedestrians not involved in interactions as the former have a significantly slower average walking speed and a lower average step frequency. A high variability in walking behavior tends to be observed in pedestrians participating in an interaction with approaching vehicles as demonstrated by the statistically significant difference in the acceleration autocorrelation.

Distracted pedestrians involved in interactions with approaching vehicles tend to have a significantly slower average walking speed and a shorter average step length compared with non-distracted pedestrians involved in interactions. However, no significant difference is found in the step frequency. For pedestrians involved in interactions with approaching vehicles, the results may suggest that those pedestrians tend to control their walking speeds by adjusting their step lengths rather than step frequencies when distracted. For all pedestrians involved in interactions with approaching vehicles, there are no significant differences in gait variability and stability measure values across the distraction state.

Discussion and Conclusions

This study demonstrated the practice of using automated video analysis to capture the influence of pedestrian distractions on their walking behavior. The study investigated the effect of different factors such as distraction state, distraction type (texting/reading and talking/listening), and pedestrian-vehicle interaction on the walking

Table 2. Features' Significance across Pedestrian-Vehicle Interaction State

Parameter	Mean [SD], distracted pedestrian by interaction state		Mean [SD], non-distracted pedestrian by interaction state		Mean [SD], all by interaction state		Mean [SD], interaction by distraction state	
	No interaction	Interaction	No interaction	Interaction	No interaction	Interaction	Non-distracted involved in interaction	Distracted involved in Interaction
Count	114	22	200	21	314	43	21	22
Step frequency (step/s)	1.87 [0.30]	1.64[a] [0.38] (<0.01)	1.91 [0.31]	1.72[a] [0.43] (0.01)	1.89 [0.31]	1.68[a] [0.40] (<0.01)	1.72 [0.43]	1.64 [0.38]
Walking speed (m/s)	1.52 [0.22]	1.35[a] [0.27] (<0.01)	1.66 [0.19]	1.64 [0.19]	1.61 [0.21]	1.49[a] [0.27] (<0.01)	1.64 [0.19]	1.35[a] [0.27] (<0.01)
Step length (m/step)	0.84 [0.19]	0.84 [0.16]	0.91 [0.26]	1.02[b] [0.30] (0.07)	0.88 [0.24]	0.93 [0.26]	1.02 [0.30]	0.84[a] [0.16] (0.02)
WR	0.96 [0.45]	1.12 [0.49]	1.05 [0.66]	1.37[a] [0.83] (0.04)	1.02 [0.60]	1.25[a] [0.68] (0.02)	1.37 [0.83]	1.12 [0.49]
Acceleration autocorrelation	0.25 [0.19]	0.25 [0.20]	0.36 [0.21]	0.22[a] [0.19] (<0.01)	0.32 [0.21]	0.24[a] [0.19] (<0.01)	0.22 [0.19]	0.25 [0.20]

Notes: Values in parentheses = *p*-value of the *t*-test; SD = standard deviation; 1 m = 3.28 ft.
[a]Statistically significant difference (at 95% confidence level) compared with the cell directly to the left.
[b]Statistically significant difference (at 90% confidence level) compared with the cell directly to the left.

behavior and gait stability of pedestrians at crosswalks. Several observations can be drawn from the study:

- Pedestrians seem to have difficulty maintaining their walking speed and tempo-spatial characteristics of gait when distracted. The slower crossing speed of distracted pedestrians increases their potential conflict with approaching vehicle traffic.
- Distracted pedestrians differ in the way they control their walking speed based on the type of distraction (e.g., talking/listening or texting/reading). Pedestrians distracted by text-messaging/reading tend to reduce their walking speed by adjusting the step length instead of step frequency. However, distracted pedestrians who are listening/ talking on a cell phone reduce their walking speed by adjusting their step frequency rather than step length.
- The difficulty of maintaining the tempo-spatial characteristics of gait for distracted pedestrians led to relatively unstable movements and disruptions in their walking behavior. Pedestrians distracted by texting/reading have less stability in walking behavior compared with pedestrians distracted by talking/listening on cell phones.
- The differences in the tempo-spatial characteristics and stability measures of gait seem to be related to the degree of complexity of the task and to the degree of visual sensory constraint, which is required to maintain gait balance. Compared with talking/listening, texting/reading while walking is a more complex task, calls for focused attention and higher cognitive ability, and leads to a higher visual sensory constraint of pedestrians.
- The walking behavior changes when involved in an interaction with an approaching vehicle. Distracted pedestrians tend to control and reduce their walking speeds by adjusting their step frequencies when involving in interactions with approaching vehicles. For non-distracted pedestrians, significant tempo-spatial changes in step length and frequency are observed. Non-distracted pedestrians tend to reduce the step frequency and increase the step length. These changes in gait tempo-spatial characteristics can be evasive actions taken by pedestrians involved in interactions with approaching vehicles. These changes can be potential indicators for pedestrian-vehicle conflicts.

Recommendations and Future Directions

Limitations of the current work are mainly related to the sample size of distracted pedestrians. Considering other data sets in different environments is important. The gait parameters can be affected by the density of pedestrians crossing and the levels of opposing flows (considering bi-directional flow analysis), which require further investigation. Inter-pedestrian conflict rate can be investigated as another metric of the distraction impact. Other distraction types (e.g., talking to another pedestrian or looking at some other objects) can be included in future research. The walking behavior of distracted pedestrians in spatial or temporal violation needs to be investigated in future research. The impact of the type of distraction (visual or auditory) on the walking behavior of pedestrians involved in interactions with vehicles can be investigated. Considering the effect of pedestrian acknowledgments to vehicles in analyzing the walking behavior of pedestrians involved in interactions with vehicles is also important. Other gait parameters can be included in future research, including gait kinematics data at various locations of the human body. This is possible through careful selection of tracked moving features. Other measures of distraction, such as path variance, starting delay, temporal and spatial violation, and head orientation, can be investigated in future research.

There are several future applications for analyzing the walking behavior of distracted pedestrians. First, this data can help in developing pedestrian safety intervention programs and legislation. Second, the distinctive features of pedestrians across distraction states, distraction type, and pedestrian-vehicle interaction state can be used in an automated video analysis system to recognize the different categories of pedestrians. This extension enables safety practitioners to conduct automated before-and-after safety studies using computer vision techniques to assess various pedestrian safety intervention programs. Third, analyzing the walking behavior of pedestrians provides data that can be used to model and calibrate the walking mechanisms of pedestrians, including distracted pedestrians, in pedestrian simulation programs. Modeling distracted pedestrians can be beneficial in evaluating the safety of pedestrians under various designs of intersections and pedestrian facilities, and signal timing strategies and plans.

The prevalence of distractions among pedestrians, especially young populations, school and university students, and at active pedestrian areas (e.g., downtown areas), may be establishing new societal norms. In that case, understanding the walking behavior of distracted pedestrians in the crossing of roads and at crosswalks becomes most important to provide pedestrian facility designs that accommodate and also protect limited attention and compromised mobility. Moreover, understanding the walking behavior and the changes in gait speed and stability not only of distracted pedestrians but also of more vulnerable pedestrians (e.g., children, elders, persons with physical, cognitive, or sensory disabilities) helps in better planning and designing of pedestrian facilities to improve their safety. Using cell phones while walking limits the visual sensory of pedestrians required for controlling gait balance, resulting in abnormal walking function, and this may lead to problems such as falling. This issue is relevant not only in the crossing of crosswalks but also in shared space areas (i.e., areas where pedestrians, bikes, and vehicles share the same routes), where a fall can lead to a contact with a moving bike or vehicle and can lead to a severe injury. Automatic identification of distracted pedestrians and pedestrians with less stability at crosswalks and shared space areas by autonomous vehicles can have potential benefits in increasing pedestrians' safety.

Author Contributions

The authors confirm contribution to the paper as follows: study conception and design: Rushdi Alsaleh, Tarek Sayed, Mohamed Zaki; data collection: Rushdi Alsaleh, Tarek Sayed, Mohamed Zaki; analysis and interpretation of results: Rushdi Alsaleh, Tarek Sayed, Mohamed Zaki; draft manuscript preparation: Rushdi Alsaleh, Tarek Sayed, Mohamed Zaki. All authors reviewed the results and approved the final version of the manuscript.

References

1. Lipovac, K., M. Đerić, M. Tešić, Z. Andrić, and B. Marić. Mobile Phone Use While Driving-Literary Review. *Transportation Research Part F: Traffic Psychology and Behaviour*, Vol. 47, 2017, pp. 132–142.
2. Smith, D. C., K. M. Schreiber, A. Saltos, S. B. Lichenstein, and R. Lichenstein. Ambulatory Cell Phone Injuries in the United States: An Emerging National Concern. *Journal of Safety Research*, Vol. 47, 2013, pp. 19–23.
3. Woollacott, M., and A. Shumway-Cook. Attention and the Control of Posture and Gait: A Review of An Emerging Area of Research. *Gait & Posture*, Vol. 16, No. 1, 2002, pp. 1–14.
4. Schabrun, S. M., W. van den Hoorn, A. Moorcroft, C. Greenland, and P. W. Hodges. Texting and Walking: Strategies for Postural Control and Implications for Safety. *PLoS One*, Vol. 9, No. 1, 2014, p. e84312.
5. Lamberg, E. M., and L. M. Muratori. Cell Phones Change the Way We Walk. *Gait & Posture*, Vol. 35, No. 4, 2012, pp. 688–690.
6. Rebold, M. J., C. A. Croall, E. A. Cumberledge, T. P. Sheehan, and M. T. Dirlam. The Impact of Different Cell Phone Functions and Their Effects on Postural Stability. *Performance Enhancement & Health*, Vol. 5, No. 3, 2017, pp. 98–102.
7. Lin, M. I. B., and Y. P. Huang. The Impact of Walking While Using a Smartphone on Pedestrians' Awareness of Roadside Events. *Accident Analysis & Prevention*, Vol. 101, 2017, pp. 87–96.
8. Ismail, K., T. Sayed, and N. Saunier. Automated Analysis of Pedestrian-vehicle: Conflicts Context for Before-and-after Studies. *Transportation Research Record: Journal of the Transportation Research Board*, 2010. 2198: 52–64.
9. Zaki, M. H., and T. Sayed. Exploring Walking Gait Features for the Automated Recognition of Distracted Pedestrians. *IET Intelligent Transport Systems*, Vol. 10, No. 2, 2016, pp. 106–113.
10. BC Injury Research and Prevention Unit. *Injury Insight*. 2017. http://www.injuryresearch.bc.ca/wp-content/uploads/2017/05/Injury-Insight-Pedestrian-Safety-May-2017-FINAL.pdf. Accessed July 22, 2017.
11. Thompson, L. L., F. P. Rivara, R. C. Ayyagari, and B. E. Ebel. Impact of Social and Technological Distraction on Pedestrian Crossing Behaviour: An Observational Study. *Injury Prevention*, Vol. 19, No. 4, 2013, pp. 232–237.
12. Schwebel, D. C., D. Stavrinos, K. W. Byington, T. Davis, E. E. O'Neal, and D. De Jong. Distraction and Pedestrian Safety: How Talking on the Phone, Texting, and Listening to Music Impact Crossing the Street. *Accident Analysis & Prevention*, Vol. 45, 2012, pp. 266–271.
13. Byington, K. W., and D. C. Schwebel. Effects of Mobile Internet Use on College Student Pedestrian Injury Risk. *Accident Analysis & Prevention*, Vol. 51, 2013, pp. 78–83.
14. Nasar, J. L., and D. Troyer. Pedestrian Injuries Due to Mobile Phone Use in Public Places. *Accident Analysis & Prevention*, Vol. 57, 2013, pp. 91–95.
15. Ferguson, R. W., Z. Xu, A. Green, and K. M. Rosentha. *Teens and Distraction: An In-depth Look a Teens' Walking Behaviors*. Safe Kids Worldwide, Washington, D.C., 2013.
16. Wells, H. L., L. A. McClure, B. E. Porter, and D. C. Schwebel. Distracted Pedestrian Behavior on two Urban College Campuses. *Journal of Community Health*, 2017, pp. 1–7.
17. Mori, H., N. M. Charkari, and T. Matsushita. On-line Vehicle and Pedestrian Detections Based on Sign Pattern. *IEEE Transactions on Industrial Electronics*, Vol. 41, No. 4, 1994, pp. 384–391.
18. BenAbdelkader, C., R. Cutler, and L. Davis. Stride and Cadence as a Biometric in Automatic Person Identification and Verification. *Proc., 5th IEEE International Conference on Automatic Face and Gesture Recognition*, Washington, D.C., IEEE, New York, 2002, pp. 372–377.
19. Saunier, N., A. El Husseini, K. Ismail, C. Morency, J. M. Auberlet, and T. Sayed. Estimation of Frequency and Length of Pedestrian Stride in Urban Environments with Video Sensors. *Transportation Research Record: Journal of the Transportation Research Board*, 2011. 2264: 138–147.
20. Zaki, M. H., and T. Sayed. A Framework for Automated Road-users Classification Using Movement Trajectories. *Transportation Research Part C: Emerging Technologies*, Vol. 33, 2013, pp. 50–73.
21. Yoo, J. H., D. Hwang, and M. S. Nixon. Gender Classification in Human Gait Using Support Vector Machine. *Advanced Concepts for Intelligent Vision systems*, Vol. 5, 2005, pp. 138–145.
22. Makihara, Y., H. Mannami, and Y. Yagi. Gait Analysis of Gender and Age Using a Large-scale Multi-view Gait

Database. *Proc., Asian Conference on Computer Vision Conference*, Berlin, Heidelberg, Springer, Vol. 6493, 2010, pp. 440–451.

23. Menz, H. B., S. R. Lord, and R. C. Fitzpatrick. Acceleration Patterns of the Head and Pelvis When Walking on Level and Irregular Surfaces. *Gait & Posture*, Vol. 18, No. 1, 2003, pp. 35–46.
24. Mizuike, C., S. Ohgi, and S. Morita. Analysis of Stroke Patient Walking Dynamics Using a Tri-axial Accelerometer. *Gait & Posture*, Vol. 30, No. 1, 2009, pp. 60–64.
25. Sekiya, N., and H. Nagasaki. Reproducibility of the Walking Patterns of Normal Young Adults: Test-retest Reliability of the Walk Ratio (Step-length/Step-rate). *Gait & Posture*, Vol. 7, No. 3, 1998, pp. 225–227.
26. Lin, M. I. B., and K. H. Lin. Walking While Performing Working Memory Tasks Changes the Prefrontal Cortex Hemodynamic Activations and Gait Kinematics. *Frontiers in Behavioral Neuroscience*, Vol. 10, 2016, p. 92.
27. Strubhar, A. J., M. L. Peterson, J. Aschwege, J. Ganske, J. Kelley, and H. Schulte. The Effect of Text Messaging on Reactive Balance and the Temporal and Spatial Characteristics of Gait. *Gait & Posture*, Vol. 42, No. 4, 2015, pp. 580–583.
28. Plummer, P., S. Apple, C. Dowd, and E. Keith. Texting and Walking: Effect of Environmental Setting and Task Prioritization on Dual-task Interference in Healthy Young Adults. *Gait & Posture*, Vol. 41, No. 1, 2015, pp. 46–51.
29. Chabrun, S. M., W. Van den hoorn, A. Moorcroft, C. Greenland, and P. W. Hodges. Texting and Walking: Strategies for Postural Control and Implications for Safety. *PLoS One*, Vol. 9, No. 1, 2014, p. e84312.
30. Nasar, J., P. Hecht, and R. Wener. Mobile Telephones, Distracted Attention, and Pedestrian Safety. *Accident Analysis & Prevention*, Vol. 40, No. 1, 2008, pp. 69–75.
31. Hatfield, J., and S. Murphy. The Effects of Mobile Phone Use on Pedestrian Crossing Behaviour at Signalised and Unsignalised Intersections. *Accident Analysis & Prevention*, Vol. 39, No. 1, 2007, pp. 197–205.
32. Ismail, K., T. Sayed, and N. Saunier, A Methodology for Precise Camera Calibration for Data Collection Applications in Urban Traffic Scenes. *Canadian Journal of Civil Engineering*, Vol. 40, No. 1, 2013, pp. 57–67.
33. Saunier, N., and T. Sayed. A Feature-based Tracking Algorithm for Vehicles in Intersections. *Proc., 3rd IEEE Canadian Conference on Computer and Robot Vision*, Quebec, Canada, IEEE, New York, 2006, pp. 59–59.
34. Hediyeh, H., T. Sayed, M. H. Zaki, and G. Mori. Pedestrian Gait Analysis Using Automated Computer Vision Techniques. *Transportmetrica A: Transport Science*, Vol. 10, No. 3, 2014, pp. 214–232.
35. John, A., and A. Ruina. Multiple Walking Speed–Frequency Relations are Predicted by Constrained Optimization. *Journal of Theoretical Biology*, Vol. 209, No. 4, 2001, pp. 445–453.
36. Zaki, M. H., and T. Sayed. Using Automated Walking Gait Analysis for the Identification of Pedestrian Attributes. *Transportation Research Part C: Emerging Technologies*, Vol. 48, 2014, pp. 16–36.
37. Tageldin, A., and T. Sayed. Developing Evasive Action-Based Indicators for Identifying Pedestrian Conflicts in Less Organised Traffic Environments. *Journal of Advanced Transportation*, Vol. 50, 2016, pp. 1193–1208.
38. Tageldin, A., M. H. Zaki, and T. Sayed. Examining Pedestrian Evasive Actions as a Potential Indicator for Traffic Conflicts. *IET Intelligent Transport Systems*, Vol. 11, 2017, pp. 282–289.

The Pedestrians and Cycles Section (ANF00) peer-reviewed this paper (18-01580).

Article

TRR
JOURNAL OF THE TRANSPORTATION RESEARCH BOARD

Transportation Research Record
2018, Vol. 2672(35) 58–68

DOI: 10.1177/0361198118773555
journals.sagepub.com/home/trr
SAGE

Making Strides: State of the Practice of Pedestrian Forecasting in Regional Travel Models

Patrick A. Singleton[1,2], Joseph C. Totten[1], Jaime P. Orrego-Oñate[1], Robert J. Schneider[3], and Kelly J. Clifton[1]

Abstract

Much has changed in the 30 years since non-motorized modes were first included in regional travel demand models. As interest in understanding behavioral influences on walking and policies requiring estimates of walking activity increase, it is important to consider how pedestrian travel is modeled at a regional level. This paper evaluates the state of the practice of modeling walk trips among the largest 48 metropolitan planning organizations (MPOs) and assesses changes made over the last 5 years. By reviewing model documentation and responses to a survey of MPO modelers, this paper summarizes current practices, describes six pedestrian modeling frameworks, and identifies trends. Three-quarters (75%) of large MPOs now model non-motorized travel, and over two-thirds (69%) of those MPOs distinguish walking from bicycling; these percentages are up from nearly two-thirds (63%) and one-half (47%), respectively, in 2012. This change corresponds with an increase in the deployment of activity-based models, which offer the opportunity to enhance pedestrian modeling techniques. The biggest barrier to more sophisticated models remains a lack of travel survey data on walking behavior, yet some MPOs are starting to overcome this challenge by oversampling potential active travelers. Decision-makers are becoming more interested in analyzing walking and using estimates of walking activity that are output from models for various planning applications. As the practice continues to mature, the near future will likely see smaller-scale measures of the pedestrian environment, more detailed zonal and network structures, and possibly even an operational model of pedestrian route choice.

A growing number of U.S. metropolitan planning organizations (MPOs) have incorporated walking and other non-motorized modes into their regional travel demand forecasting models (*1*). These models are used for long-range planning, allowing regions to analyze land use and transportation scenarios and projects for their impacts on walking and bicycling. Models that can forecast walking also have many other applications. In the short term, they are useful for prioritizing non-motorized infrastructure investments. In the long run, models that are sensitive to how the pedestrian environment influences travel behavior could better predict mode shifts and the resulting impacts on motor vehicle emissions of greenhouse gases and other air pollutants. Regional pedestrian models can also inform traffic safety analyses and health impact assessments, providing required estimates of the location and number of walk trips.

Partly in response to these new policy demands and partly due to advances in computational power, travel demand models are becoming more sensitive to finer-grained representations of travelers themselves and the environments through which people travel. The growing number of implemented activity-based models is one example of this evolution. Two other major trends support these modeling enhancements. First, archived spatial data on the built environment are becoming ubiquitous, allowing small-scale, pedestrian-relevant environmental measures to be included in models (*2*). Second, research on pedestrian travel behavior and its environmental determinants has advanced to the point of offering at least initial guidance about which relationships may be important to represent (*3, 4*).

In the face of these trends and opportunities, MPOs have taken various approaches to characterizing walk trips in their travel demand models. The purpose of this study is to assess the state of the practice with respect to how pedestrian travel is incorporated into these regional models; specifically, this paper reviews if and how the largest 48 MPOs model walk

[1]Department of Civil and Environmental Engineering, Portland State University, Portland, OR
[2]Department of Civil and Environmental Engineering, Utah State University, Logan, UT
[3]School of Architecture and Urban Planning, University of Wisconsin – Milwaukee, Milwaukee, WI

Corresponding Author:
Address correspondence to Patrick A. Singleton: patrick.singleton@usu.edu

trips. The authors assess frameworks, spatial scales, behavioral and environmental data, applications, challenges, and innovations. This paper offers agency and consultant staff a better understanding of how regional pedestrian modeling is currently done and how it might evolve in the future, so that they can adapt techniques and improve their own models. It also serves as an update to several other recent (and not-so-recent) reviews on this topic (*1, 5–8*).

A few notes about the scope of this study are warranted. First, although the focus of this paper is on walking or pedestrian travel, some of its results may also be relevant for bicycle modeling. Walking and bicycling modes have been frequently lumped into a single "non-motorized" mode for modeling purposes. This practice is starting to change, and bicycle modeling has in some ways eclipsed pedestrian modeling, especially in the area of route choice and trip assignment procedures. We encourage interested parties to examine the state of the practice of regional bicycle modeling in more detail (*9–11*), as this is beyond the scope of this study. Second, we focus on "pure" walk trips, not instances of walking linked to accessing other modes. Modeling practices for walk-access-to-transit trips are relatively well-established, although they may still benefit from some of the discussions here. Third, this paper is narrowly focused on large MPO regional forecasting models. We acknowledge that many modeling innovations are taking place at sub-regional scales and through smaller and non-MPO agencies, including the use of sketch-planning and accessibility-based tools (*12–15*). Some of these standalone pedestrian planning tools may be easier to develop, more agile to apply, or better suited to addressing certain policy questions, or both, than are regional travel demand models (*14*), yet we leave this subject for others to address. Fourth, our objective is to provide a broad overview of the state of the practice, therefore, we do not have the space to describe specific case studies in depth. A number of other pedestrian modeling resources (*8, 14, 15*) include such case studies, and we encourage interested readers to contact the MPOs we describe throughout this paper for more information on their respective practices.

The remainder of this paper begins with a Background giving a brief history of non-motorized and pedestrian modeling practices and trends. Next, a short Method section describes the state-of-the-practice survey and analysis procedures. The Results section then presents and discusses findings regarding pedestrian modeling frameworks, spatial scales and networks, and built environment measures. It also discusses barriers and challenges as well as model applications. The paper concludes by summarizing the major contributions of this work, describing innovative pedestrian modeling practices, and noting opportunities and directions for the field.

Background

Until the early 1990s, regional travel demand models did not consider walking, for several reasons. First, the conventional four-step travel forecasting process evolved out of multiple efforts in the 1950s and 1960s during a time in which regional and national transportation policies were focused on planning for and accommodating rapid growth in automobile travel and constructing the Interstate Highway System (*16*). Traffic analysis zones were too large and street networks too coarse to adequately represent walking trips. By the 1970s, "multimodal" models could also analyze travel by public transit; they were developed in part to support applications for new federal transit grant programs (*16*). Second, the travel surveys and other data collection processes necessary to construct these models either underreported or did not consider non-motorized travel (*17, 18*). In fact, many surveys in the 1980s asked for non-motorized trips to be reported only if they were home-based work (i.e., commute) trips (*19*).

As far as we have been able to document, the first regional travel demand model to include non-motorized travel was developed in 1988 at the Metropolitan Service District (now Metro) in Portland, Oregon (*20*). In preparation for a light-rail expansion project, a 1985 travel survey was used to estimate a motorized/non-motorized mode split model for home-based trips, based on trip distance and the relative number of cars and workers per household; other trip purposes used a static non-motorized mode share (*7, 17, 21*). In 1993, Sacramento's MPO was perhaps the first to separate walk and bicycle modes within a mode choice model (*20*). Despite these initial efforts, models were slow to change: none of the 10 major MPOs (not including Portland) examined in a 1992 study modeled non-motorized modes (*22*, cited in *20*). Nevertheless, by the late 1990s, regions as diverse as Baltimore, Boston, Chicago, Hampton Roads, Los Angeles, Philadelphia, Portland, Sacramento, and the San Francisco Bay Area either had or were in the late stages of incorporating non-motorized modes into their models (*7, 20*).

Contemporaneous to non-motorized modes being added to regional travel models was an increased awareness of, interest in, and research on built environment and land-use effects on travel demand (e.g., *23*). Several pedestrian-specific measures of the built environment were developed for use in travel forecasting. One of the first measures, the pedestrian and bicycle friendliness index—a combination of land-use mix, building setback, bus shelters, bicycle infrastructure, and extent of sidewalks—was developed in 1988 at the Maryland-National Capital Park and Planning Commission (*5, 17*). A second influential project during the early-to-mid 1990s—Making the Land Use, Transportation, Air Quality Connection (LUTRAQ)—developed the pedestrian environment factor (PEF) for Portland's MPO, Metro (*5, 7, 17*). The PEF, an index representing the ease of street crossings, sidewalk continuity, grid street pattern, and terrain, was used in Metro's pre-mode choice non-motorized split model (*21, 24*). The PEF or similar indexes have since been applied in many models, including in the Chicago,

Hampton Roads, Miami, Philadelphia, Portland, Sacramento, and Salt Lake City regions (*7, 20, 25–27*). In time, agencies (including Portland's Metro) have transitioned from subjectively defined indices to more objectively calculated and empirically defined pedestrian environment measures (*1*).

Advances in pedestrian modeling have continued into the 21st century as more MPOs have added non-motorized modes to their travel demand forecasting models. A 2005 TRB study found that more than half of large MPOs (54%, $N = 35$) reported modeling non-motorized travel (*28, 29*). A few years later, somewhere between almost half (41–45%, $N = 29$) and around two-thirds (68%, $N = 28$) of large MPOs included non-motorized travel in their trip-based models, with about half (53%) of those doing so in the mode choice stage (*8*). In 2012, two of the authors of the present article surveyed the 48 largest MPOs and found that nearly two-thirds (63%) modeled non-motorized travel, with about half (47%) of those distinguishing between walking and bicycling (*1*). This paper updates that study by summarizing the state of the practice of pedestrian modeling as of 2017: three-quarters (75%) of large MPOs model non-motorized travel, and more than two-thirds (69%) of those distinguish between walking and bicycling.

Method

To assess the current state of pedestrian demand modeling practice, we conducted a survey of the 48 largest MPOs, those serving populations greater than 1,000,000 people as of the 2010 U.S. Census (*30*). Table 1 lists these MPOs and their respective major cities. We decided to focus on only the largest MPOs for several reasons. First, they collectively serve over 156 million people, slightly more than half of the U.S. population; decisions made at large MPOs can impact projects and plans that affect the lives of many people. Second, large MPOs tend to have larger staffs and greater resources for investing in research and development; they can be industry leaders in travel demand modeling.

Information on how walking is addressed within regional travel forecasting models was obtained in a variety of ways. We first developed an online questionnaire for MPO modeling staff members, including questions about model structures, incorporation of walking (and bicycling) modes, spatial and network systems, built environment measures, barriers and challenges, and planning applications of walking forecasts. In May–June 2017, this survey was sent via email to staff at all 48 MPOs; 29 responses were received, a 60% response rate. Next, we obtained and analyzed model documentation for as many MPOs as possible through agency websites, survey responses, and direct correspondence. A similar process (survey and documentation review) was conducted in 2012 (*1*); 29 responses (60%) to that survey were received. While focusing on the 2017 survey, this paper also incorporates results from the 2012 survey and compares maturity of the practice over the 5-year span.

Results

Results from the MPO survey and model documentation review are presented in the sections below. First, we classify and describe six pedestrian modeling frameworks, reflecting different ways by which walking (and bicycling) can be represented in regional travel forecasting models. Next, we discuss key findings with respect to geographic and environmental attributes of the models, including spatial/zonal scales, street networks, and built environment measures. We then discuss challenges and barriers hampering further modeling of walk trips, before noting ways in which these models have been applied in practice. This section concludes with a summary of key findings.

Pedestrian Modeling Frameworks

After reviewing MPO model documentation, and informed by previous work (*1, 5, 8*), we have identified six frameworks for how walking (and bicycling, or non-motorized) travel can be incorporated into model structures. There are specific variations within these typologies, but the six pedestrian modeling frameworks provide a useful approximation of the spectrum of options that have been used by MPOs. These frameworks are generically applicable to both trip-based and activity-based model structures, as both typically still follow a four-step travel forecasting process: trips (or tours), destinations, modes, and routes. The frameworks are distinguished by the point in the process where walk trips are identified. They also suggest how much information the model provides about walk trips: trips generated, trip matrices, routes, etc. Figure 1 represents and compares these pedestrian modeling frameworks graphically. Table 1 indicates the frameworks used in 2017 and in 2012 for each large MPO. The sections that follow describe each pedestrian modeling framework and summarize both current practice and recent trends.

Option Zero. This framework represents models that exclude walk (and bicycle) trips and explicitly only consider motorized transportation. As a result, it is impossible to use these models alone to evaluate improvements to pedestrian infrastructure and corresponding effects on non-/motorized trip-making.

This framework is becoming less common as more MPOs have some representation of non-motorized modes in their models. While 18 MPOs (38%) used this framework in 2012, 12 MPOs (25%) use this framework in 2017.

Option One. MPOs with option one model frameworks forecast walking or non-motorized travel using a trip/tour generation process that is separate from that used for motorized modes, applying unique trip production/attraction rates or models. Thus, they can evaluate how at least the number of walk trips might change given future scenarios. More advanced

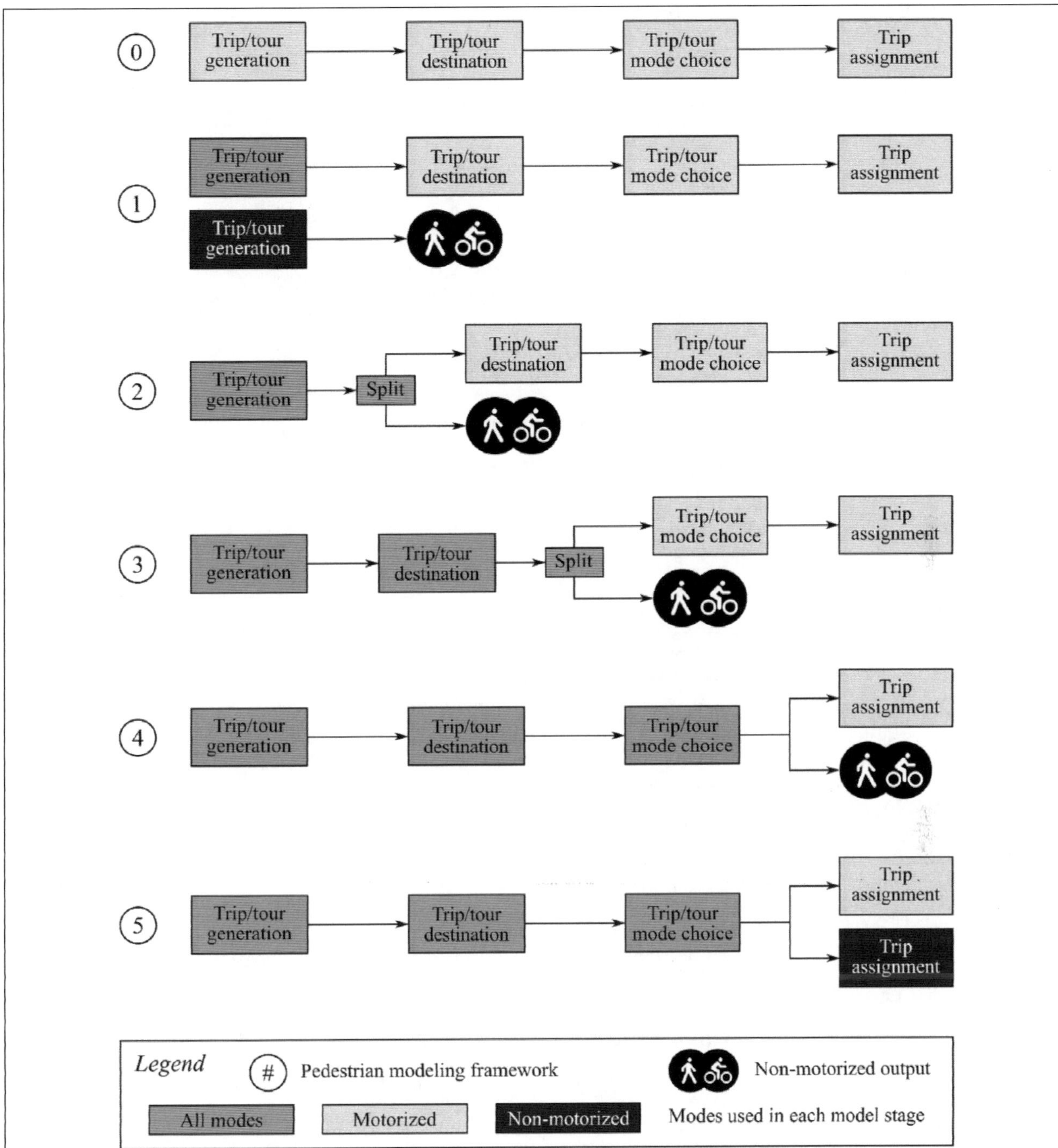

Figure 1. Pedestrian modeling frameworks.

versions of this framework could include a completely separate multi-step modeling process for walk (or bicycle or both) trips. A significant downside to this framework is the lack of interaction between the motorized and non-motorized model processes, which does not allow for representing mode shifts in response to external, transportation, or environmental changes. Nevertheless, it is a simple way for MPOs to begin modeling walk trips without having to fully update the rest of their travel demand models.

Continuing with trends identified in 2012, MPOs are moving away from using this framework, which was once relatively common (*1*). Of the two MPOs (4%) that used this

Table 1. Large MPOs and their Pedestrian Modeling Frameworks

City, State	Metropolitan Planning Organization	Model Framework 2012	Model Framework 2017
Atlanta, GA	Atlanta Regional Commission (ARC)	2	4[a]
Austin, TX	Capital Area MPO (CAMPO)	4	4
Baltimore, MD	Baltimore Regional Transportation Board (BRTB)	2	2
Boston, MA	Boston Region MPO	4	4
Buffalo, NY	Greater Buffalo-Niagara Regional Transportation Council (GBNRTC)	4	4[b]
Charlotte, NC	Charlotte Regional Transportation Planning Organization (CRTPO)	4	4[b]
Chicago, IL	Chicago Metropolitan Agency for Planning (CMAP)	2	4[a]
Cincinnati, OH	Ohio-Kentucky-Indiana Regional COG (OKI)	0	0
Cleveland, OH	Northeast Ohio Areawide Coordinating Agency (NOACA)	4	4
Columbus, OH	Mid-Ohio RPC (MORPC)	4[a]	4[a,b]
Dallas, TX	North Central Texas COG (NCTCOG)	0	0
Denver, CO	Denver Regional COG (DRCOG)	4[a]	4[a]
Detroit, MI	Southeast Michigan COG (SEMCOG)	0	2
Fort Lauderdale, FL	Broward MPO	3	4[a]
Houston, TX	Houston-Galveston Area Council (HGAC)	0	0
Indianapolis, IN	Indianapolis MPO	0	4
Jacksonville, FL	North Florida Transportation Planning Organization (TPO)	0	4[a]
Kansas City, MO	Mid-America Regional Council (MARC)	0	0
Las Vegas, NV	Regional Transportation Commission of Southern Nevada (RTC)	0	2
Los Angeles, CA	Southern California AOG (SCAG)	4	4[a]
Louisville, KY	Kentuckiana Regional Planning and Development Agency (KIPDA)	0	0
Memphis, TN	Memphis Urban Area MPO (MPO)	4	4
Miami, FL	Miami-Dade Transportation Planning Organization (TPO)	3	4[a]
Milwaukee, WI	Southeastern Wisconsin RPC (SEWRPC)	1	4
Minneapolis, MN	Metropolitan Council	4	4[a]
Nashville, TN	Nashville Area MPO	0	4[a]
New Orleans, LA	New Orleans Regional Planning Commission (RPC)	0	0[b]
New York, NY	New York Metropolitan Transportation Council (NYMTC)	3[a]	3[a]
Newark, NJ	North Jersey Transportation Planning Authority (NJTPA)	2	2
Oklahoma City, OK	Association of Central Oklahoma Governments (ACOG)	0	0[b]
Orlando, FL	MetroPlan Orlando	0	0
Philadelphia, PA	Delaware Valley RPC (DVRPC)	1	2
Phoenix, AZ	Maricopa AOG (MAG)	0	4[a]
Pittsburgh, PA	Southwestern Pennsylvania Commission (SPC)	0	0[b]
Portland, OR	Metro	4	4
Providence, RI	Rhode Island State Planning Council	0	0
Raleigh, NC	North Carolina Capital Area MPO (CAMPO)	3	3
Sacramento, CA	Sacramento Area COG (SACOG)	4[a]	4[a]
Salt Lake City, UT	Wasatch Front Regional Council (WFRC)	4	4
San Antonio, TX	Alamo Area MPO (AAMPO)	4	4
San Diego, CA	San Diego AOG (SANDAG)	4	4[a]
San Francisco Bay Area, CA	Metropolitan Transportation Commission (MTC)	4[a]	4[a]
Seattle, WA	Puget Sound Regional Council (PSRC)	4	4[a]
St. Louis, MO	East-West Gateway COG (EWG)	4	4
Tampa, FL	Hillsborough County MPO	0	0
Virginia Beach, VA	Hampton Roads Transportation Planning Organization (HRTPO)	0	0
Washington, DC	National Capital Region Transportation Planning Board (TPB)	2	2
West Palm Beach, FL	Palm Beach Transportation Planning Agency (TPA)	3	4[a]
Total, Option Zero	Models that do not include non-motorized travel	18	12
Total, Option One	Models with separate non-motorized trip generation	2	0
Total, Option Two	Models that split non-motorized trips before trip distribution	5	6
Total, Option Three	Models that split non-motorized trips after trip distribution	5	2
Total, Option Four	Models with non-motorized modes in mode choice	18	28
Total, Option Five	Models that assign walk & bicycle trips to the network	0	0
	Total activity-based models	5	17
	Total models reviewed	48	42

Note: AOG = Association of Governments; COG = Council of Governments; MPO = Metropolitan Planning Organization; RPC = Regional Planning Commission.
[a]An activity-based model.
[b]For these MPOs, survey responses were not received and recent model documentation was not found, so it was assumed that pedestrian modeling frameworks have not changed.

structure in 2012, Philadelphia moved to an option two framework, and Milwaukee moved to an option four framework. Based upon the model documentation reviewed, no large MPO currently uses this pedestrian modeling structure.

Option Two. In option two models, trips/tours by all modes are generated, but walk or non-motorized trips are removed using a special mode split model (usually binary logit) prior to trip/tour distribution or destination choice. The modal substitution patterns allowed by models of this type depend upon the variables, sensitivities, and feedback mechanisms incorporated into generation models. Therefore, this framework can yield walk trip productions/attractions but does not link trip ends together or assign destinations to walk trips. As a result, such models may not be able to determine the locations of specific barriers to walking or potential impacts of pedestrian infrastructure improvements, instead only highlighting areas that originate (and in the case of trip attractions, attract) more or fewer walk trips.

For some MPOs, the option two framework represents one step on a natural evolution to more robust pedestrian modeling capabilities. Although a similarly small number of MPOs use this framework today as did 5 years ago—five (10%) in 2012 and six (13%) in 2017—several agencies have shifted: Detroit's and Las Vegas's MPOs jumped from an option zero model, Philadelphia's MPO evolved from using an option one model, while Chicago's and Atlanta's MPOs began using option four frameworks.

Option Three. Like in option two, this framework describes models that also incorporate a special (often binary logit) mode split model, except that for option three, walk or non-motorized trips are separated after trip/tour distribution or destination choice but before mode choice. This pre-mode-choice/split model has advantages: by analyzing both trip ends, it accommodates the use of level-of-service variables (e.g., travel time or distance), and it allows for the creation of an origin–destination matrix of walk trips. It may be easier for MPOs with insufficient records of walk or non-motorized trips to estimate a binary mode split model than a full mode choice model; the downside is the ability to represent only limited modal substitution patterns.

As with the option two framework, option three seems to be used as a stepping-stone to more sophisticated pedestrian modeling frameworks. In 2012, five MPOs (10%) used this model structure. In 2017, only New York and Raleigh (2, 4%) continue to do so; the other three MPOs moved to the option four framework, described below.

Option Four. Models with an option four framework use tour/trip mode choice models to distinguish walk or non-motorized trips from trips by other modes. There is considerable variety within this framework: some models are trip-based whereas others are tour- or activity-based; some group walking and bicycling together into a single non-motorized mode whereas others keep them separate; and both multinomial and nested logit models are utilized. This framework accommodates more flexible mode shifts and produces origin–destination matrices of walk trips, but it also requires more walk (and bicycle) trip survey records for estimating the mode choice model.

As the most common pedestrian modeling framework, this is now standard practice. In 2012, 18 MPOs (38%) used this framework; over half of large MPOs (28, 58%) do so in 2017. This increase has largely corresponded with the rise of activity-based modeling; eight of the 10 MPOs that developed new option four models did so in the process of developing activity-based models. Furthermore, most (25, 89%) option four mode choice models include walking and bicycling as separate modes.

Option Five. The final pedestrian modeling framework takes walk or non-motorized trips that are estimated from a pre-/mode choice step and assigns them to a route on a non-motorized network. This may be done crudely by models utilizing the earlier frameworks when calculating walk-specific level-of-service variables (i.e., skims), but only models that go beyond simple shortest-path routing algorithms qualify. With predicted pedestrian paths, many new and more detailed model applications arise: specific non-motorized infrastructure projects, safety analysis, health impact assessments, and more (see the Application section for more details). Many of the same link-based analyses that are currently done for automobile travel could also be performed for walk trips.

For bicycle modeling, this framework represents emerging practice. Incorporating route choice models estimated with field data, forecasting models in places like Minneapolis, Portland, San Diego, San Francisco, and Seattle now route bicycle trips on a network, according to survey responses and published literature (*9, 10*). Pedestrian route choice modeling has lagged behind other aspects of pedestrian modeling and remains in the research realm. To our knowledge, no large MPOs' models currently assign pedestrian trips, although Boston and Phoenix reported planning to start doing so during model updates. Part of this gap is due to a relative lack of research on factors influencing pedestrian route choice (*31*). Another contributing factor may be the fact that link-based pedestrian capacity issues are rare. However, safety issues are a ubiquitous concern that may benefit from pedestrian trip assignment being incorporated into regional models.

Activity-Based Models. Activity-based models (ABMs), while similar to trip-based models, accommodate more complex and realistic linkages between activities and travel both for individuals and for households (*32*). Nevertheless, they still involve a similar process: activity generation, destination choice, mode choice, and network assignment; therefore, we include then in Table 1. Continuing recent trends (*28, 33*), ABMs have become more widespread among large MPOs in the last 5 years, increasing from five (10%) in 2012 to 17 (35%) in 2017. As previously mentioned, this growth in ABMs has brought with it a growth in utilization of the option four framework; only Indianapolis and Milwaukee moved to this framework without also implementing an ABM. It is reasonable to assume that the complex transition to an ABM provided many MPOs the opportunity to add more sophisticated walk or non-motorized modeling capabilities. Furthermore, most ABMs are based on a few common frameworks (*32, 34, 35*) that typically include walking as an option in mode choice models. In the 2017 survey, seven MPOs (23%)—Baltimore, Boston, Cincinnati, Cleveland, Houston, Philadelphia, and Portland—reported preparing to transition to ABMs in the near future; only three of these MPOs already utilize an option four framework. Of course, an ABM does not automatically count as an option four pedestrian modeling framework; see New York for an example.

Geographies and Environments

Spatial/zonal Scales and Networks. The 2017 questionnaire asked respondents to report the spatial/zonal system used to represent travel; there were 23 responses, with 14 responses representing trip-based models and nine from MPOs with ABMs. Many trip-based models had transportation analysis zones (TAZs) that roughly followed census geographies: tracts, block groups, or a combination of the two (with the smaller block groups more commonly used in denser parts of the region). Numbers of zones were in the 1,000–5,000 range. Several ABMs used a greater number of parcels instead. Only one MPO (San Diego) reported using specific smaller zones for non-motorized analysis: approximately five subzones for every TAZ.

The 2017 survey also asked about the networks and speeds used to represent non-motorized travel and to calculate walk/bicycle travel times for mode choice; 12 responses were received. Many networks for non-motorized travel started with the roadway network (including minor arterials or collectors, but often excluding local streets), removed limited-access highways, and added some (usually but not all) off-road paths and trails. ABMs were more likely to utilize an all-streets network. Reported walking speeds ranged from 2.5–3.0 mph and bicycling speeds were in the range of 8–12 mph; these values closely match the same speed ranges documented in the 2012 review (*1*). Denver and Boston reported that their models vary bicycling speeds to account for terrain and specific infrastructure locations.

Built Environment Measures. Both the 2017 questionnaire ($N = 27$) and the 2012 study ($N = 26$) identified specific measures of the built environment that are used to represent walking or non-motorized travel conditions. Density-type variables remain most common: 22 MPOs (85%) in 2012 and 24 MPOs (89%) in 2017 reported using population, employment, and/or activity density. Only five MPOs (19%) used a diversity variable like land-use mix in 2012 to model pedestrian travel; in 2017, this number has grown to 16 reported MPOs (59%). Among those reporting using street network design variables like block, intersection, or street network density, 10 MPOs (38%) modeled walk trips using these variables in 2012; about the same number of MPOs (13, 48%) included these design variables in their models in 2017. Currently, some MPOs are even representing small-scale measures of the pedestrian environment, including traffic control devices (5, 19%), sidewalks and other pedestrian facilities (4, 15%), and traffic calming devices (3, 11%).

Barriers and Challenges

Surveys in both 2012 ($N = 19$) and 2017 ($N = 25$) asked MPO modelers about challenges faced when modeling walking or non-motorized travel and barriers to doing so or doing so more robustly. These barriers and challenges can be grouped into the general categories of data and forecasting as well as decision-maker interest, described in the subsections below.

Data and Forecasting. Behavioral data remains among the most substantial challenges for modeling walking; one MPO modeler stated that a "Lack of data is by far our biggest issue." In 2012, 16 MPOs (84%) described limited records of walking (and bicycling) in household travel surveys as being a barrier. In 2017, all 25 MPOs (100%) reported not having enough non-motorized behavioral data for model development or validation or both, and 21 (84%) listed few resources for data collection as being a challenge. Written responses also noted that travel surveys did not always adequately represent non-motorized travel. Some MPOs reported oversampling walking and bicycling modes in order to help alleviate this problem.

Another data limitation, perhaps more relevant for forecasting, is about measures of the built environment. In the 2017 survey, 12 MPOs (48%) listed forecasting future scenarios as a challenge. This could be tied to challenges measuring or predicting pedestrian-relevant environmental and land-use attributes. Ten MPOs (40%) had insufficient information on the built environment in 2017; in 2012, 11 MPOs (58%) noted limited resources for collecting such data. A related issue (not addressed in the surveys) could be that MPOs often rely on cities and counties to report projects and land-use changes in future scenarios, and there may be variations in how their constituent municipalities collect and archive this information. Additionally, some respondents (in both surveys) noted a more fundamental issue: that a large regional model may not be the best tool for analyzing shorter-distance walk trips.

Decision-Maker Interest. Although data issues remain challenging, one barrier that may be lessening is interest in walking and non-motorized travel from decision-makers. In 2012, nine MPOs (47%) noted a lack of decision-maker interest in their rationale for not further modeling non-motorized travel. In 2017, only four (16%) MPOs reported non-motorized travel not being a priority among decision-makers, and only one (4%) said it was not a priority among the public. From this small-sample survey, it appears that there is increasing interest among decision-makers and the public in being able to understand and model walking and bicycling at a regional level.

Applications

The outputs of travel demand models—walk trips, to varying degrees of specificity—have many potential uses to MPO respondents. To examine planning applications that may be motivating improvements to pedestrian modeling practice, the 2017 survey asked respondents ($N = 23$) to report both current and desired future applications for estimates of walking travel. (The 2012 survey did not examine this issue.) The most common use of these types of non-motorized model outputs was for traditional activities like project prioritization (14, 61%) and scenario planning (10, 43%). Other more specialized applications included corridor planning (10, 43%), traffic safety analysis (8, 35%), health impact assessment (8, 35%), and infrastructure gap analysis (7, or 30%). For instance, MPOs in the Portland and San Francisco Bay areas use their travel demand models to inform public health performance measures related to physical activity (*36*). Interestingly, five MPOs (22%) reported not using estimates or forecasts of walk trips for any particular planning purpose.

Respondents were interested in using estimates of walk (and bicycle) trips for all of these applications, including project prioritization (18, 78%), scenario planning (15, 65%), corridor planning (14, 61%), and traffic safety analysis, health impact assessment, and infrastructure gap analysis (all 13, 57%). These results suggest an interest in and willingness to use the outputs of more robust pedestrian models for a variety of planning applications.

Discussion

This study analyzed the current state of the practice with respect to incorporating pedestrian travel in regional travel demand forecasting models at the 48 largest MPOs. It specifically updated and compared results with a 2012 study on the same subject (*1*), allowing for an analysis of trends in pedestrian modeling. In the paragraphs below, we summarize key findings, discuss innovative practices, and consider future directions in the areas of pedestrian modeling frameworks, behavioral and environmental data (and representations thereof), and interest in and applications for estimates of walking activity.

The last five years have seen a marked increase in the number of agencies whose models represent walk or non-motorized trips to some degree. Furthermore, pedestrian modeling capabilities have become more robust and sophisticated, moving later in the travel demand modeling process and yielding more detailed outputs of walking activity. The option four framework—modeling walking alongside other modes in the mode choice module—has become common practice. From the perspective of model frameworks, innovations will likely soon happen in the area of pedestrian route choice: assigning walk trips to the network. This task has been hampered by a relative lack of research on pedestrian route choice and other characteristics of walking behavior, but a handful of MPOs plan to develop ways to assign pedestrian trips to the network in the near future.

The growing adoption of ABMs has helped to propel pedestrian modeling practice forward. ABMs can capture the constraints and linkages among individual/household travel and often characterize walk trips within mode choice stages by default. They are also more likely to utilize more detailed spatial/zonal structures and more complete networks of streets and pedestrian paths. (Of course, many of these innovations are not limited to ABMs; even a basic four-step model could employ fine-grained built environment measures and more accurately represent walk trip distances by using disaggregate zonal and network structures.) Further diffusion of ABMs may be one of the most effective ways of increasing both the number and sophistication of pedestrian-sensitive regional travel forecasting models. However, this opportunity requires ABM developers to carefully consider the best ways to model walk trips. Representations should take into account current knowledge about pedestrian behavioral influences from research and allow for the emergence of more detailed and widely available pedestrian environment measures.

Regarding representation of the built environment, traditional measures of density, diversity, and design still predominate, although land-use mix (diversity) variables saw a significant increase in use from 2012 to 2017. Some MPOs have begun to collect and represent micro-scale attributes like sidewalks, traffic calming devices, crosswalks, and stop signs (*37*), allowing for the development of more complex and behaviorally sensitive models. A related issue is the representation of zones and networks in travel demand models. Zonal and network structures have not changed significantly: models still tend to use a few thousand zones and rough approximations of street networks. However, as computational limitations become less problematic, models may be able to utilize smaller spatial units such as "pedestrian analysis zones" (*38*), and more detailed networks, similar to ABMs detailed above. These trends may soon allow for walk trips to be modeled while being nearly unaffected by zonal and network aggregation issues.

Although obtaining (and forecasting) measures of the pedestrian environment is challenging, the more critical barrier is a lack of data on walking from both pedestrian counting programs and travel surveys. Better pedestrian models are impossible without better data, and according to our survey results, this is the biggest impediment currently faced by MPOs. Nevertheless, some MPOs are making strides to overcome the significant data barrier. For instance, during their regional household travel surveys, Portland and San Diego specifically oversampled households with higher likelihoods of walk and bicycle trip-making, ensuring that there would be a sufficient sample of trips for estimating mode choice models with these modes. As new travel surveying efforts get underway, it is important to consider how to target a representative and sufficiently large sample of these active travelers for future model estimation.

Although data challenges remain, a lack of decision-maker interest is no longer the barrier it was just 5 years ago. MPOs and their leaders are becoming increasingly aware of the value of being able to analyze, model, and predict walking (and bicycling) travel demand. There is growing interest in using model outputs, including walk trip estimates, for a variety of traditional and emerging planning applications. For instance, MPOs are beginning to consider the health and safety impacts of transportation systems in their regional planning initiatives (*36*). Numbers of walk trips along or across a corridor can be used as a measure of exposure when calculating crash rates and assessing traffic safety. Additionally, walk trips with detailed origins and destinations could be used to calculate units of physical activity (metabolic equivalents, METs), values that are critical inputs to health impact assessment models. Routing walk trips on a pedestrian-relevant network would only improve the precision of these estimates. The increased interest in non-motorized travel modeling and its applications among decision-makers and the public removes a key challenge, leaving, for many MPOs, only resources and data as remaining barriers to a more complete representation of pedestrians in travel demand models.

Conclusion

Through a thorough review of the state of the practice of modeling walk trips among large MPOs, this study contributes to travel demand modeling practice. It fills a gap in the literature in this regard. It provides an updated snapshot of how models currently account for walk and non-motorized trips and sketches broad trends about the evolution of practice. The review also identifies best-practice regional pedestrian modeling techniques and suggests opportunities for the field's next steps.

Staff members at consultancies, MPOs, and other government agencies can use this review to compare their own models to others currently in the field, identifying potential techniques to borrow. Teams looking to improve their pedestrian modeling capabilities can use the summaries contained in this paper to select the framework best suited to their organizations' planning needs and modeling capabilities. Other parties interested in obtaining estimates of walking activity can reference this study when borrowing or developing forecasting procedures of their own. Finally, researchers today and in the future can use this paper as a benchmark against which to evaluate the progress of how urban travel demand models forecast walking. Overall, pedestrian modeling practices still lag behind those for other modes, but they are making strides and moving forward.

Acknowledgements

Thanks to four anonymous reviewers for their valuable feedback. This work was made possible by a project grant from the National Institute for Transportation and Communities (NITC), a program of the Transportation Research and Education Center at Portland State University. The first and second authors' work on this study was also partially supported by graduate fellowships from the Dwight David Eisenhower Transportation Fellowship Program, a program of the Federal Highway Administration of the U.S. Department of Transportation.

Author Contributions

The authors confirm contribution to the paper as follows: study conception and design: PAS, KJC; data collection: JCT; analysis and interpretation of results: PAS, JCT, JPO, RJS, KJC; draft manuscript preparation: PAS, JCT, JPO, RJS, KJC. All authors reviewed the results and approved the final version of the manuscript.

References

1. Singleton, P. A., and K. J. Clifton. Pedestrians in Regional Travel Demand Forecasting Models: State-of-the-Practice. Presented at 92nd Annual Meeting of the Transportation Research Board, Washington, D.C., 2013.
2. Brownson, R. C., C. M. Hoehner, K. Day, A. Forsyth, and J. F. Sallis. Measuring the Built Environment for Physical Activity: State of the Science. *American Journal of Preventive Medicine*, Vol. 36, No. 4, 2009, pp. S99–S123.
3. Saelens, B. E., and S. L. Handy. Built Environment Correlates of Walking: A Review. *Medicine and Science in Sports and Exercise*, Vol. 40, No. 7 Suppl, 2008, pp. S550–556.
4. Ewing, R., and R. Cervero. Travel and the Built Environment: A Meta-analysis. *Journal of the American Planning Association*, Vol. 76, No. 3, 2010, pp. 265–294.
5. Replogle, M. Integrating Pedestrian and Bicycle Factors into Regional Transportation Planning Models: Summary of the State-of-the-art and Suggested Steps Forward. *Urban Design, Telecommuting and Travel Forecasting Conference*, 1997, pp. 75–94. https://trid.trb.org/view/540125.
6. Porter, C., J. Suhrbier, and W. L. Schwartz. Forecasting Bicycle and Pedestrian Travel: State of the Practice and Research Needs. *Transportation Research Record: Journal of the Transportation Research Board*, 1999. 1674: 94–101.
7. Rossi, T. F. Modeling Non-Motorized Travel. Presented at 79th Annual Meeting of the Transportation Research Board, Washington, D.C., 2000.
8. Liu, F., J. Evans, and T. Rossi. Recent Practices in Regional Modeling of Nonmotorized Travel. *Transportation Research Record: Journal of the Transportation Research Board*, 2012. 2303: 1–8.
9. Broach, J., J. Dill, and J. Gliebe. Where do Cyclists Ride? A Route Choice Model Developed with Revealed Preference GPS

Data. *Transportation Research Part A: Policy and Practice*, Vol. 4, No. 10, 2012, pp. 1730–1740.
10. Hood, J., E. Sall, and B. Charlton. A GPS-Based Bicycle Route Choice Model for San Francisco, California. *Transportation Letters*, Vol. 3, No. 1, 2011, pp. 63–75.
11. Zimmermann, M., T. Mai, and E. Frejinger. Bike Route Choice Modeling using GPS Data Without Choice Sets of Paths. *Transportation Research Part C: Emerging Technologies*, Vol. 75, 2017, pp. 183–196.
12. Bomberg, M., L. Zorn, and E. Sall. Incorporating User Based Perspective of Livability Projects in SF-CHAMP Mode Choice Models. *Transportation Letters*, Vol. 5, No. 2, 2013, pp. 83–95.
13. Hood, J., G. Erhardt, C. Frazier, and A. Schenk. Estimating Emissions Benefits of Bicycle Facilities with Stand-Alone Software Tools: Incremental Nested Logit Analysis of Bicycle Trips in California's Monterey Bay Area. *Transportation Research Record: Journal of the Transportation Research Board*, 2014. 2430: 124–132.
14. Kuzmyak, J. R., J. Walters, M. Bradley, and K. M. Kockelman. *NCHRP Report 770: Estimating Bicycling and Walking for Planning and Project Development: A Guidebook*. Transportation Research Board of the National Academies, Washington, D.C., 2014.
15. Aoun, A., J. Bjornstad, B. DuBose, M. Mitman, and M. Pelon. *Bicycle and Pedestrian Forecasting Tools: State of the Practice*. Pedestrian and Bicycle Information Center, Chapel Hill, NC., 2015.
16. Weiner, E. *Urban Transportation Planning in the United States: History, Policy, and Practice*. Springer, New York, NY, 2013.
17. Cambridge Systematics, Inc., and Barton Aschman Associates. *Short-Term Travel Model Improvements*. Final Report, DOT-T-95-05. Travel Model Improvement Program, Federal Highway Administration, U.S. Department of Transportation, Washington, D.C., 1994.
18. 1000 Friends of Oregon. *Analysis of Alternatiaves: LUTRAQ Report*. Cambridge Systematics, Inc., and Parsons Brinckerhoff Quade & Douglas, Inc., Portland, Oregon, 1996.
19. Clifton, K., and C. Muhs. Capturing and Representing Multimodal Trips in Travel Surveys: Review of the Practice. *Transportation Research Record: Journal of the Transportation Research Board*, 2012. 2285: 74–83.
20. Purvis, C. *Review of the Literature: Home-Based Work Mode Choice Models: Technical 5*. Memorandum HBWMC #1. Memo. Metropolitan Transportation Commission, Oakland, CA, 1997.
21. 1000 Friends of Oregon. *Model Modifications: Volume 4 (revised). Making the Land Use Transportation Air Quality Connection*. Cambridge Systematics, Inc., Parsons Brinckerhoff, Quade & Douglas, and S.H. Putman Associates, Inc., Portland, Oregon, 1996.
22. Institute of Transportation Engineers. *Travel Demand Forecasting Processes Used by Ten Large Metropolitan Planning Organizations*. Institute of Transportation Engineers, Washington, D.C., 1994.
23. Cervero, R., and K. Kockelman. Travel Demand and the 3Ds: Density, Diversity, and Design. *Transportation Research Part D: Transport and Environment*, Vol. 2, No. 3, 1997, pp. 199–219.
24. 1000 Friends of Oregon. *The Pedestrian Environment: Volume 4A. Making the Land Use Transportation Air Quality Connection*. Parsons Brinckerhoff Quade & Douglas, Inc., Cambridge Systematics, Inc., and Calthorpe Associates, Portland, Oregon, 1993. http://www.friends.org/resources/reports#LUTRAQ.
25. Nourzad, F. *Modeling of the Non-Motorized Travel Interregional Travel Model Development*. (Phase I-Task 3). Memo. Wasatch Front, Regional Council, Salt Lake City, UT, 2000.
26. Eash, R. Incorporating Urban Design Variables in Metropolitan Planning Organizations' Travel Demand Models. *Urban Design, Telecommuting and Travel Forecasting Conference*, 1997, pp. 167–186. https://trid.trb.org/view/540125.
27. Eash, R. Destination and Mode Choice Models for Nonmotorized Travel. *Transportation Research Record: Journal of the Transportation Research Board*, 1999. 1674: 1–7.
28. *Special Report 288*: *Metropolitan Travel Forecasting: Current Practice and Future Direction*. TRB, National Research Council, Washington, D.C., 2007. http://onlinepubs.trb.org/onlinepubs/sr/sr288.pdf.
29. Vanasse Hangen Brustlin, Inc. *Determination of the State of the Practice in Metropolitan Area Travel Forecasting: Findings of the Surveys of Metropolitan Planning Organizations*. TRB, National Research Council, Washington, D.C., 2007. http://onlinepubs.trb.org/onlinepubs/reports/VHB-2007-Final.pdf.
30. Transportation Planning Capacity Building Program. *Metropolitan Planning Organization (MPO) Database*. Federal Highway Administration, U.S. Department of Transportation, Washington, D.C., 2014. https://www.planning.dot.gov/mpo.asp.
31. Broach, J., and J. Dill. Using Predicted Bicyclist and Pedestrian Route Choice to Enhance Mode Choice Models. *Transportation Research Record: Journal of the Transportation Research Board*, 2016. 2564: 52–59.
32. Castiglione, J., M. Bradley, and J. Gliebe. *SHRP 2 Report S2-C46-RR-1: Activity-Based Travel Demand Models: A Primer*. Transportation Research Board of the National Academies, Washington, D.C., 2015.
33. Donnelly, R., G. D. Erhardt, R. Moeckel, and W. A. Davidson. *NCHRP Synthesis 406: Advanced Practices in Travel Forecasting: A Synthesis of Highway Practice*. TRB, National Research Council, Washington, D.C., 2010.
34. Bowman, J. L., and M. E. Ben-Akiva. Activity-based Disaggregate Travel Demand Model System with Activity Schedules. *Transportation Research Part A: Policy and Practice*, Vol. 35, No. 1, 2001, pp. 1–28.
35. Vovsha, P., J. Freedman, V. Livshits, and W. Sun. Design Features of Activity-based Models in Practice: Coordinated Travel-Regional Activity Modeling Platform. *Transportation Research Record: Journal of the Transportation Research Board*, 2011. 2254: 19–27.
36. Singleton, P. A., and K. J. Clifton. Considering Health in US Metropolitan Long-range Transportation Plans: A Review of Guidance Statements and Performance Measures. *Transport Policy*, Vol. 57, 2017, pp. 79–89.
37. Singleton, P. A., R. J. Schneider, C. D. Muhs, and K. J. Clifton. *The Pedestrian Index of the Environment (PIE): Representing*

the Walking Environment in Planning Applications. Presented at 93rd Annual Meeting of the Transportation Research Board, Washington, D.C., 2014.
38. Clifton, K. J., P. A. Singleton, C. D. Muhs, and R. J. Schneider. Representing Pedestrian Activity in Travel Demand Models: Framework and Application. *Journal of Transport Geography*, Vol. 52, 2016, pp. 111–122.

The Pedestrians and Cycles Section (ANF00) peer-reviewed this paper (18-00963).

Article

TRR
JOURNAL OF THE TRANSPORTATION RESEARCH BOARD

Transportation Research Record
2018, Vol. 2672(35) 69–78

DOI: 10.1177/0361198118790645
journals.sagepub.com/home/trr
SAGE

Exploration of Pedestrian Assertiveness and Its Association with Driver Yielding Behavior at Uncontrolled Crosswalks

Mohammad Razaur Rahman Shaon[1], Robert J. Schneider[2], Xiao Qin[1], Zhaoxiang He[1], Aida Sanatizadeh[1], and Matthew Dreis Flanagan[1]

Abstract

Improving driver yielding to pedestrians at crosswalks may help prevent pedestrian fatalities, which have increased over the last decade in the United States. The level of assertiveness exhibited by pedestrians when they arrive at a crosswalk may have a significant impact on driver yielding behavior, but assertiveness is not defined clearly or studied thoroughly in the literature. This study defined three levels of pedestrian assertiveness and collected naturalistic video data at two uncontrolled crosswalks in Madison and Milwaukee, Wisconsin to explore the relationship between pedestrian assertiveness and driver yielding behavior. Driver yielding rates were 71% for pedestrians exhibiting Level 1 (high), 30% for Level 2 (moderate), and 3% for Level 3 (low) assertiveness. The pedestrian assertiveness definitions were also used to assess the potential impact of a high-visibility enforcement (HVE) program in the communities where the study took place. Observations taken after the HVE program showed a significantly higher rate of driver yielding to pedestrians exhibiting a moderate level of assertiveness. This result is promising, since a moderate level of assertiveness may be reasonable for pedestrians to adopt, especially if supported by educational messages for pedestrians to clearly indicate their intent to cross within a crosswalk. This exploratory study provides a framework for future analysis and highlights the need for additional research on the relationship between pedestrian assertiveness and driver yielding behavior.

According to the National Highway Traffic Safety Administration (NHTSA), there was a 12.1% increase in annual pedestrian fatalities between 2006 and 2016 (*1*). In 2016, the Governors Highway Safety Association estimated that nearly 6,000 pedestrians died while crossing the street in the United States (*2*). For such a seemingly simple task, this number is far too high. These recent increases in pedestrian fatalities underscore the need to protect pedestrians from crashes with motor vehicles and reduce their risk on roadways. Further, fatalities only represent the most harmful outcomes of these crashes – serious pedestrian injuries number more than 10 times the number of fatalities each year, which can have life-time impacts on the people involved, in addition to their family, friends, and co-workers (*1*).

An improved understanding of how motorists and pedestrians interact at pedestrian crossings can lead to better engineering, education, and enforcement strategies to improve pedestrian safety. Most state traffic laws require drivers to yield the right-of-way to pedestrians crossing within a crosswalk provided the driver has sufficient time to do so. Yet, these laws are often violated throughout North America, resulting in pedestrians being struck (*3*). The failure of drivers to yield to pedestrians at crosswalks was the most common contributing circumstance cited in Wisconsin pedestrian crash reports between 2011 and 2015 (*4*). Similarly, between 2008 and 2012, 37% of pedestrian crashes reported in California involved motorists violating the "pedestrian right-of-way" law (*5*). Therefore, it is important to observe driver yielding behavior and gather better information about driver and pedestrian interactions. Note that Wisconsin State Statute 346.24 requires drivers to yield the right-of-way to pedestrians rather than stop for pedestrians (*4*), which is required in several other states.

This study explores the relationship between pedestrian assertiveness and driver yielding behavior. Previous

[1]Department of Civil and Environmental Engineering, University of Wisconsin-Milwaukee, Milwaukee, WI
[2]Department of Urban Planning, University of Wisconsin-Milwaukee, Milwaukee, WI

Corresponding Author:
Address correspondence to Mohammad Razaur Rahman Shaon: mshaon@uwm.edu

studies have explored pedestrian characteristics such as age, gender, and race, but pedestrian assertiveness has only been examined by a few researchers. The video data collected in this study made it possible to define different categories of pedestrian assertiveness. Pedestrian assertiveness data were also compared with vehicle speed and vehicle distance to crosswalk when the pedestrian arrived, to provide a complete understanding of the complex interactions between drivers and pedestrians at uncontrolled crosswalks. As one component of a larger Wisconsin Department of Transportation (Wisconsin DOT) project, the definitions of assertiveness developed for this study were also applied to assess the impact of a high-visibility enforcement (HVE) program on pedestrian and driver interactions (*4*).

Factors Affecting Driver Yielding Behavior

Although roadway geometric design features and treatments – such as the number of roadway lanes and crossing distance, posted and actual speeds, and signs, markings, and beacons – have been found to have a significant relationship with driver yielding (*3*, *6–17*), only a small number of studies have examined how pedestrians and drivers interact and whether or not this interaction might also influence driver yielding behavior (*18–21*). Several studies identify relationships between pedestrian characteristics and driver yielding, hinting at the value of examining the role of pedestrian behavior directly. For example, Stapleton et al. found that driver yielding rates increase with higher pedestrian volume (*10*). This suggests that drivers tend to yield more when the intensity of pedestrians crossing a roadway increases or when pedestrians cross in groups. Personal pedestrian characteristics such as race (*22*, *23*), clothing color (*20*), and holding a cane (*24*) have also been associated with driver yielding behavior. Results of these studies suggest that drivers yield more often to white pedestrians, pedestrians wearing brighter clothing, and those holding a cane while crossing. Neighborhood socioeconomic characteristics may also be related to driver yielding behavior. Coughenour et al. noted that black pedestrians needed to wait for twice the number of cars to pass the crosswalk before being able to cross in high-income areas compared to lower-income neighborhoods (*23*).

Further, relatively few driver factors related to yielding behavior have been identified in the literature. Piff et al. noted that drivers of more expensive automobiles may yield less often to pedestrians than drivers of less expensive automobiles (*25*). A study conducted in Alberta, Canada found that more than twice as many men stopped for pedestrians as women (*19*). However, the authors noted that this conclusion may not apply in all contexts. Driver factors that could influence yielding behavior but which have not been explored in the literature include, but are not limited to: speed preferences, their perceived value of their time, vehicle constraints, social attitudes, economic considerations, and awareness of law enforcement. Traffic characteristics may also contribute to the decision to yield to pedestrians. Schroeder and Rouphail noted that when vehicles are in a platoon, they are less likely to yield (*20*).

Pedestrian–vehicle interaction is a topic that has gained traction in recent years following advancements in computer-aided modeling techniques. Chen et al. discussed the importance of using vehicle and pedestrian characteristics along with genetic-based modeling parameters to attain an accurate representation of pedestrian–vehicle interactions (*26*). A novel contribution was made when the sociological attribute of herd mentality was applied to computer-based pedestrian–vehicle modeling. This refers to the tendency of pedestrians to follow a more aggressive pedestrian when crossing. Understanding what role factors such as these play in pedestrian–vehicle interactions is key to developing more accurate modeling, allowing for increased efficiency and safety for both drivers and pedestrians. Jin et al. proposed guidelines to model pedestrian–vehicle interactions and developed a modified visual angle car-following model with consideration of the lateral effect of waiting pedestrians (*27*). The authors noted that considering pedestrian crossing behavior in a car-following model may help to make better-informed decisions about traffic flow efficiency and safety.

Concept of Assertiveness

Pedestrian assertiveness has been suggested as an important factor in a theoretical framework representing driver yielding (*3*), but relatively few studies have examined this behavior (Table 1) (*18–21*). Assertiveness is a characteristic that the pedestrian shows before or during the crossing activity (*3*, *20*). Pedestrian assertiveness may impact driver yielding behavior by increasing visibility, making the driver more aware of the presence of a pedestrian at the crosswalk, or by clearly indicating an intention to cross (as opposed to other actions, such as waiting for a bus or waiting to meet friends on the sidewalk).

The small number of studies that have examined pedestrian assertiveness do not use a common definition of this behavior (*18–21*, *28*). For example, Harrell sought to understand the impact of pedestrian assertiveness (or pedestrian boldness) on the yielding rates in pedestrian–vehicle interactions (*19*). Harrell's study proposed a method to classify assertive pedestrians as a function of where they stood before crossing. Assertive (bold) pedestrians were those who waited in the street before attempting to cross (stood "just off the curb"). Passive (meek)

Table 1. Summary of Literature on Assertiveness

Authors	Assertiveness definition	Use of assertiveness in analysis	Observation method	Intersection type	Study area
Crowley-Koch et al. (*21*)	Pedestrians either extended their arm or raised their arm. These two variables were compared to the control (pedestrian making no arm movements)	Yes	Manual field observations	Uncontrolled intersections	Chicago and Western Michigan
Harrell (*19*)	Bold – pedestrian stood just off the curb and in the passing zone. Meek – pedestrian stood at least 15 cm from the edge of the curb	Yes	Manual field observations	Unsignalized intersections	Alberta, Canada
Schroeder and Rouphail (*20*)	Pedestrian assertiveness was defined as binary variable, defined as whether or not the pedestrian walked fast when crossing the road	Yes	Video observations	Unsignalized midblock crossings	North Carolina
Schroeder (*18*)	Pedestrian with faster speed and lower gap acceptance	Yes	Video with a speed gun	Unsignalized midblock crosswalk	North Carolina
Mitman et al. (*28*)	How many times pedestrian looked before crossing and how big a gap the pedestrian required to cross	Yes	Manual field observations	Marked, unmarked, and midblock crossings	San Francisco and Lake Tahoe

pedestrians were those who waited on the curb before crossing ("stood 15 cm from the edge of the curb"). Their findings showed that pedestrians who were assertive and wore brightly colored clothing received the highest driver yielding rate (75%), while pedestrians who were meek and wore muted (drab) clothing received the lowest driver yielding rate (39%).

Body or arm gestures have also been used as an indication of pedestrian assertiveness. Crowley-Koch et al. found that pedestrians who used assertive arm motions (i.e., extended arm or raised hand) experienced higher yielding rates than pedestrians who made non-assertive motions (*21*). Pedestrian pace has also been used to define assertiveness. Pedestrians who approach crosswalks quickly may generate a more alarming visual cue and are yielded to more often than pedestrians approaching in a slow manner (*18*, *20*).

Pedestrian assertiveness may also vary significantly between individuals and communities (*3*). A public survey conducted for a separate part of the Wisconsin DOT study suggested that many pedestrians are cautious or afraid to assert their legal right to cross because they do not trust drivers to yield (*4*). Of approximately 390 online survey respondents, 39% perceived that pedestrians in their local community would not cross the street in an uncontrolled, marked crosswalk on a residential two-lane road with a motor vehicle approaching (assuming that the driver had time to stop), and 70% said that local pedestrians would not cross the street in a similar situation on a major four-lane road (*4*).

Video Data Processing

To explore the relationship between pedestrian assertiveness and driver yielding behavior, video data were collected at one uncontrolled intersection in Milwaukee (Kenwood Boulevard and Farwell Avenue) and one uncontrolled intersection in Madison (Dayton Street and Charter Street) near the University of Wisconsin-Milwaukee and the University of Wisconsin-Madison, respectively. The study sites in Milwaukee and Madison have traffic volumes of 5,000 and 6,000 vehicles per day and speed limits of 30 mph and 25 mph, respectively. Both parking and bike lanes are present upstream and downstream of the crossing location at the Milwaukee study site. In Madison, a bike lane is present, but there is no on-street parking near the study location. The crosswalk lengths at the Milwaukee and Madison sites are 71 feet and 40 feet, respectively. The Milwaukee crossing is longer because it includes a raised median. At each site, pedestrians cross one travel lane in each direction. Both intersections have high pedestrian volumes during

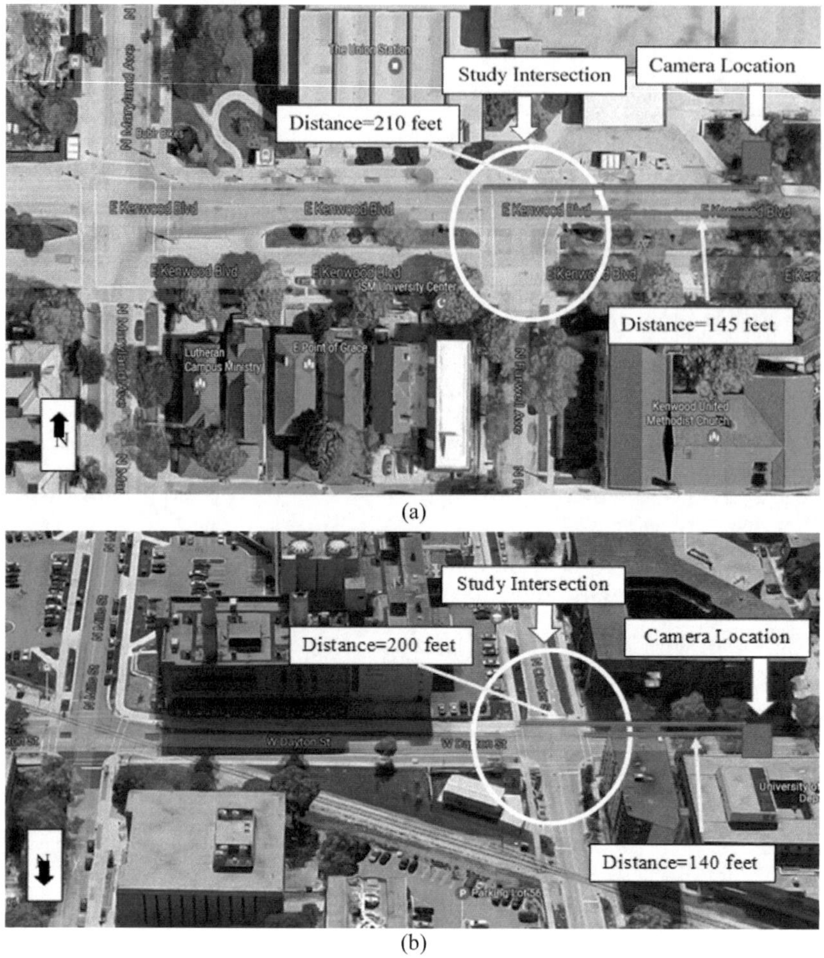

Figure 1. (*a*) Milwaukee video observation intersection (E. Kenwood Boulevard and N. Farwell Avenue); (*b*) Madison video observation intersection (W. Dayton Street and N. Charter Street).

the day because they are close to university campuses. Video data were collected during midday periods at both sites (approximately 9:30 a.m. to 2:30 p.m.) before and after HVE efforts were conducted in Milwaukee and Madison. Figure 1 shows the study intersections and video camera locations at (a) Milwaukee and (b) Madison.

The research team analyzed the recorded video to collect detailed pedestrian assertiveness and driver yielding observations for both study sites. Importantly, the video observations were naturalistic (i.e., they did not use a staged pedestrian). This allowed the research team to document naturally occurring interactions between drivers and pedestrians in both communities.

Several other pedestrian variables were also collected, including demographic characteristics (age, gender, ethnicity), waiting location (e.g., curb versus street), and pedestrian crossing group size. Characteristics of the first automobile with an opportunity to yield to the pedestrian were observed, including vehicle type, vehicle distance from crosswalk when the pedestrian arrived, estimated vehicle speed, and whether or not the driver yielded. Driver yielding behavior was classified as:

- Hard yield (HY) – the driver yielding scenario in which the pedestrian is assertive, and the driver is located closer to the intersection forcing the driver to yield. In this case, the driver decelerated quickly to allow the pedestrian to cross, demonstrated by the vehicle hood going down visibly;
- Soft yield (SY) – the driver decelerated comfortably to allow the pedestrian to cross;
- No yield (NY) – the driver did not yield to the pedestrian who was waiting to cross.

Table 2. Descriptive Statistics of Variables Collected from Video Data

Characteristics	Variables	Type	Total	Yielding	Yielding percentage
Pedestrian	Gender	Male	120	52	43.33
		Female	35	15	42.86
		Both[a]	24	7	29.17
	Race	Black	5	4	80.00
		White	174	70	40.23
	Group size[b]	1	136	61	44.85
		>1	43	13	30.23
	Standing location	In street	41	23	56.10
		On curb	138	51	36.96
Vehicle	Vehicle type	Car	103	45	43.69
		SUV	25	8	32.00
		Van	15	7	46.67
		Bus	5	2	40.00
		Truck	31	12	38.71
	Vehicle location[c]	<20 m	35	21	60.00
		20–50 m	91	35	38.46
		>50 m	53	18	33.96
	Vehicle speed[d]	10 mph	10	6	60.00
		15 mph	48	25	52.08
		20 mph	68	29	42.65
		25 mph	48	13	27.08
		30 mph	5	1	20.00

[a]The "both" gender category was used when the pedestrian group size was more than one and at least one male and female were each present in the group.
[b]The "group size" is defined as the total number of pedestrians waiting to cross at one time, as long as at least one person in the group is intending to enter the crosswalk. If additional pedestrians arrived after the initial pedestrian or pedestrian group passed the roadway centerline, this additional pedestrian (or group) was recorded as a separate observation.
[c]The "vehicle location" is the distance of the vehicle from the crosswalk when the pedestrian first arrived at the curb at the beginning of the crosswalk. This variable was collected in three categories (<20 m, 20–50 m and >50 m).
[d]The "vehicle speed" is the vehicle's actual approach speed, estimated at the same time the pedestrian arrived at the crosswalk and intended to cross the intersection. Vehicle speed was rounded to the nearest 5 mph for simplicity.

The research team was not able to collect driver demographic characteristics because drivers were not clearly visible in the video. To reduce errors of subjectivity, two individuals each made observations separately and then cross-validated their data. Adjustments were made where disagreements were found. Overall, there were disagreements on only 10 observations (less than 6%) out of the 179 records, indicating the video observation protocol generally worked well.

Data Analysis

Before analyzing the association between pedestrian assertiveness and driver yielding behavior, a preliminary analysis was conducted to explore the relationship between several pedestrian- and driver-related characteristics and driver yielding behavior (Table 2).

The descriptive statistics provided in Table 2 show that the driver yielding rate is almost the same for both male and female pedestrians. There were not enough observations to draw conclusions about yielding rates for pedestrians with different racial backgrounds. Interestingly, the yielding rate was lower for pedestrians who waited in groups rather than as individuals. The result is statistically significant at the 95% confidence level ($p = 0.044$). One possible reason could be pedestrians crossing in larger groups are less likely to be assertive while crossing. Pedestrians could also be more cautious because they are concerned about the safety of other members of the group while crossing. It is also possible that when a larger group is waiting to cross, pedestrians do not want to start crossing on their own and stand out from the group. For example, several groups of children attempted to cross the study intersection. The video showed the group waiting until they find the largest safe gap to cross together. Pedestrians standing in the street had a higher yielding rate (56%) than pedestrians standing on the curb (37%). A *Z*-test between the two yielding rates for pedestrians' standing location shows that the difference is statistically significant at the 95% confidence level ($p = 0.014$). Different vehicle types had different yielding rates: SUVs and trucks were less likely to yield to pedestrians (yielding rate = 32% and 39%, respectively). However, the yielding rate for SUVs

Table 3. Categorization of Pedestrian Assertiveness

	Assertiveness Level 1 (high)	Assertiveness Level 2 (moderate)	Assertiveness Level 3 (low)
Pedestrian action	No stopping before crossing (A3)	• Body gesture (A1) • Hand gesture (A2) • Observing traffic from street (A4) • Observing traffic from curb edge (A5)	• Inattentive in street (A6) • Inattentive on curb edge (A7) • Waiting far from curb edge (A8)

($p = 0.144$) and trucks ($p = 0.312$) is not statistically different than for passenger car at the 95% confidence level. As seen in other studies, yielding rates tended to be higher when vehicle speeds were lower: Only 20% of vehicles traveling 30 mph yielded, but 60% of vehicles traveling 10 mph yielded (*7*, *29*). A *Z*-test shows that vehicles traveling 25 mph have a statistically significant lower yielding rate than vehicles traveling at 20 mph at the 95% confidence interval ($p = 0.043$).

Defining Assertiveness

The main analysis focused on pedestrian assertiveness. This variable may have a significant impact on driver behavior, but it is not clearly defined or thoroughly studied in the literature.

For the purposes of this study, pedestrian assertiveness was generally defined as an additional hand or body gesture shown by a pedestrian to improve his or her detection by the driver at the crosswalk. Further, whether or not pedestrians looked at drivers and how closely to the edge of the roadway pedestrians waited were also important indicators of assertiveness. To capture these nuanced behaviors, eight specific pedestrian actions were identified from video data:

- A1: pedestrian uses body gesture to cross;
- A2: pedestrian uses hand gesture to cross;
- A3: pedestrian does not stop before crossing (his or her waiting time was less than 1 s);
- A4: pedestrian stands on street and observes oncoming traffic for an opportunity to cross;
- A5: pedestrian stands on curb (< 50 cm from edge of pavement) and observes oncoming traffic for a minimum of 3 s for an opportunity to cross;
- A6: pedestrian stands in street but is inattentive (does not observe oncoming traffic);
- A7: pedestrian stands on curb (< 50 cm from edge of pavement) but is inattentive (does not observe oncoming traffic);
- A8: pedestrian stands behind curb (>50 cm from edge of pavement) waiting to cross.

Initially, pedestrian assertiveness was categorized into these eight actions. However, some assertiveness categories had similar data, so they likely represented similar concepts. Therefore, the pedestrian assertiveness categories were combined into three levels of assertiveness. This three-level categorization is provided in Table 3:

- Assertiveness Level 1 (high) – crossing without stopping;
- Assertiveness Level 2 (moderate) – stopping but includes making a body gesture or hand gesture to indicate a desire to cross the street, standing in the street and observing (looking at) the traffic, and standing on curb (< 50 cm) and observing the traffic. Among the observations, body gesture (A1) or hand gesture (A2) were always associated with pedestrians who stood on the curb (< 50 cm) and observed the traffic (A5), thus the count of A1 or A2 actions was not counted when summarizing the number of pedestrians in Level 2.
- Assertiveness Level 3 (low) – standing in the street or curb but being inattentive (not looking at the traffic) as well as standing further than 50 cm from the curb.

Relationship between Pedestrian Assertiveness and Driver Yielding Behavior

The pedestrian assertiveness and driver yielding video observations are organized into matrices (Table 4). The top third of the table shows all observations in Madison and Milwaukee; the middle third shows observations made before the HVE program was implemented in each community; the bottom third shows observations made after the HVE program in each community. Several general patterns can be found from all observations (top third of the table):

- Regardless of their vehicle distance from the crosswalk when the pedestrian arrived, drivers tended to yield more often to more assertive pedestrians than less assertive pedestrians: 71% (48 of 68) of drivers yielded to pedestrians exhibiting Assertiveness Level 1; 30% (25 of 82) yielded to pedestrians exhibiting Assertiveness Level 2;

Table 4. Relationship between Pedestrian Assertiveness and Driver Yielding Behavior

Study period	Driver behavior			Pedestrian assertiveness			Total	
	Vehicle location group	Yielding type[a]	Average speed (MPH)	Level 1 (A3)	Level 2 (A1 + A2 + A4 + A5)	Level 3 (A6 + A7 + A8)	Count	Percentage
All	<10 m	HY[b]	18.00	4	6	0	10	28.57
		NY[c]	18.57	4	7	3	14	40.00
		SY[d]	16.82	9	2	0	11	31.43
	10–20 m	HY	–	0	0	0	0	0.00
		NY	20.36	11	35	10	56	61.54
		SY	18.00	24	11	0	35	38.46
	>20 m	HY	–	0	0	0	0	0.00
		NY	21.71	5	15	15	35	66.04
		SY	20.88	11	6	1	18	33.96
	Total		19.72	68	82	29	179	100
Before HVE	<10 m	HY	18.13	3	5	0	8	27.59
		NY	18.50	3	4	3	10	35.71
		SY	17.50	9	1	0	10	35.71
	10–20 m	HY	–	0	0	0	0	0.00
		NY	22.06	7	24	3	34	66.67
		SY	18.82	12	5	0	17	33.33
	>20 m	HY	–	0	0	0	0	0.00
		NY	21.82	1	6	4	11	84.62
		SY	17.50	1	1	0	2	15.38
	Total		20.11	36	46	10	92	100
After HVE	<10 m	HY	17.50	1	1	0	2	28.57
		NY	18.75	1	3	0	4	57.14
		SY	10.00	0	1	0	1	14.29
	10–20 m	HY	–	0	0	0	0	0.00
		NY	17.73	4	11	7	22	55.00
		SY	17.22	12	6	0	18	45.00
	>20 m	HY	–	0	0	0	0	0.00
		NY	21.67	4	9	11	24	60.00
		SY	21.33	10	5	1	16	40.00
	Total		19.30	32	36	19	87	100

Note: 10 mph = 16.1 km/h

[a]The levels in the "yielding type" variable were based loosely on research by Schroeder (*18*).

[b]In the "hard yield (HY)" category, the driver started to decelerate <30 feet in advance of the crosswalk but still stopped or slowed to let the pedestrian cross.

[c]In the "no yield (NY)" category, the driver did not slow down or stop to accommodate the pedestrian crossing.

[d]In the "soft yield (SY)" category, the driver started to decelerate > 30 feet in advance of the crosswalk and either stopped or slowed to let the pedestrian cross.

and 3% (1 of 29) yielded to pedestrians exhibiting Assertiveness Level 3.

- Drivers only made a HY when the pedestrian arrived when they were close to the crosswalk (within 10 m). This suggests that some drivers were braking quickly, just in case the pedestrian continued across the street without stopping. Drivers likely did not make HYs when the pedestrian arrived when they were farther away from the crosswalk because they had more time to choose between yielding or not.
- Driver yielding rates ((HY + SY)/yielding opportunities) were highest (60%) when the pedestrian arrived and the driver was close to the crosswalk (within 10 m). This may seem counterintuitive, but the difference between yielding from a closer distance and further distances is almost entirely accounted for by HYs (likely caused by drivers reacting quickly and cautiously, as described in the previous bullet). The proportion of SYs was similar for all distance categories (31–38%).
- Average vehicle speeds were slightly slower when pedestrians arrived when they were closer to the crosswalk. While these speeds are only slightly different, it is possible that the faster speeds could lead to more focus toward the center of the roadway (a narrower "vision cone") and less recognition of pedestrians waiting at the end of a crosswalk.

Application: Driver Yielding Before and After HVE

Previous studies showed that HVE programs can change driver yielding behaviors and increase pedestrian safety over time (*30*). HVE programs combine enforcement of driver yielding and other laws along with media outreach to increase awareness of pedestrian safety. Wisconsin DOT supported an HVE program near three major university campuses in Milwaukee and Madison, Wisconsin in fall 2016. The program involved police issuing warnings and tickets to drivers and pedestrians for 2–3 weeks in each location. Efforts at each of these three sites were also covered by local media (*4*). These programs were generally viewed as successful by local law enforcement agencies, though the enforcement and media outreach time frames were shorter than the year-long HVE program that resulted in significant increases in driver yielding in Gainesville, Florida (*29*, *31*). This case study applies the pedestrian assertiveness framework from this study to evaluate the potential impact of the HVE program. One possible outcome of HVE is that pedestrians become more confident and more likely to trust drivers to stop when attempting to cross the street, which could be reflected in a higher level of assertiveness. Ultimately, this could help increase driver yielding rates, creating a positive feedback loop.

The assertiveness matrix (Table 4) compares data collected before HVE program implementation (middle third of the table) and after HVE program implementation (bottom third of the table). The overall driver yielding rate (including both HY and SY) increased slightly from 40.2% (37 of 92) before to 42.5% (37 of 87) after the HVE program. To conduct a more detailed statistical test of whether drivers were deciding to yield or not, SY behavior was examined in isolation. This is because HY was likely a reaction to a potentially hazardous situation rather than a more conscious choice to yield or not to yield. Overall, the SY rate increased from 31.5% (29 of 92) to 40.2% (35 of 87) after the HVE program activities. However, according to a one-tailed Z-test of the difference between these two proportions, this increase was not statistically significant at a 90% confidence level ($p = 0.11$). This means that the HVE program activities may have improved the yielding rate, but this improvement could also result from expected random fluctuations in driver yielding rates.

Importantly, the assertiveness framework made it possible to explore how the HVE program may have influenced driver yielding behavior in response to different levels of pedestrian assertiveness. A t-test was conducted on the proportion of drivers who yielded to pedestrians exhibiting Assertiveness Levels 1 and 2 (data were not sufficient to analyze Level 3). For Level 1, 61.1% (22 of 36) drivers did an SY before and 68.8% (22 of 32) drivers did an SY after the HVE program. As expected (caused by pedestrians being so assertive that they "force" drivers to yield), there was no significant difference in driver yielding at this level ($p = 0.26$). For Level 2, 15.2% (7 of 46) drivers did an SY before and 33.3% (12 of 36) drivers did an SY after the HVE program. This increase was significant at the 95% confidence level ($p = 0.027$). While further research is needed, HVE was associated with a higher rate of driver yielding for pedestrians exhibiting a moderate level of assertiveness. This is a promising result. A moderate level of assertiveness may be reasonable for pedestrians to adopt, especially if supported by educational messages to clearly indicate an intent to cross within a crosswalk.

Conclusions and Future Research

Understanding how motorists and pedestrians interact at pedestrian crossings can lead to better engineering, education, and enforcement strategies to improve pedestrian safety. This exploratory study defined three levels of pedestrian assertiveness at uncontrolled crosswalks and classified pedestrians into these categories using video data. These levels of pedestrian assertiveness were then linked to different driver yielding behaviors. Drivers were more likely to yield to pedestrians exhibiting more assertive behaviors. A case study comparing pedestrian assertiveness before and after an HVE program showed that the program may have helped increase driver yielding rates to pedestrians exhibiting moderate levels of assertiveness.

In this study, video data were collected at two intersections: one in Milwaukee and the other in Madison, both near university campuses. More sites should be included in the future to generate more statistically rigorous results. Furthermore, there were some challenges with the video data collection procedures. A tree blocked the view of pedestrians and vehicles on one side of the intersection in Milwaukee, limiting the number of observations at this site. Future studies should carefully scout camera locations before field data collection.

In addition, video data collection equipment should be adjusted carefully to capture images of vehicles and vehicle speeds and images of pedestrians and pedestrian speeds at greater distances from the crosswalk. The vehicle speed data collected for this study were only estimates because of the limitations of the video camera perspective. Since vehicle and pedestrian acceleration and deceleration are important for evaluating yielding behavior, specific measures of these characteristics should be developed in the future to create more representative classifications of pedestrian and driver assertiveness.

This study was exploratory, so further research should be conducted on the relationship between pedestrian assertiveness on driver yielding. Larger samples of video observations should be collected to increase the overall number of pedestrian and driver interactions available for analysis, represent a wider range of uncontrolled crosswalks (e.g., multi-lane crossings, crossings with facilities such as flashing beacons and curb extensions), cover more time periods (e.g., nighttime), and include data from multiple communities with different urban contexts and different pedestrian and driver behavioral norms. Specifically, it would be useful to conduct this study in a state where drivers are required to stop for pedestrians at crosswalks (rather than just yield to pedestrians) and in communities with high levels of police enforcement of pedestrian rights of way. In addition, future studies should also explore how driver yielding behavior may adjust over time as pedestrians behave more or less assertively.

Acknowledgments

The authors would like to acknowledge and thank the Wisconsin Department of Transportation for providing grant funding and guidance for this study.

Author Contributions

The authors confirm contribution to the paper as follows: study conception and design: RRS, RJS, XQ; data collection: RRS, AS, ZH; literature review: MDF, ZH, AS; analysis and interpretation of results: RRS, RJS, ZH; draft manuscript preparation: RRS, RJS, XQ. All authors reviewed the results and approved the final version of the manuscript.

References

1. *Pedestrian Safety*. National Highway Traffic Safety Administration, Washington, D.C. https://www.nhtsa.gov/road-safety/pedestrian-safety. Accessed July 1, 2017.
2. Retting, R. *Pedestrian Traffic Fatalities by State: 2016 Preliminary Data*. Governors Highway Safety Association, Washington, D.C., 2017.
3. Schneider, R. J., and R. L. Sanders. Pedestrian Safety Practitioners' Perspectives of Driver Yielding Behavior across North America. *Transportation Research Record: Journal of the Transportation Research Board*, 2015. 2519: 39–50.
4. Schneider, R. J., X. Qin, M. R. R. Shaon, A. Sanatizadeh, Z. He, P. Wkyhuis, B. Block, K. Santiago, and A. Bill. *Evaluation of Driver Yielding to Pedestrians at Uncontrolled Crosswalks*. Prepared by University of Wisconsin-Milwaukee and University of Wisconsin-Madison Traffic Operations and Safety (TOPS) Laboratory, for the Wisconsin Department of Transportation, 2017.
5. *Transportation Injury Mapping System (TIMS)*. Safe Transportation Research and Education Center, Berkeley, CA, 2014.
6. Fitzpatrick, K., S. Turner, M. Brewer, P. Carlson, B. Ullman, N. Trout, E. S. Park, J. Whitacre, N. Lalani, and D. Lord. *TCRP Report 112/NCHRP Report 562: Improving Pedestrian Safety at Unsignalized Crossings*. Transportation Research Board of the National Academies, Washington, D.C., 2006.
7. Bertulis, T., and D. Dulaski. Driver Approach Speed and Its Impact on Driver Yielding to Pedestrian Behavior at Unsignalized Crosswalks. *Transportation Research Record: Journal of the Transportation Research Board*, 2014. 2464: 46–51.
8. Lindheimer, T. *Will You Stop for Me? Roadway Design and Traffic Control Device Influences on Drivers Yielding to Pedestrians in a Crosswalk with a Rectangular Rapid-Flashing Beacon*. Final Report. Center for Transportation Safety, Texas A&M Transportation Institute, College Station, TX, 2016.
9. Huang, H., C. Zegeer, R. Nassi, and B. Fairfax. *The Effects of Innovative Pedestrian Signs at Unsignalized Locations: A Tale of Three Treatments*. Report FHWA-RD-00-098. U.S. Federal Highway Administration, Washington, D.C., 2000.
10. Stapleton, S., T. Kirsch, T. J. Gates, and P. T. Savolainen. "Factors Affecting Driver Yielding Compliance at Uncontrolled Midblock Crosswalks on Low-Speed Roadways." *Transportation Research Record: Journal of the Transportation Research Board*, 2017. 2661: 95–102.
11. Bennett, M. K., and R. Van Houten. Variables Influencing Efficacy of Gateway In-Street Sign Configuration on Yielding at Crosswalks. *Transportation Research Record: Journal of the Transportation Research Board*, 2016. 2586: 100–105.
12. Houten, R., J. Malenfant, and D. McCusker. Advance Yield Markings: Reducing Motor Vehicle–Pedestrian Conflicts at Multilane Crosswalks with Uncontrolled Approach. *Transportation Research Record: Journal of the Transportation Research Board*, 2001. 1773: 69–74.
13. Van Houten, R., R. Ellis, and E. Marmolejo. Stutter-Flash Light-Emitting-Diode Beacons to Increase Yielding to Pedestrians at Crosswalks. *Transportation Research Record: Journal of the Transportation Research Board*, 2008. 2073: 69–78.
14. Shurbutt, J., and R. Van Houten. *Effects of Yellow Rectangular Rapid-Flashing Beacons on Yielding at Multilane Uncontrolled Crosswalks*. U.S. Federal Highway Administration, Research, Development, and Technology, Turner-Fairbank Highway Research Center, McLean, VA, 2010.
15. Ross, J., D. Serpico, and R. Lewis. *Assessment of Driver Yielding Rates Pre-and Post-RRFB Installation, Bend, Oregon*. Oregon Department of Transportation, Salem, OR, 2011.
16. Fitzpatrick, K., S. T. Chrysler, R. Van Houten, W. W. Hunter, and S. Turner. *Evaluation of Pedestrian and Bicycle Engineering Countermeasures: Rectangular Rapid-Flashing Beacons, HAWKs, Sharrows, Crosswalk Markings, and the Development of an Evaluation Methods Report*. Report FHWA-HRT-11-039. Research, Development, and Technology, Turner-Fairbank Highway Research Center, McLean, VA, 2011.
17. Houten, R., R. Retting, C. Farmer, and J. Houten. Field Evaluation of a Leading Pedestrian Interval Signal Phase at Three Urban Intersections. *Transportation Research Record: Journal of the Transportation Research Board*, 2000. 1734: 86–92.

18. Schroeder, B. J. *A Behavior-Based Methodology for Evaluating Pedestrian–Vehicle Interaction at Crosswalks*. North Carolina State University, Raleigh, NC, 2008.
19. Harrell, W. A. The Impact of Pedestrian Visibility and Assertiveness on Motorist Yielding. *The Journal of Social Psychology*, Vol. 133, No. 3, 1993, pp. 353–360.
20. Schroeder, B. J., and N. M. Rouphail. Event-Based Modeling of Driver Yielding Behavior at Unsignalized Crosswalks. *Journal of Transportation Engineering*, Vol. 137, No. 7, 2010, pp. 455–465.
21. Crowley-Koch, B. J., R. Houten, and E. Lim. Effects of Pedestrian Prompts on Motorist Yielding at Crosswalks. *Journal of Applied Behavior Analysis*, Vol. 44, No. 1, 2011, pp. 121–126.
22. Goddard, T., K. B. Kahn, and A. Adkins. Racial Bias in Driver Yielding Behavior at Crosswalks. *Transportation Research Part F: Traffic Psychology and Behaviour*, Vol. 33, 2015, pp. 1–6.
23. Coughenour, C., S. Clark, A. Singh, E. Claw, J. Abelar, and J. Huebner. Examining Racial Bias as a Potential Factor in Pedestrian Crashes. *Accident Analysis & Prevention*, Vol. 98, 2017, pp. 96–100.
24. Harrell, W. A. Driver Response to a Disabled Pedestrian Using a Dangerous Crosswalk. *Journal of Environmental Psychology*, Vol. 12, No. 4, 1992, pp. 345–354.
25. Piff, P. K., D. M. Stancato, S. Côté, R. Mendoza-Denton, and D. Keltner. Higher Social Class Predicts Increased Unethical Behavior. *Proceedings of the National Academy of Sciences*, Vol. 109, No. 11, 2012, pp. 4086–4091.
26. Chen, P., C. Wu, and S. Zhu. Interaction Between Vehicles and Pedestrians at Uncontrolled Mid-Block Crosswalks. *Safety Science*, Vol. 82, 2016, pp. 68–76.
27. Jin, S., X. Qu, C. Xu, and D. H. Wang. Dynamic Characteristics of Traffic Flow with Consideration of Pedestrians' Road-Crossing Behavior. *Physica A: Statistical Mechanics and its Applications*, Vol. 392, No. 18, 2013, pp. 3881–3890.
28. Mitman, M., D. Cooper, and B. DuBose. Driver and Pedestrian Behavior at Uncontrolled Crosswalks in Tahoe Basin Recreation Area of California. *Transportation Research Record: Journal of the Transportation Research Board*, 2010. 2198: 23–31.
29. Salamati, K., B. Schroeder, D. Geruschat, and N. Rouphail. Event-Based Modeling of Driver Yielding Behavior to Pedestrians at Two-Lane Roundabout Approaches. *Transportation Research Record: Journal of the Transportation Research Board*, 2013. 2389: 1–11.
30. Van Houten, R., L. Malenfant, R. D. Blomberg, B. E. Huitema, and S. Casella. *High-Visibility Enforcement on Driver Compliance with Pedestrian Right-of-Way Laws*. Report No. DOT HS 811 786. National Highway Traffic Safety Administration, Washington, D.C., 2013.
31. Van Houten, R., L. Malenfant, R. D. Blomberg, and B. E. Huitema. The Effect of High-Visibility Enforcement on Driver Compliance with Pedestrian Right-of-Way Laws: 4-Year Follow-Up. Report No. DOT HS 812 364. National Highway Traffic Safety Administration, Washington, D.C., 2017. https://www.nhtsa.gov/sites/nhtsa.dot.gov/files/documents/812364_highvisibilityenfdrivercomppeds4yearfollowup.pdf. January 2017.

The Pedestrians and Cycles Section (ANF00) peer-reviewed this paper (18-05396).

Article

TRR
JOURNAL OF THE TRANSPORTATION RESEARCH BOARD

Transportation Research Record
2018, Vol. 2672(35) 79–89

DOI: 10.1177/0361198118796961
journals.sagepub.com/home/trr
SAGE

Equitable Access to Services in Large-Scale Pedestrian Facilities

Khaled Abdelghany[1], Ahmed Abdelghany[2], Hani Mahmassani[3], and Isam Kaysi[4]

Abstract

This paper presents a modeling framework for the planning of spatially distributed public service facilities. The framework assists in determining the locations and capacities of public facilities in different contexts including urban planning, architectural design, and special events. Given a preliminary set of feasible facility locations, the framework solves for the optimal capacity of each facility and the associated pedestrian-facility distribution pattern. The capacities are allocated such that the total time spent by all pedestrians is minimized. The framework assumes informed pedestrians where each pedestrian is minimizing the combined access and service times at the facility. The framework is first illustrated using a hypothetical area with few service facilities. Then, the results are presented for the application of the framework to plan the amenity services in a major pilgrimage site in Makkah, Saudi Arabia. The results illustrate the capabilities of the framework in determining the optimal service area for each facility, which is essential for the design of wayfinding schemes that ensure service efficiency. Furthermore, it provides a plan for prioritizing capacity expansion of the facilities such that pedestrians' benefits are maximized.

Planning spatially distributed public service facilities is a fundamental location optimization problem with numerous applications. The problem involves the selection of the optimal locations and capacities of service facilities to maximize the offered level of service and ensure equity among pedestrians. For a pedestrian to receive a service at a facility, she/he has to travel from her/his origin location to a facility of her/his choice. If the facility is busy on arrival, the pedestrian is expected to wait for the service to become available. The waiting time at the facility is a function of the number of pedestrians and the available service capacity at the facility. Figure 1 shows two different public service facilities and their planning represents the problem considered in this paper. Figure 1*a* shows a food distribution center in a refugee camp. Depending on the size of the camp, multiple centers with different service capacities could be planned to provide the refugees with their daily needs. An efficient distribution of these centers in the camp area would significantly reduce the refugees' suffering in obtaining their needs from these centers. Figure 1*b* shows a public restroom facility that serves the pilgrims in the city of Makkah during the pilgrimage season. As one of the largest special events worldwide, the city hosts more than three million visitors during the season. Public restrooms with adequate capacities should be distributed around the city to ensure the convenience of the pilgrims while practicing their rituals. Other examples of public service facilities include post offices, health-care centers, vote-casting stations, banks, and chain-based services, to name a few. Furthermore, in the context of architectural design, the problem involves determining the locations and capacities of restrooms, elevators, and information desks in large-scale public buildings (e.g., business centers, libraries, stadia, etc.).

As there are numerous versions of the public facility planning problem, it has been intensively studied in the literature. Early research work concentrated on developing measures to evaluate service equity among pedestrians is presented by Teitz (*1*) and McAllister (*2*). Marsh and Schilling (*3*) provided a review of equity measures in

[1]Department of Civil and Environmental Engineering, Southern Methodist University, Dallas, TX
[2]College of Business, Embry-Riddle Aeronautical University, Daytona Beach, FL
[3]Department of Civil and Environmental Engineering, Northwestern University, Evanston, IL
[4]Department of Civil and Environmental Engineering, American University of Beirut, Beirut, Lebanon

Corresponding Author:
Address correspondence to Khaled Abdelghany: khaled@lyle.smu.edu

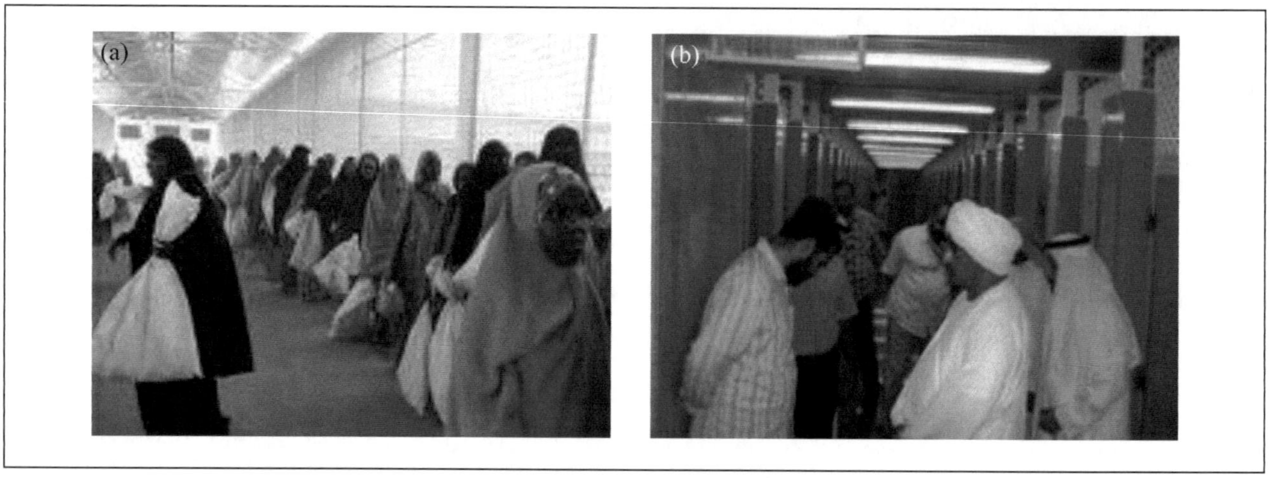

Figure 1. Examples of pedestrians waiting at a public service facility: (a) refugees waiting at a food distribution center in a refugee camp (source: http://www.voanews.com); (b) pilgrims waiting at a restroom facility in Makkah, KSA (source: Center of Research Excellence in Hajj and Omrah).

facility planning and discussed approaches that can be used to select an appropriate measure of equity. Developed models that solve the optimal location problem can be generally categorized, in terms of how the pedestrians' level of service is treated, into two main categories. The first category determines the optimal locations of the facilities such that the pedestrians' travel distance (time) is minimized. The second category minimizes the combined travel and waiting times at the facilities to better represent the pedestrians' level of service.

Hakimi (*4*) and ReVelle and Swain (*5*) are among the first to define the *p*-median problem which is an example of the first category. Given a set of demand centers and a set of potential sites for the service facilities, the problem is to assign the pedestrians of each demand center to one of the sites such that the total travel cost is minimized and the number of open sites is equal to *p*. The problem is shown to be NP-hard, and several heuristic-based approaches are proposed to provide an efficient solution. Reese (*6*) provides a review of main techniques that are used to solve this problem. For example, Christofides and Beasley (*7*) integrated relaxation techniques and a tree search algorithm to solve the problem. Hribar and Daskin (*8*) adopted a dynamic programming heuristic approach in which the capacities of the facilities are ignored and therefore the optimal solutions have the so-called single assignment and closest assignment properties, as described by Krarup and Pruzan (*9*). To consider the capacity of the facilities, the formulation is modified by introducing constraints to define the maximum and minimum capacity of each facility. Examples of the capacity-based formulations are provided by Carreras and Serra (*10*) and Verter and Lapierre (*11*). In addition to considering the capacity constraints, Texeira and Antunes proposed different pedestrian-facility assignment constraints that ensure practical representation of the obtained solution (*12*). For example, a constraint is introduced to ensure that pedestrians from the same origin are assigned to the same facility.

Hansen and Thisse compared the solutions of the Condorcet, Weber, and Rawls locations, which optimally solve for a certain objective (*13*). For example, the Condorcet location is the closest location to the majority of pedestrians. The Weber location problem minimizes the total distance traveled by all pedestrians to go to their facilities. The Rawls location problem minimizes the largest distance traveled by a pedestrian. The maximum coverage location problem is another variation of the p-median problem where constraints on the maximum travel distance/time between any demand center and facility are considered. If the demand center is not within this maximum distance from any of the facilities, the demand of this center is marked as not covered. Different formulations of and solution approaches to this problem can be found in the literature (*14–17*).

As an example of the second model category, a mathematical formulation for the facility optimal location problem with an objective function that minimizes the combined travel and waiting times is presented by Wang et al (*18*). The facilities are modeled as M/M/1 queuing systems and pedestrians are assumed to travel to the closest facility. The model considers an upper bound on the number of facilities that may be opened and the allowable expected waiting time at a facility. Different heuristics are developed to efficiently solve the problem. Berman and Drezner (*19*) extended the model presented by Wang et al (*18*) by allowing multiple servers at the facilities which are modeled as M/M/k queuing systems.

A similar model in which additional constraints are considered on the amount of lost demand due to insufficient coverage and congestion is presented by Berman et al (*20*). The objective is to minimize the number of facilities. An alternative objective function for the formulation which minimizes the maximum time spent by any pedestrian is proposed by Aboolian et al (*21*).

One limitation of the above-mentioned models is that they assume that the service is provided by the nearest facility, which could contribute to unrealistic results predicted by these models. An excellent discussion on the limitations of the proximity assumption adopted in these models is presented by Drezner and Drezner (*22*). For instance, the pedestrians' assignment to facilities follows an unrealistic all-or-nothing behavior especially in the case of high congestion. As such, a small shift in the location of a facility could result in a shift of its entire demand to another closer facility. In addition, service facilities could have different levels of attractiveness. Pedestrians could be willing to travel to a farther facility if they are receiving a higher level of service at that facility. Furthermore, pedestrians usually have difference perception of the actual travel distance/time when deciding on the closest service facility. In other words, they might travel to a farther facility under the perception that it is the closest to their origin location. Models in which the distribution of the demand among the facilities is governed by the gravity rule are presented by Drezner and Drezner (*22*). Following this rule, the probability that a pedestrian selects a facility is assumed to be proportional to the facility's level of attractiveness (e.g., waiting time) and inversely proportional to the travel cost to reach this facility (e.g., travel time).

This paper presents an equilibrium-based formulation for the facility location problem. Pedestrians are assumed to select their facilities such that their total service time (access time plus service time) is minimized, and to have perfect information on the level of congestion at the different facilities. Under these two assumptions, pedestrians are expected to distribute themselves among the different facilities such that a state of equilibrium is achieved. At equilibrium, no pedestrian can improve her/his total service time by unilaterally changing her/his facility. A facility could be equipped by multiple servers. The presented framework consists of two models along with their solution algorithms. The first model solves for the pedestrian distribution pattern at equilibrium for any given facility configuration. The second model solves for the optimal allocation of the capacity among the facilities such that the total service time for all pedestrians is minimized. The framework integrates a simulation-based model to estimate the performance of the general-type queuing systems at the facilities. The framework can be used in determining the service area for each facility, which is essential for designing wayfinding schemes that achieve equity among pedestrians. In addition, the framework could be used to provide a plan for prioritizing capacity expansion at the facilities such that the pedestrians' benefits are maximized. Accordingly, the paper contributes primarily through (1) providing a mathematical formulation for the pedestrian-facility assignment problem and showing that the formulation yields the equilibrium conditions; (2) presenting an equilibrium-based mathematical formulation for the facility location and capacity allocation problem; (3) developing a solution methodology that integrates a simulation model for the queuing systems at the facility which enable solving problems with general-type queuing systems; and (4) illustrating a real-world application of the developed methodology in which the optimal locations and capacities of the restroom facilities are determined for a large-scale special event that is organized as part of the pilgrimage season in Makkah, Saudi Arabia.

Problem Definition and Formulation

Given is an area A which is defined in terms of the possible locations of a set of facilities J that are planned to offer a public service in this area. A service facility at location $j \in J$ is assumed to operate in the form of a queuing system with multiple identical servers. The service interval of a single server at location $j \in J$ is assumed to be a random variable and to follow the probability distribution $g_j(a_j)$, where a_j is the set of parameters of this probability distribution including the average service rate μ_j. Similarly, the pedestrians' inter-arrival interval at location $j \in J$ is assumed to be a random variable and to follow the probability distribution $h_j(b_j)$, where b_j is the set of parameters of this probability distribution including the average arrival rate λ_j. In addition, we define for each location $j \in J$ the maximum number of servers N_{j-max} that can be installed, which depends on the available space at each location. The entire area is divided into a set of demand zones I. Zone $i \in I$ is defined by its location and the total number of pedestrians d_i per unit of time who originate from this zone. Define t_{ij} as the travel (access) time from zone $i \in I$ to facility $j \in J$, which is assumed to be fixed for all $i - j$ pairs. Also, define w_j as the average service time at facility $j \in J$. This service time is assumed to be a function of the number of servers at facility n_j and the pedestrians' arrival rate at the facility λ_j.

Consider the case in which the number of servers n_j at each site $j \in J$ is known. The problem is to determine the demand distribution pattern among the different facilities and the corresponding level of service for each facility. We assume a scenario in which all pedestrians have perfect information on the system in terms of access times to all facilities and the average waiting time at each facility. Pedestrians are assumed to acquire this information based on their day-to-day learning experience or through an advanced information system that feeds this information to the pedestrians. We further assume that all pedestrians are choosing their facility such that each pedestrian minimizes her/his total service times. Considering these assumptions, pedestrians are expected to distribute themselves among the different facilities resulting in a state of equilibrium. At equilibrium, no pedestrian can improve her/his total service time by unilaterally changing her/his facility. Define λ_{ij} as the demand contribution from zone $i \in I$ to facility $j \in J$. The waiting time for each facility is considered to be a function of the arrival rate at the facility, which is computed as the sum of the arrival rates from all demand zones in the area $w_j\left(\sum_i \lambda_{ij}\right)$.

The mathematical program (MP-1) solves for the demand distribution among the facilities at equilibrium. The objective function minimizes the sum of the areas under the performance function for all facilities. Constraints in (2) are the demand conservation constraints. Constraints in (3) define the total demand (arrival rate) for each facility which is equal to the sum of the demand portions that are generated from all zones. Finally, constraints in (4) are the non-negativity constraints.

Mathematical Program MP-1:

$$\text{Minimize } z1 = \sum_j \int_0^{\lambda_j} w_j(v_j)dv + \sum_i \sum_j t_{ij} \cdot \lambda_{ij} \quad (1)$$

Subject to:

$$\sum_j \lambda_{ij} = d_i \qquad \forall i \in I \qquad (2)$$

$$\sum_i \lambda_{ij} = \lambda_j \qquad \forall j \in J \qquad (3)$$

$$\lambda_{ij} \geqslant 0 \qquad \forall i \in I \text{ and } \forall j \in J \qquad (4)$$

To show the equivalency between MP-1 and the equilibrium conditions, an equivalent Lagrangian minimization formulation is derived as given in (5).

$$L(\lambda, u) = \sum_j \int_0^{\lambda_j} w_j(v_j)dv + \sum_i \sum_j t_{ij} \cdot \lambda_{ij} + \sum_i u_i \cdot \left(d_i - \sum_j \lambda_{ij}\right) \qquad (5)$$

where, u_i is the Lagrangian multiplier for the demand conservation constraint for each origin zone $i \in I$, and v_j denotes the demand share of facility j.

The service time function $w_j(.)$ can generally be described as a continuous monotonically increasing function. Thus, the objective function $L(\lambda, u)$ can be assumed to be convex and differentiable in the valid range of λ_{ij}. Solving for the optimal demand split λ^*_{ij} by differentiating (5) with respect to λ_{ij}, two cases can be encountered. First, the minimum value of $L(.)$ occurs at some positive value for λ_{ij}. Second, this minimum value occurs at $\lambda_{ij} = 0$. Equation 6 and the relation in (7) describe these two cases mathematically. Equation 6 shows that at equilibrium the demand portion from zone $i \in I$ to the facility $j \in J$ is more than zero as the service time at j plus the access time from zone i to location j (total service time) is equal to the minimum total service time u_i for any pedestrian that originates from zone i. As illustrated in (7), if the total service time for pedestrians that originate from zone i and target zone j is greater than the minimum total service time u_i for any pedestrian that originates from zone i, the demand share λ_{ij} is equal to zero.

$$\lambda_{ij}.\left((w_j + t_{ij}) - u_i\right) = 0 \qquad \forall i \in I \text{ and } \forall j \in J \qquad (6)$$

$$\left((w_j + t_{ij}) - u_i\right) \geqslant 0 \qquad \forall i \in I \text{ and } \forall j \in J \qquad (7)$$

The problem above solves for the equilibrium pedestrian-facility assignment pattern for a given service configuration (i.e., the locations and capacities of the facilities are given) while ensuring that all pedestrians are receiving the minimum service time. Providing a solution to this problem is essential in determining the expected service area for each facility. Such information is valuable in designing wayfinding schemes that ensure balanced distribution of the pedestrians among the facilities. The presented solution framework can also be used to examine several what-if scenarios related to changes in the facilities' configuration and/or the demand pattern in the study area as presented hereafter.

Another problem that is equally important is to determine the optimal service configuration such that the total time required by pedestrians to receive the service is minimized. In other words, the problem is to determine the optimal locations and corresponding capacities of the facilities such that pedestrians' total service time is minimized. The problem could be solved to develop new facilities and/or to expand the capacities of existing facilities. In both cases, there is a maximum available budget that determines the number of new service units that can be provided. Define N_{max} as the maximum number of service units that can be provided for a given budget. The mathematical program given below (MP-2) formally describes this problem. It solves for the number of servers n_j at each facility $j \in J$ such that the total time spent by the pedestrians in the system is minimized.

Mathematical Program MP-2:

$$\text{Minimize } z2 = \sum_i \sum_j \left(w_j\left(n_j, \sum_i \lambda_{ij}\right) + t_{ij}\right) \cdot \lambda_{ij} \quad (8)$$

Subject to:
(6) and (7)

$$\sum_j n_j \leqslant N_{max} \quad (9)$$

$$n_j \leqslant N_{j-max} \quad \forall j \in J \quad (10)$$

$$\lambda_{ij} \geqslant 0 \quad \forall i \in I \text{ and } \forall j \in J \quad (11)$$

$$n_j \text{ is integer} \quad \forall j \in J \quad (12)$$

The objective function in (8) minimizes the total service time for all pedestrians. The expected waiting time at facility $j \in J$ is assumed to be a function of the number of servers at the facility n_j and the expected arrival rate λ_j at this facility. The number of servers n_j at each facility site is treated in MP-2 as a decision variable. All potential facility locations are considered as an input to the problem. If one or more locations are assigned no servers (i.e., $n_j = 0$), these locations are considered to be not optimal for developing service facilities. While the assumption of the availability of a list of potential locations for facility development significantly reduces the search space of the problem, it follows most current practice as most projects start by scanning the study area to determine an initial set of feasible locations for developing these facilities.

In addition, the demand share λ_j of facility $j \in J$ is expected to depend on the available capacity at the facility. For any service configuration, pedestrians are assumed to be distributed among the different facilities following the equilibrium pattern. Constraints (6) and (7) state that the pedestrian distribution pattern among the facilities follows an equilibrium pattern as described in MP-1. Constraint (9) ensures that the total number of newly added service units is less than the budget limit. Constraints in (10) are required to ensure that the number of servers assigned to a facility is less than the maximum number of servers that can be included at this facility. Finally, constraints in (11) and (12) ensure that λ_{ij} is non-negative and n_j is integer.

Solution Approach

Two solution algorithms are described in this section. The first algorithm is designed to solve MP-1. It determines the pedestrian-facility assignment pattern at equilibrium. The second algorithm solves MP-2 which determines the optimal locations and capacities of the facilities. The developed solution algorithm for MP-1 is based on the Frank–Wolfe decomposition algorithm, which consists of two main steps that are solved iteratively: (a) finding the optimal search direction; (b) determining the optimal step size along this direction (*23*). The following describes the main steps of this algorithm.

Algorithm A1:

Step 0: Initialization

Determine the average access time from all zones to all potential facility locations.

Set iteration counter $itr = 0$.

Generate an initial feasible solution ($\lambda_{ij}^{itr}, \forall i,j$).

Based on the demand share of each facility $\sum_i \lambda_{ij}^{itr}$, simulate the queuing systems at the different facilities to determine their average service times $w_j \forall j$.

Step 1: Direction finding

Determine the auxiliary demand shares φ_{ij}^{itr} $\forall i,j$ by assigning the entire demand of each zone to the facility with the least total service time (i.e., all-or-nothing assignment).

Step 2: Solution update

Update the pedestrian-facility assignment pattern using the method of successive averages.

$$\lambda_{ij}^{itr+1} = \lambda_{ij}^{itr} + \frac{1}{itr+1}\left(\varphi_{ij}^{itr} - \lambda_{ij}^{itr}\right)$$

Determine the average service time at each facility w_j.

Step 3: Convergence checking

Terminate, if $\frac{|z1^{itr} - z1^{itr-1}|}{z1^{itr}} \leqslant \eta$, and percentage of origin-facility pairs that satisfy $\left|\lambda_{ij}^{itr} - \lambda_{ij}^{itr-1}\right| \leqslant \varepsilon$ is greater than M, where η, ε and M are predefined parameters. Otherwise, update iteration counter $itr = itr + 1$ and go to Step 1.

The algorithm starts by finding an initial feasible solution. For instance, a feasible initial solution could be obtained by assigning the entire demand of each zone to the nearest facility. Based on the number of pedestrians that are assigned to each facility, the average service time at each facility is computed. In the current implementation of the algorithm, a simulation routine of a multi-server queuing system is developed to estimate the average waiting time at each facility as described below. In Step 2, an efficient descending direction along the objective function is determined. At the current iteration, the number of pedestrians that are assigned to each facility is assumed to no longer be a function of the waiting time at the facility resulting in a linear approximation of the objective function. This linear approximation of the problem can be solved following an all-or-nothing assignment procedure. Pedestrians originating from each zone are assigned to the facility with the least total service time. The auxiliary demand shares $\varphi_{ij}^{itr}\ \forall i,j$ are then determined. Considering a moving step that is equal to the inverse of the iteration counter, the solution is then updated using the method of successive averages. The average waiting time at each facility is again determined as a function of the number of pedestrians assigned to the different facilities at the current iteration. The algorithm is checked for termination. It terminates if the percentage change in the objective function $z1$ in two successive iterations is within a predefined threshold η, and the demand share for a predefined number of the zone-facility pairs M stabilizes within two successive iterations.

Very few queuing systems have closed analytical forms that can be used to estimate their performance (*24*). Thus, simulation emerges as a plausible alternative approach that can be used to evaluate a wide variety of queuing systems with different arrival and service patterns. A simulation routine is implemented to estimate the average waiting time at each facility. The developed routine follows an event based logic that tracks the number of pedestrians in the queue and their waiting times. A probability distribution function g_j with an average of $1/\sum_i \lambda_{ij}^{itr}$ is assumed to represent the time interval between two successive arrivals at facility $j \in J$. Without loss of generality, if all servers are assumed to be identical, a probability distribution function h_j with an average of $1/\mu_j$ is used to represent the time interval between two service completions by one of the servers at facility $j \in J$. Pedestrians are sorted based on their arrival times. For the earliest pedestrian to arrive at the system, the pedestrian is assigned to any of the available servers and this server is marked as busy. If no servers are available, the pedestrian is kept in the queue. Once the service is completed, the total time spent in the system is recorded for the pedestrian. The process is repeated until all pedestrians complete the service.

Algorithm A2:

while $\sum_j n_j \leq N_{max}$ ***do***
Determine the facility j' with the maximum $\delta_{j'}$, where
$\delta_{j'} = \frac{z2(n_{j'} + \Delta) - z2(n_{j'})}{\Delta}$ such that $\lambda_{ij}\ \forall i,j$ are determined using A1 and
$n_{j'} + \Delta \leq N_{j-max}$
For facility j', update the number of servers $n_{j'} \rightarrow n_{j'} + \Delta$
Record $z2(n_{j'} + \Delta)$ as the overall performance at the current iteration
end

The second algorithm solves MP-2 which determines the optimal capacities of the service facilities. The algorithm adopts an iterative gradient-based search technique. At each iteration, the facility with the maximum marginal improvement in the overall system performance (i.e., the total service time for all pedestrians) is determined. The marginal improvement for a facility δ_j is computed as the improvement in the system performance associated with adding Δ servers at this facility. To determine the marginal system improvement for a facility, the algorithm A1 is used to determine the pedestrian-facility assignment pattern before and after incrementing the capacity at this facility. The marginal improvement is then computed as the difference in the total system performance for the two cases. In each iteration, the marginal improvement for all facilities is computed and the facility with the maximum value is marked as a candidate for capacity expansion. If its expansion capacity allows, the number of servers at this facility is updated. Otherwise, the facility with the next highest marginal improvement and a capacity of expansion is selected. The process is repeated until the maximum available number of service units N_{max} is allocated.

Experiments and Results

The models described above were applied to examine the optimal configuration of a set of service facilities in a hypothetical area. As illustrated in Figure 2, a rectangular area of 8 km length and 4 km width is considered. The area is divided into eight equal demand zones. A set of service facilities is assumed to be distributed in the study area. As illustrated in the figure, two types of facility sites are considered: (a) existing sites and (b) potential sites where new facilities can be developed. Each site is defined in terms of its location, existing number of servers, maximum number of servers that can be included, and the average service rate per server. Each zone is

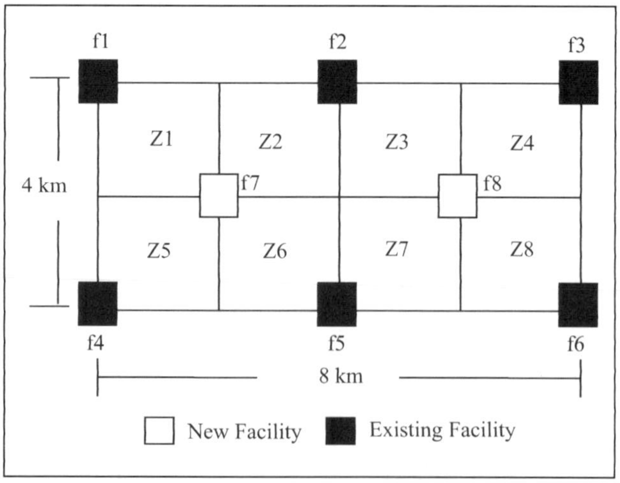

Figure 2. A hypothetical study area.

Table 1. The Pedestrian-Facility Distribution Pattern for the Base Case

Facility	No. of servers	Avg. service time (hour)	Total number of pedestrians	Demand from zones	Travel time to facility (hour)
f1	5	0.52	50	Z1: 50	0.28
f2	45	0.12	420	Z1: 20	0.63
				Z2: 200	0.28
				Z3: 200	0.28
f3	15	0.12	130	Z4: 100	0.28
				Z8: 30	0.63
f4	15	0.12	130	Z1: 30	0.63
				Z5: 100	0.28
f5	45	0.12	420	Z6: 200	0.28
				Z7: 200	0.28
				Z8: 20	0.63
f6	5	0.52	50	Z8: 50	0.28
Total time spent in the system (pedestrian-hours)		555			

assumed to generate a number of pedestrians who seek service at one of the facilities. The demand of each zone is assumed to be generated from an imaginary centroid that is located at the center of the square area representing each zone. The travel time from a zone to a facility is given as the distance between the centroid of the zone and the facility divided by the pedestrians' average walking speed. A base case scenario is considered for this example. In this scenario, zones Z1, Z4, Z5, and Z8 are assumed to generate 100 pedestrians/hour while zone Z2, Z3, Z6, and Z7 are assumed to generate 200 pedestrians/hour. Facilities f1 and f6 are assumed to have 5 servers. Facilities f3 and f4 are assumed to have 15 servers. Facilities f2 and f5 are the largest with 45 servers each. Facilities f7 and f8 are assumed to have no servers in the base case scenario. All servers are assumed to operate at an average server rate of 10 pedestrians/hour. All pedestrians in the study area are assumed to be able to access any of the facilities with an average walking speed of 5.0 km/hour. Finally, the criteria for convergence of algorithm A1 are set such that η, ε, and M are equal to 1.0%, 1.0%, and 95.0%, respectively. Without loss of generality, a Poisson distribution is used to describe the arrival and service processes at all facilities.

Figure 3 illustrates the convergence pattern of Algorithm A1 for the base case scenario. As shown in the figure, the solution quickly stabilizes within 20 iterations. Table 1 illustrates the pedestrian-facility distribution pattern at equilibrium. As shown in the table, the demand of Z1 splits among facilities f1 (50% of the demand), f2 (20% of the demand), and f4 (30% of the demand). Similarly, the demand of Z8 splits among facilities f6 (50% of the demand), f5 (20% of the demand), and f3 (30% of the demand). Considering the equilibrium assumption, all pedestrians of Z1 are shown to have almost the same total service time (walking time

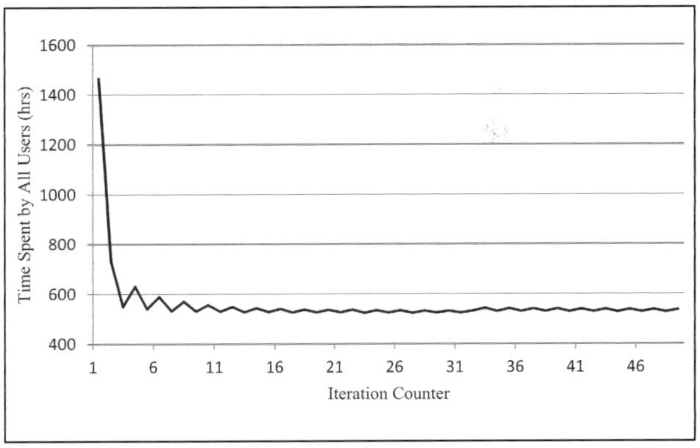

Figure 3. The convergence pattern of Algorithm A1.

plus service time). As shown in the table, the total service time for pedestrians from Z1 that receive service at facility f1 is 0.80 hour. The total service time for pedestrians from zone Z1 that receive service at facilities f2 or f4 is 0.75 hour. The same service time pattern is also obtained for the pedestrians of Z6. Facility f2 (f5) serves all pedestrians of zones Z2 and Z3 (Z6 and Z7) with a total service time of 0.40 hour. Finally, facility f3 (f4) serves all the pedestrians of Z4 (Z5) with a total service time of 0.40 hour. As mentioned earlier, the obtained solution could be used to determine the service area for each facility to ensure that all pedestrians are getting the minimum total service time. Furthermore, defining these service areas could be used in designing wayfinding schemes that ensure that pedestrians are experiencing the least possible total service time. For example, signage could be used to distribute the pedestrians according to the split ratios obtained in the solution. According to the solution in Table 1, wayfinding signage could be distributed in Z1 such that 50% of the demand is directed to f1, 20% of

Table 2. The Pedestrians Distribution Pattern with 5% Demand Increase in Z2, Z3, Z7, and Z8

Facility	No. of servers	Avg. service time (hour)	Total number of pedestrians	Demand from zones	Travel time to facility (hour)
f1	5	0.52	50	Z1: 50	0.28
f2	45	0.14	434	Z1: 14	0.63
				Z2: 210	0.28
				Z3: 210	0.28
f3	15	0.14	136	Z4: 100	0.28
				Z8: 36	0.63
f4	15	0.14	136	Z1: 36	0.63
				Z5: 100	0.28
f5	45	0.14	434	Z6: 210	0.28
				Z7: 210	0.28
				Z8: 14	0.63
f6	5	0.52	50	Z8: 50	0.28
Total time spent in the system (pedestrian-hours)		594			

Table 3. The Pedestrian-Facility Distribution Pattern After Distributing 20 New Servers Among the Facilities

Facility	No. of servers	Avg. service time (hour)	Total number of pedestrians	Demand from zones	Travel time to facility (hour)
f1	15	0.10	100	Z1: 100	0.28
f2	45	0.10	400	Z2: 200	0.28
				Z3: 200	0.28
f3	15	0.10	100	Z4: 100	0.28
f4	15	0.10	100	Z5: 100	0.28
f5	45	0.10	400	Z6: 200	0.28
				Z7: 200	0.28
f6	15	0.10	100	Z8: 100	0.28
Total time spent in the system (pedestrian-hours)		456			

the demand is directed to f2, and the remaining 30% is directed to f4, respectively.

Table 2 gives the pedestrian-facility assignment pattern at equilibrium for the case in which the demand of the inner zones Z2, Z3, Z6, and Z7 is increased by 5%. As illustrated in the table, as the demand of the inner zones increases, facilities f2 and f5 are shown to serve less of the demand of zones Z1 and Z8, respectively. These portions of the demand of Z1 and Z8 are shifted to facilities f4 and f3, respectively. For example, facility f2 serves only 14% of the demand of Z1 compared with 20% in the base case. The demand not served by f2 shifts to facility f4. In the 5% demand increase case, facility f4 serves 36% of the demand of Z1. In this scenario, the demand is distributed such that the equilibrium conditions are satisfied. For example, all pedestrians of Z1 are receiving almost the same total service time at the different facilities (0.80 hour at f1 and 0.77 hour at f2 and f4).

As described above, program MP-2 is designed to determine the optimal allocation of the service units among the facilities. The objective is to distribute a given number of service units among existing and new sites such that the total time spent by all pedestrians is minimized. In this set of experiments, we assume that 20 new server units are to be used. These 20 new servers are assigned to the sites using an increment of five servers in each iteration ($\Delta = 5$). As illustrated in Table 3, the new units are equally split between facilities f1 and f6 with a total of 15 service units at each of them. The addition of these units results in a new pedestrian-facility assignment pattern. Comparing the pedestrian-facility distribution patterns in Tables 1 and 3, adding new service units at facilities f1 and f6 allows these two facilities to serve the entire demand of zones Z1 and Z8, respectively. Consequently, facility f2 serves only the demands of Z2 and Z3, and facility f5 serves only the demand of Z6 and Z8. Increasing the number of service units is associated with an improvement in the total service time. In the base case, the total service time is recorded at 555 hours, which is reduced to 456 hours after the 20 service units are optimally distributed among the facilities.

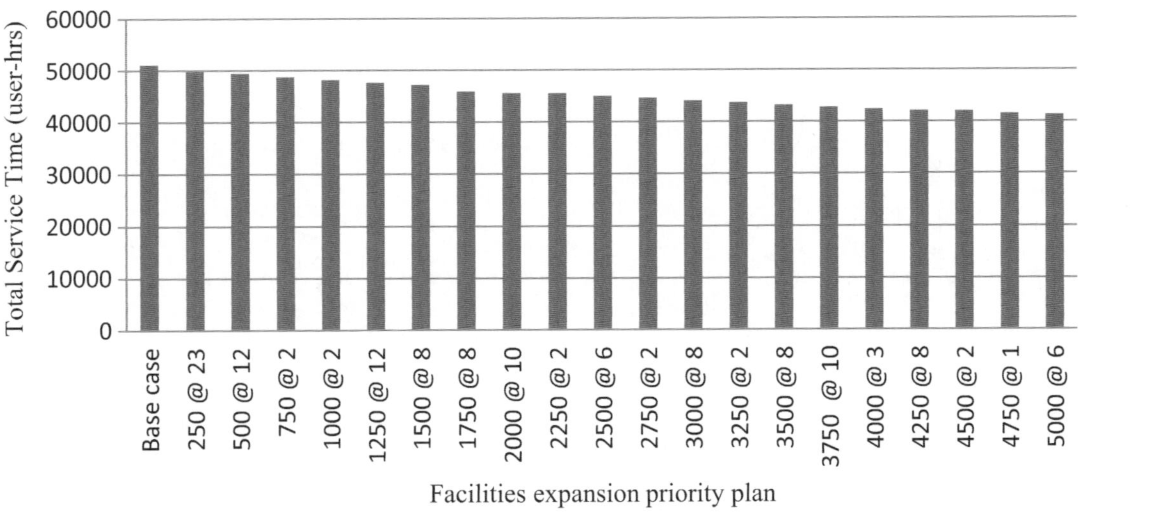

Figure 4. Facility expansion plan and corresponding overall system performance for Arafat Camp.

Real-World Application

In this section, we present a real-world application of the modeling framework described above. The application is related to the planning of the locations and capacities of the amenity service (e.g., restroom/ablution) facilities in the mountain of Arafat, which is a major pilgrimage site in Makkah, Saudi Arabia. All visitors to the city of Makkah are required to spend one day at the mountain of Arafat as part of the pilgrimage rituals. The site is planned in the form of an open camp which is currently capable of hosting about three million pilgrims with a plan to expand its capacity to accommodate five million pilgrims by the year 2040. The area is divided into 42 demand zones with different sizes. The boundaries of these zones are determined mainly by existing streets within the area. The number of pedestrians that are generated from a zone is assumed to be proportional to the area of each zone. Considering the forecasted demand for the year 2040, a homogenous density of one person per two square meters is assumed for all zones. In addition, it is assumed that 5% of the people in a zone seek service at one of the amenity facilities in each hour. Thus, the total number of pedestrians is estimated at 250,000 per hour. Those pedestrians will be distributed among the existing service facilities in the camp. The developed master plan for the camp suggests the construction of 37 service facilities. All peripheral facilities are assumed to have an initial capacity of 500 service units (e.g., restrooms) per facility. Facilities in the central area are assumed to have an initial capacity of 1,000 restrooms per facility. The service rate of all service units is 10 persons/hour. The model is used to determine the optimal allocation of 5,000 additional restrooms among the existing facilities. For the purpose of construction efficiency, if a site is selected for expansion, a minimum of 250 units must be allocated to this site. All facility sites are assumed to have no constraints on expansion.

Figure 4 provides the expansion plan for the facilities in the camp and the expected overall performance after each expansion phase. The x-axis lists the facilities in the order in which they appear in the solution in the successive iterations. An increment of 250 restroom units is used in each iteration. For example, the solution suggests that 250 units should be first added at facility 23. In the next phase, facility 12 is expanded by adding 250 units. Facility 6 is the last facility to be considered for expansion. As illustrated in the figure, a systematic improvement in the overall system performance is recorded after each expansion phase. The total service time is reduced from 51,039 pedestrian-hours in the base case to 41,975 pedestrian-hours after the distribution of the 5,000 new restrooms. Figure 5 illustrates the model's convergence pattern to obtain the pedestrian-facility distribution at equilibrium for the base case and after expansion.

Conclusion

The paper presents a modeling framework for the planning of spatially distributed public service facilities. Pedestrians are assumed to choose their facilities such that a state of equilibrium is achieved. At equilibrium, no pedestrian can improve her/his total service time by unilaterally changing her/his facility. Two mathematical models are presented as part of this framework. The first

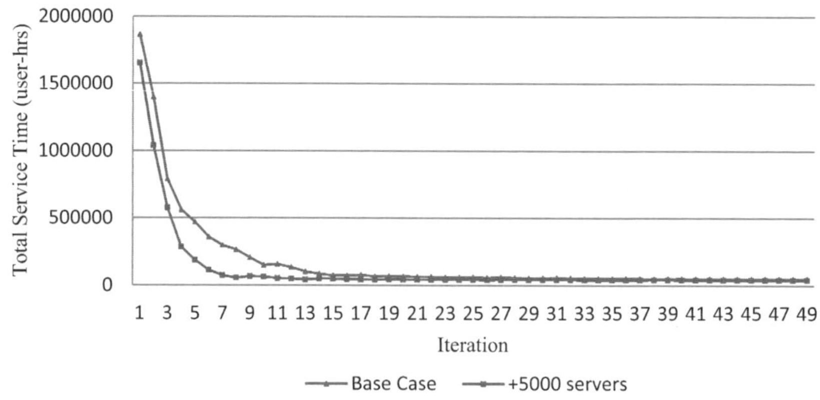

Figure 5. Conversion patterns for the base case and the expansion scenarios.

model determines the pedestrian-facility distribution pattern at equilibrium. The second program solves for the optimal allocation of the capacity among the different facilities such that the total service time for all pedestrians is minimized. The framework is first illustrated using a hypothetical area with simplified configuration. Then, a real-world application is presented related to planning the locations and capacities of amenity services in a major pilgrimage site in Makkah, Saudi Arabia. The results illustrate the capability of the presented framework to determine the service area for each facility, which is essential for the design of wayfinding schemes that ensure that the pedestrians are receiving the best possible service. In addition, the framework provides a scheme for prioritizing capacity expansion at the facilities that maximizes the pedestrians' benefits. Several extensions are considered for this research work. For example, the pedestrian equilibrium approach could be extended to consider the stochastic equilibrium version of the problem where pedestrians do not have perfect information on the facilities' level of service. In addition, solving the problem considering multiple service priority levels and pedestrian classes is another extension. Finally, the locations of the facilities are assumed to be pre-determined in the presented models. These models can be extended to consider the locations of the facilities as additional decision variables to be solved for.

Author Contributions

The authors confirm contributing to the paper as follows: study conception and design: Khaled Abdelghany, Ahmed Abdelghany, Hani Mahmassani and Isam Kaysi; data collection: Khaled Abdelghany and Ahmed Abdelghany; analysis and interpretation of results: Khaled Abdelghany, Ahmed Abdelghany and Hani Mahmassani; draft manuscript preparation: Khaled Abdelghany and Isam Kaysi. All authors reviewed the results and approved the final version of the manuscript.

References

1. Teitz, M. Locational Strategies for Competitive Systems. *Journal of Regional Science*, Vol. 8, 1968, pp. 135–148.
2. McAllister, D. Equity and Efficiency in Public Facility Location. *Geographical Analysis*, Vol. 8, 1976, pp. 47–63.
3. Marsh, M., and D. Schilling. Equity Measurement in Facility Location Analysis: A Review and Framework. *European Journal of Operational Research*, Vol. 74, No. 1, 1994, pp. 1–17.
4. Hakimi, S. Optimum Distribution of Switching Centers in a Communication Network and Some Related Graph Theoretic Problems, *Operations Research*, Vol. 13, No. 3, 1965, pp. 462–475.
5. ReVelle, C., and R. Swain. Central Facilities Location. *Geographical Analysis*, Vol. 2, 1970, pp. 30–42.
6. Reese, J. *Methods for Solving the p-Median Problem: An Annotated Bibliography*. Technical Report. Department of Mathematics, Trinity University, San Antonio, Texas, 2005.
7. Christofides, N., and J. Beasley. A Tree Search Algorithm for the p-Median Problem. *European Journal of Operational Research*, Vol. 10, 1982, pp. 196–204.
8. Hribar, M., and M. Daskin. A Dynamic Programming Heuristic for the p-Median Problem. *European Journal of Operational Research*, Vol. 101, No. 3, 1997, pp. 499–508.
9. Krarup, J., and P. Pruzan. The Simple Plant Location Problem: Survey and Synthesis. *European Journal of Operational Research*, Vol. 12, No. 1, 1983, pp. 36–81.
10. Carreras, M., and D. Serra. On Optimal Location with Threshold Requirements. *Socio-Economic Planning Sciences: The International Journal of Public Sector Decision Making*, Vol. 33, No. 2, 1999, pp. 91–103.

11. Verter, V., and S. Lapierre. Location of Preventative Healthcare Facilities. *Annals of Operations Research*, Vol. 110, 2002, pp. 123–132.
12. Teixeira, J., and A. Antunes. A Hierarchical Location Model for Public Facility Planning. *European Journal of Operational Research*, Vol. 185, No. 1, 2008, pp. 92–104.
13. Hansen, P., and J. Thisse. Outcomes of Voting and Planning Condorcet, Weber and Rawls Location. *Journal of Public Economics*, Vol. 16, 1981, pp. 1–15.
14. Chung, H., D. Schilling, and R. Barbone. The Capacitated Maximal Covering Problem: A Heuristic. *Proc., Fourteenth Annual Pittsburgh Conference on Modeling and Simulation*, Pittsburgh, PA, 1983, pp. 1423–1428.
15. Current, J., and J. Storbeck. Capacitated Covering Models. *Environment and Planning B: Planning and Design*, Vol. 15, 1988, pp. 153–164.
16. Karasakal, O., and E. Karasakal. A Maximal Covering Location Model in the Presence of Partial Coverage. *Computers and Operations Research*, Vol. 31, No. 9, 2004, pp. 1515–1526.
17. Pirkul, H., and D. Schilling. The Maximal Covering Location Problem with Capacities on Total Workload. *Management Science*, Vol. 37, 1991, pp. 233–248.
18. Wang, Q., R. Batta, and C. Rump. Algorithms for a Facility Location Problem with Stochastic Customer Demand and Immobile Servers. *Annals of Operations Research*, Vol. 111, 2002, pp. 17–34.
19. Berman, O., and Z. Drezner. The Multiple Server Location Problem. *Journal of the Operational Research Society*, Vol. 58, 2007, pp. 91–99.
20. Berman, O., D. Krass, and Z. Drezner. Locating Service Facility to Reduce List Demand. *IIE Transactions*, Vol. 38, 2006, pp. 933–946.
21. Aboolian, R., O. Berman, and Z. Drezner. The Multiple Server Center Location Problem. *Annals of Operations Research*, Vol. 167, 2009, pp. 337–352.
22. Drezner, T., and Z. Drezner. The Gravity Multiple Server Location Problem. *Computers and Operations Research*, Vol. 38, 2011, pp. 694–701.
23. Frank, M., and P. Wolfe. An Algorithm for Quadratic Programming. *Naval Research Logistics Quarterly*, Vol. 3, No. 1–2, 1956, pp. 95–110.
24. Gross, D., and C. Harris. Fundamentals of Queuing Theory. In *Wiley Series in Probability and Statistics*, 3rd ed., Wiley-Interscience, Hoboken, NJ, 1998.

The Standing Committee on Pedestrians (ANF10) peer-reviewed this paper (18-04531).

Article

TRR
JOURNAL OF THE TRANSPORTATION RESEARCH BOARD

Transportation Research Record
2018, Vol. 2672(35) 90–100

DOI: 10.1177/0361198118793496
journals.sagepub.com/home/trr
SAGE

Influence of the Built Environment on the Spatiotemporal Variation of Pedestrian Collision in Seoul: An Application of Big Data

Dohyung Kim[1] and Hina Maheshbhai Chanchlani[1]

Abstract

Although it is reasonable to hypothesize that the spatial pattern of pedestrian collisions changes by the time of day, there has been a lack of research that addresses this spatiotemporal variation because of a shortage of pedestrian volume data by time of day. The goal of this paper is to examine the contribution of the built environment to the spatiotemporal variations by employing a big data set of pedestrian count data that is based on cell phone signals on the streets. This paper constructs eight spatial error models that are based on a 3-h time period and a 300 m by 300 m (984 feet) grid cell. Of the twelve built environmental factors, the employment and restaurant factors show consistent correlation with pedestrian collisions. Local roads, office land uses, and bars do not present a correlation. Although the outputs of the models do not show distinct patterns of correlation between built environment and pedestrian collisions by time period, they indicate that the built environment contributes to pedestrian collisions more during nighttime than daytime. This distinction is primarily caused by transportation factors associated with pedestrian collisions when people journey back home during the night period. They include bus stops, rail transit stations, and arterial roads. School is a factor that positively correlates with pedestrian collisions during the school exit time, while population density negatively correlates with collisions during periods with lower pedestrian and automobile volumes. Employment density shows a negative correlation with pedestrian collisions, regardless of the time period.

In discussions about recent trends in transportation planning, walking has been identified as a valuable transportation mode. Walking can be a sustainable transportation option by replacing automobile traffic for short-distance trips. It contributes to more healthy and sustainable communities by promoting active lifestyles and social interaction. However, the danger of unsafe roads is a significant barrier to the promotion of pedestrian activities. Although recent trends show a decline in the number of crashes and fatalities, the perception of danger is still a significant deterrent to pedestrian activities. In the U.S., pedestrians were one of the few groups of road users to experience an increase in fatalities in 2015, totaling 5,376 deaths and 70,000 injured (*1*).

For this reason, a large volume of research has focused on the factors contributing to pedestrian collisions. The contributing factors are broken down into the social characteristics of residents; the physical and land use characteristics of neighborhoods; roadway/traffic characteristics; and the behavior of drivers and pedestrians (*2*). Some research has found the influence of socio-demographic factors on pedestrian collisions. These factors include: low income, high youth population density, high unemployment rate, lower senior population, more linguistically isolated households, and more people taking public transit to work (*3–5*).

Furthermore, a growing body of research has paid attention to the physical environment factors that contribute to pedestrian collisions. In relationship to pedestrian safety, the physical environment includes a wide range of factors ranging from urban form to roadway design. In terms of urban form, research identifies the influence of the dynamics of urban space on pedestrian collisions. The dynamics include land use factors such as population density, commercial and retail strip, and big box stores (*6*, *7*). Some studies also associate pedestrian

[1]Department of Urban and Regional Planning, California State Polytechnic University, Pomona, Pomona, CA

Corresponding Author:
Address correspondence to Dohyung Kim: dohyungkim@cpp.edu

collisions with some pedestrian-scale trip destinations such as schools, retail shops, public transit stops, and bars (*5*, *8*).

A large volume of research also identifies roadway design as a contributing factor in pedestrian collisions. Another study identifies street layout, road networks, sidewalk conditions, and speed limits as influential urban features (*9*). Other studies point out that pedestrian-oriented design factors such as sidewalks, signalized crosswalks, marked crosswalks, roundabouts, speed humps, and cushions contribute to improving pedestrian safety (*10*, *11*). There have been studies that relate pedestrian collisions to behavioral factors such as the yield behavior of drivers, pedestrians' crossing behavior, dart-out, jaywalking, and alcohol consumption (*12–14*). In many cases, roadway design associates with behavioral issues. For example, pedestrians' behavior and vehicles' movements at intersections are different from those at mid-blocks. Thus, much research separately addresses pedestrian collisions at mid-blocks from ones at intersections (*15*, *16*).

Although a large volume of research has addressed the factors contributing to pedestrian collision, it is mostly cross-sectional research. The research was conducted under the assumption that the spatial distribution of pedestrian collisions is consistent, regardless of the time of day. However, it is reasonable to hypothesize that the spatial pattern of pedestrian collisions changes by the time of day. The built environment contributes to the variation of daily pedestrian collision patterns. For example, the areas with the concentration of pedestrian collisions during morning peak hours are probably different from the areas with concentrations during the mid-day period. Public transit is more likely to be a significant contributing factor to the former collisions rather than the latter. There has been a lack of research that addresses the spatiotemporal variations of the daily pedestrian collision patterns. Limited research has shed light on spatiotemporal analysis, with a particular focus on nighttime collisions and their relationship to alcohol consumption (*17*, *18*). With the visualization of the spatiotemporal variation of pedestrian collision by time of day, one study found that pedestrian crashes were most common around 8–9 a.m. and 3–6 p.m. (*19*). Another study also indicates that pedestrian collisions involving children occur frequently during the same time period (*20*). However, this finding is because of the high volume of pedestrians on the street during this time period. It is desirable to measure collision rates with the collision density and the number of pedestrian collisions per pedestrian counts. Conventional pedestrian volume data, which typically counts the volume for only limited time periods of the day, does not support a spatiotemporal analysis by time of day. Therefore, the spatiotemporal research on pedestrian safety is limited.

The purpose of this paper is to examine the spatiotemporal variations of pedestrian collision by time of day and the contribution of the built environment to the variations in the City of Seoul. This paper overcomes the lack of pedestrian volume data by employing a big data set: pedestrian count data based on cell phone signals. Big data refers to the large volume of data collected from the combination of new data collection mechanisms based on ubiquitous digital devices, greatly enhanced storage capacity, and automated systems, as well as enhanced sensing and communication technologies (*21*). By this definition, the pedestrian volume data can be categorized as big data. These data allow the accurate measurement of the numbers of pedestrian collisions per 1,000 pedestrians by hour. After summarizing the built environment factors including transportation, land use, and destination by 300 m by 300 m grid cells, this paper tests the correlation between the built environment and pedestrian collisions, by constructing eight spatial error models by 3 h time periods for 24 h.

Spatiotemporal Analysis

The study area of this paper is the City of Seoul, the capital of South Korea. With a population of 10,297,138, the Seoul was ranked the 14th largest city in the world in 2015. Of 26 "Gus" (which are the largest administrative boundaries in Seoul), this paper selected five Gus because of the availability of pedestrian count data. The Gus selected include: Jung-Gu, Jongro-Gu, Gangnam-Gu, Seocho-Gu, and Songpha-Gu. As of 2016, the population of 1,984,894 residents in these Gus occupies 154.26 square kilometers of land. They are equivalent to about 19.5% and 25.5% of the population and area in Seoul, respectively. They represent various elements of the built environment and land use in Seoul, from the central business district (CBD) of Seoul to high-density residential areas and from the historic downtown to new office-commercial centers.

The spatial unit of analysis for this study is a uniform grid, which is captured with 300 m by 300 m (984 feet), set over the five Gus. The size of the grid cell was selected as this size can generally capture the areas with prototypical land use and built environment patterns in the study area. This methodological approach provides insight into overall local pedestrian safety and built environment (*22*). The grid, which is smaller than other spatial units such as a census tract, facilitates a micro-level analysis and provides a way to include the collisions located along arterials that typically become the border for the units, without giving special consideration. Using the grid cell as the unit of analysis, this paper builds a data set of analysis variables for a total of 1,703 cells within the Gus' boundary. After defining the geographical unit, this study specified a time block of 3 h as a temporal setting. In other words, this study defines eight periods (3 h for each period), sequentially from Period 1 (midnight to 2:59 a.m.) to Period 8 (9 p.m. to 11:59 p.m.).

After setting these spatiotemporal units, this paper summarizes the weighted number of pedestrian collisions per 1,000 pedestrians by the units, employing pedestrian collision data from the Seoul Police Department (SPD) and the pedestrian count data from SK Telecommunication (SKT). SKT, a cell phone service provider, measures pedestrian counts by counting cell phone signals on the streets captured within an area defined by a 50 m by 50 m (159 feet) cell. SKT counts the number of cell phones within the area of the cell (50 m by 50 m) by cross-validating the strengths of the cell phone signals on the street with the database that contains the strength of cell phone signals from each cell. Based on the counts collected, SKT screens out pedestrian signals based on the movements of signals (e.g., speed). SKT finalizes the estimation of pedestrian counts by taking into consideration the rates of cell phone services' market share and any cell phones which were turned off. The data consist of the average hourly pedestrian counts for each month.

The pedestrian collision data within the study area from 2010 to 2014 were available from the SPD. Accompanied by the time of the collision, the data provided not only the accurate locations of the collisions, but also the time of the collisions. For the five years from 2010 to 2014, a total of 4,043 pedestrian collisions occurred in the study area including 85 fatalities and 1,932 severe injury collisions. Instead of counting the number of pedestrian collisions, this study computed the weighted collision counts by injury severity in order to incorporate the severity of collision into the analysis. The collision data from the SPD classify collisions into four categories including: fatality, severe injury, minor injury, and no injury. This study computed the weighted pedestrian collision counts (WPCC) by assigning the values of 2, 1.5, 1.25, and 1 to fatality, severe injury, minor injury, and no injury collisions, respectively. In other words, a fatal collision was counted as 2 rather than 1. Whereas pedestrian collision counts are weighted, the overall trends of the weighted counts are comparable to those of the unweighted collision counts (Figure 1). This indicates that this weighting method does not distort the collision count data.

The WPCC were normalized by the pedestrian count data from SKT. This paper collected the data of 4 months including April, June, September, and December. Representing four seasons in Korea, these data sets can incorporate the seasonal variations of pedestrian volume. Computing the average of pedestrian counts of the 4 months, this study was able to calculate average hourly pedestrian counts (AHPCs). In total, about 5.2 million pedestrian counts were observed. By dividing WPCC by AHPC, this study computed the weighted pedestrian collisions per 1,000 pedestrian (WPCC1K).

The temporal trends of pedestrian counts and weighted collisions are comparable, but the trend of WPCC1K is different from the others. While the pedestrian volume consistently increases from Period 2 to Period 7 and then decreases, pedestrian collisions keep increasing from Period 4 to the end of day. However, the trend of the normalized line, weighted pedestrian

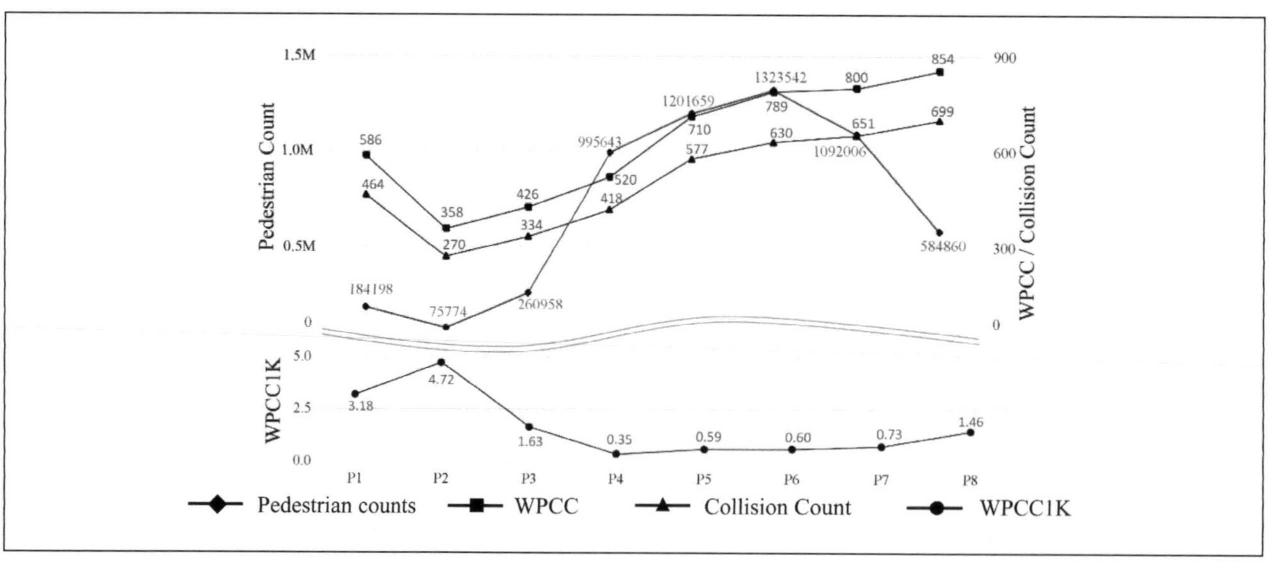

Figure 1. Temporal trend of pedestrian volume and collision.

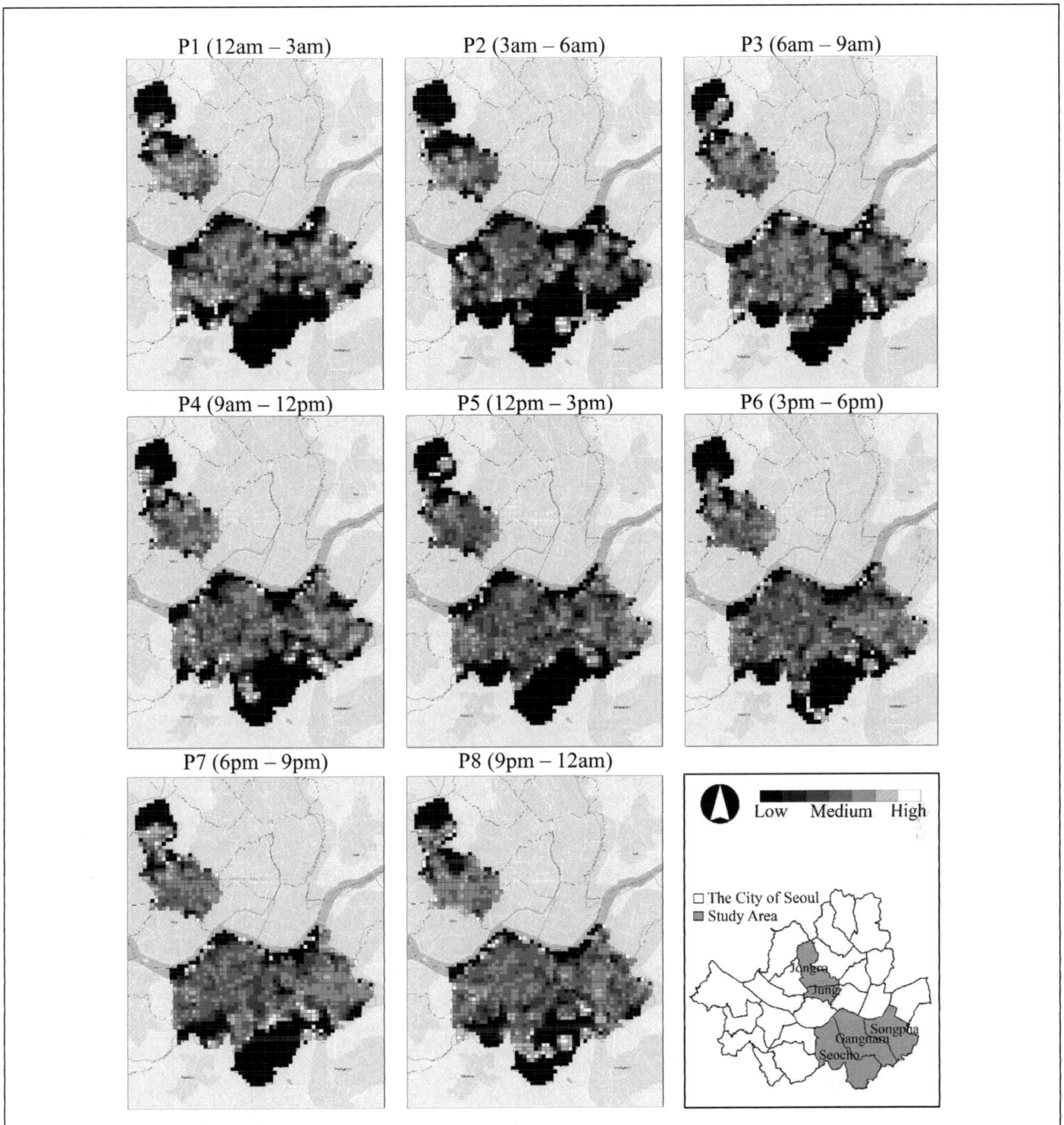

Figure 2. The distribution of WPCC1K by a 3 h time period.

collisions per 1,000 pedestrians, presents a contrasting trend. During the 9 h from 12 a.m. to 9 a.m., represented by periods 1, 2, and 3, the normalized pedestrian collisions are much higher than those during the rest of the day, as a result of the relatively small number of pedestrians.

By computing the value of WPCC1K for eight time periods by the grid of 0.09 square km (0.04 square miles), this paper illustrates the spatiotemporal distribution of WPCC1K (Figure 2). The spatial patterns of WPCC1K differ by the time period. In general, while the patterns within Jung-Gu and Jongro-Gu (located to the north in

Table 1. List of Independent Variables

Variables	Definitions	Descriptions
Transportation variables		
ARTRD	Arterial road length	The length of arterial roads (miles)
LOCRD	Local road length	The length of secondary roads (miles)
PTBUS	Bus stop density	The total number of bus stops (count)
PTRAIL	Subway station accessibility	Dummy, if the grid is within 0.25 miles from a station, 1, otherwise 0
INTRSCT	Intersection density	The number of intersections (count)
Land Use variables		
POPDEN	Population	The number of residents (count)
EMPDEN	Employment	The number of employees (count)
LUOFF	Office land use	The percent of office land use (percent)
LUCOM	Commercial land use	The percent of commercial land use (percent)
Destination variables		
RSTRT	Restaurant	The total number of restaurants (count)
BAR	Bar density	The total number of bars that sell alcohol (count)
SCHL	School density	The total number of K-12 schools (count)

the maps) are relatively consistent regardless of the time period, the patterns in the other Gus vary by time period. Consistent patterns are rarely found by time period.

Spatial Regression Analysis

After identifying the spatiotemporal variations of pedestrian collisions in the study area, this paper attempts to test built environment factors that influence the variations. In order to do so, it employs a spatial regression methodology, which is a spatial error model (SE). Traditional ordinary least square (OLS) regression is not an ideal method for this paper, as it does not take into account spatial heterogeneity by incorporating a global coefficient. As one of the OLS regression assumptions is the independence of observations, if this does not hold, the regression obtains error terms that contain spatial dependencies. Similar to spatial dependence, spatial autocorrelation indicates the observations that are closer to each other in space have related values. The related values contribute to violate the fact that observations are independent from one another.

Unlike OLS regression, spatial regression considers spatial dependences. Spatial regression particularly addresses spatial autocorrelation. The SE model treats the spatial correlation primarily as a nuisance, much like how statistical approaches often treat temporal serial correlation as something to be eliminated and solely regarded as an estimation problem (*23*). Therefore, this paper employs the SE model. The SE model can be expressed with the following formula.

$$y = X\beta + \varepsilon \text{ with } \varepsilon = \lambda W\varepsilon + u$$

Where y = a vector of observations on the dependent variable

W = the spatial weights matrix

X = a matrix of observations on the explanatory variables

ε = a vector of spatially autocorrelated error terms,

u = a vector of errors, and

λ, β = parameters

Using the spatiotemporal setting described above, eight SE models were developed. The dependent variable is the logged value of WPCC1K. As the dependent variable does not follow a normal distribution, the number was logged, and the logged values provide better fit. Independent variables consist of a variety of built environmental factors including transportation, land use, and travel destination factors (Table 1). The transportation variables incorporate various components of transportation factors such as automobile traffic, public transit, and active transportation. In order to measure the influence of automobile traffic on pedestrian collision, this paper measures the lengths of arterial and collector roads (ARTRD) and local roads (LOCRD). Excluding highways and expressways that have minimal interactions between automobiles and pedestrians, this study attempted to measure the influence of roadway types on pedestrian collision. This paper also includes two main public transit systems in the City of Seoul, bus and subway. The variables, PTBUS and PTRAIL, indicate bus stop density and accessibility to rail transit station, respectively. The variable, INTRSCT, refers to intersection density, which is well known for a measurement of walkability. The higher number of intersections symbolizes a walkable environment that attracts pedestrians (*24*).

The land use factors include the types of land use associated with pedestrian activities. In order to measure pedestrian activities in residential areas, this study includes population density (POPDEN). The variable,

EMPDEN, referring to employment density, targets the analysis of pedestrian collisions near people's workplaces. The density of office (LUOFF) and commercial (LUCOM) land uses are also included, in order to measure the relationship between pedestrian collisions and typical travel destinations. In addition to the land use factors, pedestrian destinations that have a strong correlation with pedestrian collisions are included in this study. The destinations include restaurants, (RSTRT), bars (BAR), and schools (SCHL). The variable RSTRT examines the influence of alcohol on pedestrian collisions. The variable SCHL particularly targets the analysis of children's collisions around schools.

This paper computed and aggregated the independent variables by the grid cell, conducting a series of geospatial analyses. Point data were summed by a count of the elements in each cell, the line data were summed for total length, and the polygon data were summed proportional to the percentage of the polygon's area that was within a grid cell. The data for the independent variables were obtained from the City of Seoul, Seoul Institute (SI), Korea National Spatial Data Infrastructure Portal (KNSDIP), and the Korean Population and Housing Census (KPHC).

Results

Before spatial regression modeling, this paper summarized the descriptive statistics for the variables (Table 2). The majority of the transportation values present relatively small standard deviation values, except bus stop density (PTBUS), office land use (LUOFF), and commercial land use (LUCOM). This implies that the variables are closely distributed around the mean value. In other words, the variables are consistent in the study area with less variation. The relatively large standard deviations of PTBUS, LUOFF, and LUCOM indicate the uneven spatial distribution of bus stops and types of land use including commercial and office. For example, bus stops in the study area are concentrated along major arterials. The large standard deviations of LUOFF, and LUCOM represent the dynamic land use types within grid cells.

As indicated in the higher values of R-squared, the SE models of P2 and P5 present the better general model fit, but the R-squared values from the SE models ranging from 0.22 and 0.32 present a consistent general model fit. All of the models present the variation of the correlation of the dependent variable with the independent variables. The P2 and P8 models have five out of the twelve independent variables that present a statistically significant correlation with the dependent variable. It is noteworthy that the time of day represented by both of the time periods are nighttime and early morning. In addition, four independent variables of the P1 and P5 models present statistically significant correlations. While P5 represents the time from noon to 3 p.m., P1 fills the gap between P2 and P8. This indicates the significant contribution of the built environment to pedestrian collisions during nighttime.

A statistically significant correlation is found between two independent variables in the other four models. The correlations between the independent and dependent variables in the models are comparable. Especially, the models in morning times, P3 and P4, present the identical two independents variables that correlates with the dependent variables. They include employment density (EMPDEN) and restaurant (RSTRT). While EMPDEN continuously presents a negative correlation with

Table 2. Descriptive Statistics of Variables

	Min.	Max.	Mean	Std. D.
Transportation variables				
ARTRD	0.00	1.13	0.14	0.18
LOCRD	0.00	3.60	0.51	0.59
PTBUS	0.00	562.68	45.06	70.13
INTRSCT	0.00	138.00	10.72	16.24
Land Use variables				
POPDEN	0.00	72,811.56	10,524.12	12,504.51
EMPDEN	0.00	145,057.97	10,595.29	19,340.41
LUOFF	0.00	91.66	2.24	7.42
LUCOM	0.00	23.67	1.72	4.96
Destination variables				
RSTRT	0.00	193.27	19.57	28.24
BAR	0.00	21.57	1.07	2.58
SCHL	0.00	0.21	0.03	0.04

Note: Min = Minimum; Max = Maximum; and Std. D. = standard deviation. The variable, PTRAIL, is excluded from the descriptive statistics summary as the variable is a dummy variable.

pedestrian collision in the models P6 and P7, RSTRT does not present a statistically significant correlation with the dependent variable in model P6. Instead, school (SCHL) shows a positive correlation with the dependent variable in model P6.

Although some of the independent variables show and do not show a correlation with the dependent variable depending on the time periods, two variables, employment density (EMPDEN) and restaurant (RSTRT), consistently present a negative and positive correlation, respectively, with the dependent variable regardless of the time periods. While EMPDEN consistently presents a negative correlation with the dependent variable for all eight periods, RSTRT indicates a positive correlation for all of the time periods except P6. In contrast, three variables do not present a correlation with pedestrian collision at all regardless of the time periods. They include local roads (LOCRD), office land use (LUOFF), and bars (BAR).

Discussion

The output of the models suggests interesting findings and discussion topics. The output does not show systemic spatiotemporal dynamics of pedestrian collisions by the time period (Table 3). Although there are some variations of the built environmental factors correlating with pedestrian collision by the period, the correlation is not significant enough to explain the daily patterns of pedestrian safety by 3 h. For example, the output of P3, which intends to measure the correlation during morning peak hours, is not different from the output of P4. Two variables, EMPDEN and RSTRT, demonstrate a consistent correlation with pedestrian collisions regardless of the time period. The only built environment factor that indicates a profound correlation by the time period is school (SCHL), which shows a positive correlation with pedestrian collisions during the time period, P6 (from 3 p.m. to 6 p.m.). This implies that students are vulnerable in the vicinity of schools during school exit times. It is noteworthy that P3, school entry times, does not present a significant correlation with pedestrian collisions. This probably reflects the fact that there is more congestion during school exit times than entry time, as all of the schoolchildren are released at the same time. The congestion is confusing for the schoolchildren and affects their decision-making.

Although the output does clearly suggest the systemic configuration of pedestrian collisions by time period, the output confirms a dominant spatiotemporal phenomenon, the distinct influence of the built environment on pedestrian collisions between nighttime and daytime. The relationship between the built environment and pedestrian collisions becomes significant during late nighttime. The models of P1, P2, and P8 that depict the relationship from 9 p.m. to 6 a.m. indicate four or five independent variables that present a correlation with pedestrian collisions. In comparison with the other models (from 6 a.m. to 9 p.m.) that mostly have two independent variables correlating to pedestrian collision, three models for 9 p.m. to 6 a.m. imply a stronger relationship between the built environment and pedestrian collisions during the late nighttime.

Transportation factors primarily influence the distinction between the late nighttime and the rest of the day. Positive correlations of pedestrian collisions with three out of four transportation variables are identified from the models of P1, P2, and P8. The variables include major arterial and collector roads (ARTRD), bus stops (PTBUS), and adjacency to subway stations (PTRAIL). P1, P2, and P8 models confirm the impact of ARTRD and PTRAIL, ARTRD, and PTBUS and PTRAIL on pedestrian collisions, respectively. The outputs reflect the lifestyle of the people in Seoul who enjoy active nightlife. Like many other metropolitan cities, Seoul offers a variety of cultural events and entertainment. People are willing to take advantage of these opportunities during the nighttime.

The output also strongly correlates with people's trips back home at the end of a day. The correlation between bus stops and pedestrian collisions from the P8 model (9 p.m. to 12 a.m.) is aligned with the schedule of metro bus service in the City of Seoul. In general, the city's metro bus daily provides service until midnight. Therefore, the correlation implies that there are frequent pedestrian collisions among the bus transit riders who want to utilize the late-night bus service. The same explanation can apply to the correlation between pedestrian collisions and PTRAIL. The daily subway service in the city typically lasts until 1 a.m. The positive correlation of PTRAIL from P1 and P8 indicates that there is a high probability for late-night subway riders to become victims of pedestrian collision.

The positive correlation of arterial and collector roads with pedestrian collisions is found in P1, P2, and P5. In general, major roads positively associate with pedestrian collisions, because of their higher automobile speed, wider lanes, and heavy automobile traffic (*6*, *25*). However, it is interesting to observe that the correlation is dominant during the late nighttime from midnight to 6 a.m.. This probably relates to taxis. Taxis are one of the most popular public transportation modes in Seoul. People can hail taxis everywhere within the city. It becomes a valuable transportation mode, especially during the late night after other public transit service hours. During the late night, people tend to hail taxis on major roads, where taxi traffic is concentrated. It is commonly observed that people mingle with automobile traffic in

Table 3. The Outputs of Spatial Regression

Model	P1		P2		P3		P4		P5		P6		P7		P8	
Time	12 a.m.–3 a.m.		3 a.m.–6 a.m.		6 a.m.–9 a.m.		9 a.m.–Noon		Noon–3 p.m.		3 p.m.–6 p.m.		6 p.m.–9 p.m.		9 p.m.–12 a.m.	
Variable	Coef.	z-value	Coef.	z-value	Coef.	z-value	Coef.	z-value	Coef.	z-value	Coef.	z-value	Coef.	z-value	Coef.	z-value
Constant	–4.911	–22.846	–3.142	–13.008	–4.044	–19.523	–5.236	–27.698	–4.790	–25.470	–4.887	–31.676	–4.931	–29.221	–3.924	–26.650
Transportation variables																
ARTRD	0.726	**1.982	0.629	*1.700	0.451	1.281	0.489	1.496	0.541	*1.798	–0.157	–0.530	0.134	0.403	0.211	0.778
LOCRD	0.082	0.318	0.063	0.242	–0.008	–0.031	–0.035	–0.151	0.014	0.065	–0.023	–0.109	–0.172	–0.740	–0.176	–0.919
PTBUS	0.000	0.129	0.000	0.526	0.001	0.896	0.000	0.543	0.001	0.967	0.001	0.772	0.000	0.347	0.001	*1.783
PTRAIL	0.422	***2.740	0.103	0.632	–0.100	–0.656	0.004	0.031	–0.048	–0.375	–0.109	–0.890	0.019	0.138	0.195	*1.709
INTRSCT	–0.001	–0.085	0.003	0.317	–0.002	–0.216	0.009	1.065	0.002	0.249	0.003	0.438	0.011	1.271	0.009	1.262
Land use variables																
POPDEN	0.000	–0.126	0.000	**–1.982	0.000	–0.487	0.000	–1.088	0.000	**–2.119	0.000	0.242	0.000	0.953	0.000	*–1.871
EMPDEN	0.000	***–3.605	0.000	**–1.962	0.000	**–2.102	0.000	*–1.845	0.000	**–2.437	0.000	**–2.522	0.000	**–2.308	0.000	**–2.158
LUOFF	0.019	1.356	0.006	0.464	0.007	0.475	0.005	0.388	0.007	0.578	0.007	0.652	0.005	0.360	0.008	0.773
LUCOM	0.016	0.889	–0.035	*–1.785	0.009	0.509	–0.027	–1.565	–0.001	–0.047	0.001	0.079	0.010	0.626	–0.010	–0.738
Destination variables																
RSTRT	0.395	***3.894	0.211	***2.067	0.218	**2.146	0.187	**1.975	0.183	***2.096	0.122	1.458	0.237	**2.535	0.146	*1.841
BAR	–0.041	–0.048	0.534	0.628	0.301	0.350	0.296	0.369	–0.080	–0.109	0.708	0.993	–0.206	–0.259	0.586	0.867
SCHL	0.401	0.190	–3.535	–1.593	1.207	0.579	–0.762	–0.392	–1.044	–0.588	2.740	*1.795	–0.100	–0.052	–1.397	–0.861
Lambda	0.618	18.899	0.661	21.051	0.649	21.609	0.645	21.488	0.686	24.332	0.578	17.410	0.569	17.101	0.624	20.635
R-squared	0.29		0.32		0.28		0.29		0.32		0.23		0.22		0.27	
Log likelihood -	2289.33		–1983.33		–2459.36		–2423.67		–2255.37		–2345.21		–2514.00		–2360.85	
Akaike info criterion	4604.67		3992.66		4944.71		4873.34		4536.74		4716.42		5054.01		4747.7	

Note: *, **, *** Correlations are significant at the 0.10, 0.05, and 0.01 levels, respectively (2-tailed) Coeff. = Coefficient.

order for them to hail taxis and for taxis to pick up customers. The positive correlation identified from P1 and P2 reflects this phenomenon.

Another finding is the negative correlation between the number of employees and pedestrian collisions. The correlation is found in all eight models, regardless of the time period. The areas with the highest concentration of employment includes the Seoul CBD and are characterized by high-rise office buildings and dense, intense land use. These areas provide a pedestrian-oriented environment with proper pedestrian infrastructure. Automobiles are banned on some streets in the area. Automobile speeds are low, and drivers are more cautious about pedestrians in this environment (*22*). The normalization of the dependent variable also influences the negative correlation. The areas with a concentration of employment is close to the areas with a significantly high volume of pedestrians. The dependent variable normalized by pedestrian counts becomes small in these areas. As a consequence, the consistent negative correlation can be formed.

However, it is noteworthy that the land use characteristics in the study area probably dilute the correlation between individual land use type and pedestrian collisions. The outputs do not provide plausible findings on the relationships between pedestrian collisions and two land use types, office and commercial. This possibly happens as a result of the dense, intense, and mixed land uses that are concentrated in the unit of analysis. Further study is necessary to analyze the relationships.

Unlike previous research (*6*, *7*), pedestrian collision presents a negative correlation with population in P2, P5, and P8. This reflects the unique characteristics of residential settlements in Seoul. The areas with high population concentrations in Seoul are occupied by large condominium complexes. Many of the complexes consist of 10- to 50-story condominium buildings on large parcels ranging from 0.36 to 2.07 square km (89–512 acres). The complexes typically provide a pedestrian-friendly environment while controlling automobile traffic. For example, they prohibit through traffic; enforce strict speed limits; and implement traffic calming devices. These three models indicate the time periods that may have lower automobile and pedestrian traffic in these areas, such as the early morning, mid-day, and late night. The reduced chance of pedestrian and automobile interactions under the built environment contributes to the reduction of pedestrian collisions in the areas and during the time periods.

It was hypothesized that pedestrian collisions during the nighttime are influenced by alcohol consumption, but a direct spatial correlation between pedestrian collision and bars during the nighttime is not identified. However, it is observed that the correlation between transportation factors and pedestrian collisions increases during the late night. This may suggest the probability for pedestrians to become the victims of collisions during their trips home after the consumption of alcohol, although collisions do not occur directly adjacent to bars. Further study is necessary to confirm the probability.

Conclusion

Unlike previous pedestrian safety research, this paper demonstrates a systemic approach that includes a spatiotemporal analysis on the relationship between pedestrian safety and the built environment. A big data set, with pedestrian count data based on cell phone signals, allows normalizing pedestrian collisions by the number of pedestrians on the streets. Owing to a lack of consistent, comprehensive pedestrian volume data, much of the previous pedestrian safety research did not attempt to normalize collision counts, or employed proxy variables such as population and transit ridership (*22*, *26*, *27*).

The big data, the SKT data, becomes alternative pedestrian volume data that replace conventional pedestrian count data and proxy variables. The SKT data supported a spatiotemporal analysis at the micro level. The data provided in the format of geographic information systems (GIS) data was thoroughly aligned with the other datasets employed by this paper. The data do not require intensive data processing and/or special software or data tools. Therefore, it is highly possible for the data to be widely adopted by transportation research, including safety research. Despite these advantages of the data, however, the SKT data do not completely disclose the algorithm used for distinguishing pedestrians' signals from others. This makes it hard to confirm the accuracy, reliability, and validity of the data. The undisclosed technical details of the data is problematic not only to the SKT data, but also to much of the data from the private sector. With a growing availability of big data, especially from the private sector, it is important to expand the discussion about the balance between the protection and transparency of data and technologies.

The outputs from the spatial regression analysis reveal the complicated spatiotemporal relationship between pedestrian collisions and the built environment. Although the clear influence of the built environment on the changes of pedestrian collisions by the time period of three hours is not detected, the outputs imply the impact of the built environment on the contrasts of pedestrian collisions between nighttime and daytime. The built environment, especially transportation factors, positively correlates with pedestrian collisions during late nighttime in the City of Seoul. In order to reduce pedestrian collisions during the nighttime, therefore, it would be beneficial to increase police patrols around transit stations/

stops and on arterial/collector roads during the nighttime hours. Implementing signage and pedestrian facilities that are easy to see in a dark environment can be also considered. The outputs also point out that schoolchildren become vulnerable to pedestrian collisions during school exit times. Thus, it is important to raise schoolchildren's safety awareness through education programs that emphasize their behavior during school exit times. Providing intensive police patrols, traffic light control, and crossing guards during this time will also help to protect them from pedestrian collisions.

Overall, the outputs confirm that pedestrian collisions significantly decrease in the areas with a pedestrian-friendly environment, regardless of the time of day. These areas with a concentration of employment and population are not only equipped with pedestrian-oriented facilities, but are also dominated by a large volume of pedestrians. Automobiles drive slowly and with caution, and pay more attention to pedestrians in these areas. However, this paper did not include transportation infrastructure variables such as traffic volume and lighting, mainly because of the lack of data availability. It is necessary for future studies to improve this research by incorporating these variables.

Author Contributions

The authors confirm contribution to the paper as follows: study conception and design: Dohyung Kim; data collection: Dohyung Kim; analysis and interpretation of results: Dohyung Kim, Hina Maheshbhai Chanchlani; draft manuscript preparation: Dohyung Kim, Hina Maheshbhai Chanchlani. All authors reviewed the results and approved the final version of the manuscript.

References

1. National Highway Traffic Safety Administration. *Traffic Safety Facts 2015*. National Highway Traffic Safety Administration, Washington, D.C., 2015.
2. *A Review of Pedestrian Safety Research in the United States and Abroad*. FHWA-RD-03-042. FHWA, https://ntl.bts.gov/lib/24000/24700/24702/Ped_Synthesis_Report.pdf. Accessed July 17, 2017.
3. Bejleri, I., R. Steiner, and D. Kim. GIS Methods and Tools for Bicycle and Pedestrian Crash Mapping and Analysis. Presented at 86th Annual Meeting of of the Transportation Research Board, Washington, D.C., 2007.
4. Rothman, L., C. Macarther, T. To, R. Buliung, and A. Howard. Motor Vehicle-Pedestrian Collisions and Walking to School: The Role of the Built Environment. *Pediatrics*, Vol. 133, No. 5, 2014, pp. 1–11. https://doi.org/10.1542%2Fpeds.2013-2317d.
5. Dai, D., and D. Jaworski. Influence of Built Environment on Pedestrian Crashes: A Network-based GIS Analysis. *Applied Geography*, Vol. 73, 2016, pp. 53–61. http://dx.doi.org/10.1016/j.apgeog.2016.06.005.
6. Dumbaugh, E., and W. Li. Designing for the Safety of Pedestrians, Cyclists, and Motorists in Urban Environments. *Journal of the American Planning Association*, Vol. 77, No. 1, 2010, pp. 69–88. https://doi.org/10.1080/01944363.2011.536101.
7. Wedagama, P., R. Bird, and A. Metcalfe. The Influence of Urban Land-use on Non-motorized Transport Casualties. *Accident Analysis and Prevention*, Vol. 38, No. 6, 2006, pp. 1049–1057. https://doi.org/10.1016/j.aap.2006.01.006.
8. Schuurman, N., J. Cinnamon, V. A. Crooks, and S. M. Hameed. Pedestrian Injury and the Built Environment: An Environmental Scan of Hotspots. *BMC Public Health*, Vol. 9, No. 1, 2009, p. 233. https://doi.org/10.1186/1471-2458-9-233
9. Dumbaugh, E., and R. Rae. Safe Urban Form: Revisiting the Relationship between Community Design and Traffic Safety. *Journal of the American Planning Association*, Vol. 75, No. 3, 2009, pp. 309–330. https://doi.org/10.1080/01944360902950349.
10. Loukaitou-Sideris, A., R. Liggett, and H. Sung. Death on the Crosswalk, A Study of Pedestrian- Automobile Collisions in Los Angeles. *Journal of Planning Education and Research*, Vol. 26, No. 3, 2007, pp. 338–351.
11. Hubbard, S. M. L., D. M. Bullock, and F. L. Mannering. Right Turn on Green and Pedestrian Level of Service: Statistical Assessment. *Journal of Transportation Engineering*, Vol. 135, No. 4, 2009, pp. 153–159.
12. Gårder, P. E. The Impact of Speed and Other Variables on Pedestrian Safety in Maine. *Accident Analysis and Prevention*, Vol. 36, No. 4, 2004, pp. 533–542. http://dx.doi.org/10.1016/S0001-4575(03)00059-9.
13. de Carvalho Ponce, J., D. R. Muñoz, G. Andreuccetti, G. Gonçalves de Carvalho, and V. Leyton. Alcohol-related Traffic Accidents with Fatal Outcomes in the City of Sao Paulo, *Accident Analysis and Prevention*, Vol. 43, No. 3, 2011, pp. 782–787. http://dx.doi.org/10.1016/j.aap.2010.10.025.
14. Davis, G. Relating Severity of Pedestrian Injury to Impact Speed in Vehicle-Pedestrian Crashes: Simple Threshold Model. *Transportation Research Record: Journal of the Transportation Research Board*, 2001. 1773: 108–113.
15. Quistberg, D. A., E. J. Howard, B. E. Ebel, A. V. Moudon, B. E. Saelens, P. M. Hurvitz, J. E. Curtin, and F. P. Rivara. Multilevel Models for Evaluating the Risk of Pedestrian–Motor Vehicle Collisions at Intersections and Mid-Blocks. *Accident Analysis and Prevention*, Vol. 84, 2015, pp. 99–111. https://doi.org/10.1016/j.aap.2015.08.013.
16. Diogenes, M. C., and L. A. Lindau. Evaluation of Pedestrian Safety at Midblock Crossings, Porto Alegre, Brazil. *Transportation Research Record: Journal of the Transportation Research Board*, 2010. 2193: 37–43.
17. Griswold, J., B. Fishbain, S. Washington, and D. R. Ragland. Visual Assessment of Pedestrian Crashes. *Accident Analysis and Prevention*, Vol. 43, No. 1, 2011, pp. 301–306. https://doi.org/10.1016/j.aap.2010.08.028.
18. Sullivan, J. M., and M. J. Flannagan. Differences in Geometry of Pedestrian Crashes in Daylight and Darkness. *Journal of Safety Research*, Vol. 42, No. 1, 2011, pp. 33–37. 10.1016/j.jsr.2010.11.005

19. Plug, C., J. Xia, and C. Caulfield. Spatial and Temporal Visualization Techniques for Crash Analysis. *Accident Analysis and Prevention*, Vol. 43, No. 6, 2011, pp. 1937–1946. http://dx.doi.org/10.1016/j.aap.2011.05.007.
20. Yiannakoulias, N., K. E. Smoyer-Tomic, J. Hodgson, D. W. Spady, B. H. Rowe, and D. C. Voaklander. The Spatial and Temporal Dimensions of Child Pedestrian Injury in Edmonton. *Canadian Journal of Public Health*, Vol. 93, No. 6, 2002, pp. 447–451.
21. Batty, M. Big Data, Smart Cities and City Planning. *Dialogues in Human Geography*, Vol. 3, No. 3, 2013, pp. 274–279. 10.1177/2043820613513390
22. Gladhill, K., and C. M. Monsere. Exploring Traffic Safety and Urban Form in Portland, Oregon. Presented at 91st Annual Meeting of the Transportation Research Board, Washington, D.C., 2012.
23. Cressie, N. A. C. *Statistics for Spatial Data*. John Wiley & Sons, Chichester, 1993.
24. Frank, L., J. Sallis, T. Conway, J. Chapman, B. Saelens, and W. Bachman. Many Pathways from Land Use to Health: Associations between Neighborhood Walkability and Active Transportation, Body Mass Index, and Air Quality. *Journal of the American Planning Association*, Vol. 72, No. 1, 2006, pp. 75–87. https://doi.org/10.1080/01944360608976725.
25. Quistberg, D. A., E. J. Howard, B. E. Ebel, A. V. Moudon, B. E. Saelens, P. M. Hurvitz, J. E. Curtin, and F. P. Rivara. Multilevel Models for Evaluating the Risk of Pedestrian–Motor Vehicle Collisions at Intersections and Mid-blocks. *Accident Analysis and Prevention*, Vol. 84, 2015, pp. 99–111. http://dx.doi.org/10.1016/j.aap.2015.08.013.
26. Kim, D., and D. Rojas. Does Transit-Oriented Development Create a Bicycle and Pedestrian Safe Environment? Presented at 94th Annual Meeting of the Transportation Research Board, Washington D.C., 2015.
27. Stoker, P., A. Garfinkel-Castro, M. Khayesi, O. Wilson, M. Mwangi, M. Peden, and R. Ewing. Pedestrian Safety and the Built Environment: A Review of the Risk Factors. *Journal of Planning Literature*, Vol. 30, No. 4, 2013, pp. 377–392. https://doi.org/10.1177/0885412215595438.

The Standing Committee on Pedestrians (ANF10) peer-reviewed this paper (18-01270).

Article

TRR
JOURNAL OF THE TRANSPORTATION RESEARCH BOARD

Transportation Research Record
2018, Vol. 2672(35) 101–117

DOI: 10.1177/0361198118790623
journals.sagepub.com/home/trr
SAGE

What Are the Most Important Factors for Pedestrian Level-of-Service Estimation? A Systematic Review of the Literature

Nowar Raad[1] and Matthew I. Burke[1]

Abstract

Pedestrian environments are becoming more important in an urbanizing world where walking is increasingly being encouraged. Engineers and planners determine a link's performance using measures codified as level-of-service, which provide guidance on acceptable or desirable standards. A range of approaches are used in determining pedestrian level-of-service (PLOS), incorporating a wide variety of factors and with much debate as to what should or should not be used. There has been no systematic attempt to synthesize this research and provide an over-arching perspective. Our review surveyed PLOS models from the peer-reviewed literature using a systematic quantitative literature review method based on the protocol developed by the Preferred Reporting Items for Systematic Review Recommendations (PRISMA). PLOS models are increasingly being developed; 22 of the 58 studies surveyed were produced since 2013. Earlier work adapted approaches used to determine automobile LOS to PLOS. Later approaches use a much wider range of factors but with very little consistency across the studies surveyed. Collectively these factors can be grouped in themes of: comfort, safety, and mobility. The most used factors were, in order: footpath width; obstructions to pedestrian flow; motor vehicle speeds and volumes; shoulder widths; and buffers such as on-street parking. However, many of the factors being included have not been empirically studied and almost none of the tools in use have been tested for such matters as inter-rater reliability.

Standards and performance measures often guide the planning and design of pedestrian environments. Pedestrian level-of-service (PLOS) is one of the more important measures used. For almost 50 years transport engineers, urban designers, and planners have developed and applied PLOS to determine how a pedestrian environment may function in terms of a pedestrian's comfort, safety, and amenity. As will be shown, there is dispute as to how PLOS should be calculated, and what factors should be considered. This paper systematically reviews the literature from 1971 to 2016 to identify the main approaches being used to determine PLOS and to identify and theme for the first time the most common factors used in assessments. This should assist practitioners in moving towards more robust and standardized approaches and to identify problems with PLOS measurement that require further investigation.

Level-of-service can be defined as an assessment of the quality of service or an overall measure of current conditions on any piece of transport infrastructure, covering aspects such as the facilities themselves, the contextual situation, equipment levels, and ancillary infrastructures. The level-of-service is defined as "a quantitative stratification of a performance measure or measures that represent quality of service" (*1*). Gallin defined PLOS as "an overall measure of walking condition on route, path and facility" and suggested PLOS reflects users' perceptions in terms of sense of comfort and safety (*2*).

Street designs should meet the needs of all users, including pedestrians, especially in urban areas (*3*). The US *Highway Capacity Manual* (HCM) developed a six-level scale for PLOS that ranges from *A*, which represents the "best" level-of-service, through to *F*, which represents the "worst" (*1*). The HCM provides possible methods for evaluating level-of-service in specific conditions (*4*). However, while vehicle LOS was quite advanced as far

[1]Cities Research Institute, Griffith University, Nathan, QLD, Australia

Corresponding Author:
Address correspondence to Nowar Raad:
nowarraad.abdulmajeed@griffithuni.edu.au

back as the 1980s, underpinned by measures of volume and capacity, means to produce PLOS were poorly developed by comparison (*5*). Conceptually, to measure PLOS generally requires quite different and more complex approaches than for vehicle LOS. Ideally, one might wish to capture users' perceptions across different groups of pedestrians, including the young, seniors, and the less physically able, given that footpaths must be designed to create accessibility for the majority of users. Footpaths should also have attractive features for those individuals who may choose to walk rather than those with no other mode options (*6*).

Increasingly it is qualitative factors, along with traditional volume and capacity factors, that are important in the measurement of PLOS. This recognizes that the pedestrian experience cannot be seen only as a physical movement from point A to point B (*7*). PLOS is therefore being tied to agendas to make streets and roadway environments more liveable (*5*).

PLOS dates back to the 1970s and Fruin's model, based on footpath capacity and pedestrian volume, which ignored other built environment concerns (*5*). In subsequent years many researchers adapted Fruin's approach with some additions or modifications (*8–12*). Others suggested alternative approaches based on alternative measures of pedestrian comfort, safety, and convenience (*13–17*). Many of these focused on the micro-level conditions of comfort for pedestrians (i.e., shade, lighting, greenery) but used only a limited set of factors to measure PLOS. A third set of researchers used a combination of Fruin-style volume/capacity and other qualitative street characteristics, attempting to gain the best of both approaches, as well as applying PLOS to a wider set of facility types (*5*, *18*). Researchers such as Gallin (*2*) went a little further, seeking to group and theme measures into categories, such as physical characteristics, location factors, and user factors (*19*). However, very few researchers used field studies to calibrate user perceptions, with notable exceptions including the work of Petritsch et al. (*20*). Asadi-Shekari and Zaly-Shah (*21*) grouped factors derived from more than 20 other studies and guidelines in an attempt at being more robust. Other researchers, such as Hidayat et al. (*22*), aimed to develop PLOS measures more suited to developing countries, where street vendor activities proliferate.

As will be shown, across all these different approaches and methods, there is very wide variety in the choice of factors suggested for PLOS evaluations. To help practitioners understand the different options and to try to identify, group, and theme the more commonly suggested factors, a systematic review was performed. The review gathered PLOS methods from the literature and categorized them by input data, model type, analysis method, factors included, and use of reliability testing. The next section of this paper outlines the methodology used to perform the systematic review and the analysis; the following section presents the results; and finally a discussion is provided, identifying gaps and suggesting options for future enquiry to create more reliable and comprehensive PLOS measures.

Approach and Method

A systematic quantitative literature review was conducted, systematically searching and categorizing the relevant literature. We sought to categorize and tabulate what researchers have suggested are the most important factors in generating PLOS outputs.

Systematic Review Protocol

We used the Preferred Reporting Items for Systematic Review Recommendations (PRISMA) protocol (*23*) following the steps outlined in Pickering and Byrne (*24*), as used successfully in a range of past studies (*25–28*). The approach seeks to be both robust and replicable, minimizing potential biases that can occur in narrative reviews, and by being transparent in selecting and categorizing the papers according to specific eligibility criteria (*29*). The approach documents (*i*) where, when, and by whom research was published; (*ii*) the geographical spread of the literature; (*iii*) types of methods used; (*iv*) types of subjects examined; (*v*) types of variables measured; (*vi*) different disciplines assessing the topic; and (*vii*) the types of results obtained (*24*, *26*, *27*). The approach begins with the identification of a topic, in this case: "factors affecting PLOS and/or the evaluation of PLOS". The second step involves database search and priority selection. The third step is content assessment and study selection. Finally, data collection processes are performed, extracting and handling data from within each paper for use in the review.

Research Questions, Search Terms, and Information Sources "Database"

We developed five research questions, as shown in Figure 1. Electronic databases were searched to identify original research papers related to "pedestrian level of service." We sought original research papers published in English language academic journals and conferences with the search conducted progressively from January to November 2016. We began with only the term "pedestrian level of service", but we progressively added other search terms. Each paper was scanned to check if it met the eligibility criteria developed for the review. Only peer-reviewed academic papers were included. Book chapters, guidelines, and non-academic reports were used

Review step	Information collected
Title	Determination of the most important factors affecting pedestrian level of service on footpaths using systematic quantitative literature review
Research questions	Where is there research on pedestrian level of service (PLOS), where and when was the research undertaken, who has undertaken the research? Where was the research published? Which type of facilities has been assessed? What methods and evaluation techniques have been developed? What were the most important factors used?
Search terms used in electronic databases	("Pedestrian Level of Service") AND ("Sidewalks* OR Footpaths* OR Pedestrian perceptions* OR Safety* OR Comfort* OR Quality of walking environment* OR Physical Characteristics* OR Obstructions* OR Aesthetics and Amenities* OR Pedestrian Traffic* OR Adjacent Traffic* OR PLOS Models*)
Electronic databases searched	Google Scholar, Web of Science, TRID, ScienceDirect, SCOPUS
Search strategy and Initial selection	Author, Year, Title, Abstract, Keywords, Journal
Content assessment & Eligibility criteria	**Group 1 (all answers yes):** Is the publication an original research paper'? (yes/no) Are the objectives of the article clear? Are the results related to the specific topic and research questions of this study? (yes/no) The main findings of the study are clear? (yes/no) **Group 2 (at least one answer yes):** Does the study assessing PLOS? Does the study assessing quality of walking environment? Does the study present a methodology for assessing PLOS? Does the study present method used to identify factors affecting PLOS? Does the study present the most important factors affecting PLOS on footpaths? Does the study present the most important factors affecting quality of walking environment on footpaths?
Data collection process, cross-check and synthesis of results	**Basic data on the paper** Paper titles, Author, Journal, Year of publication, Study location, Type of work **Methodological Information** Type of Facility, Sampling Methods, Data Type, Analysis Method, Model Type **Results** Factors /variables affecting PLOS, classification of factors, the relative importance of each factor

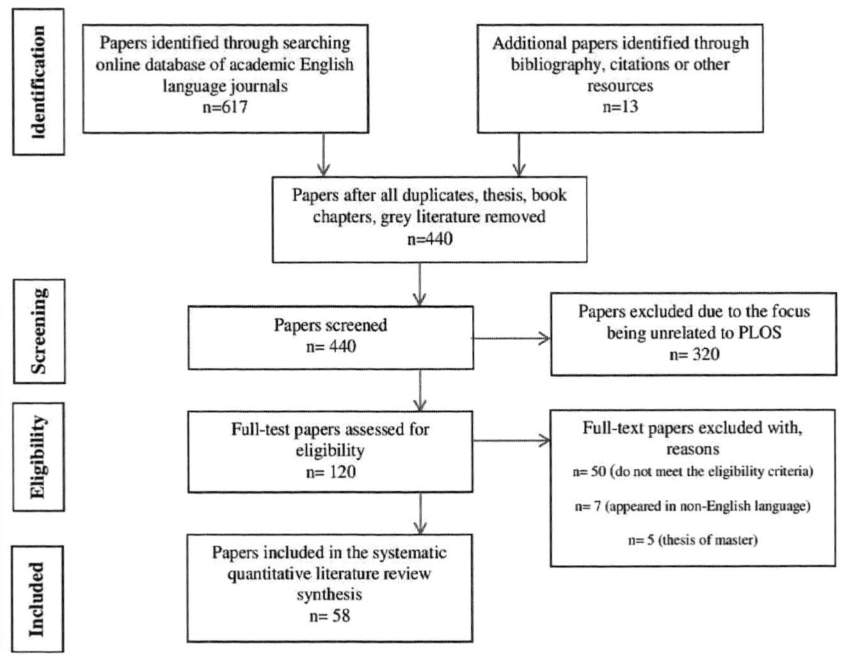

Figure 1. Review protocol (left) and papers obtained after the identification, screening, and eligibility phases of the review, using the PRISMA technique (right).

to help identify additional academic papers. We excluded evaluations of level-of-service for motorized vehicles, bicycles, and public transport.

Search Strategy and Initial Selection

A total of 617 papers including many duplicates were initially obtained. Because the concept of PLOS first emerged with the work of Fruin (*30*) the period selected for searching the papers required was set from 1971 to 2016. Zotero, EndNote X7.7, Mendeley, Jabref, and Excel software was used to support the screening stages. Zotero was used to transfer data obtained from Google Scholar to Endnote. EndNote was used for organizing references and identifying duplicates. Further filtering was undertaken using Mendeley and Jabref software and for transfer to Excel. Concurrently, all papers considered relevant to the research objective were recorded and classified using Excel spreadsheets at each of these stages. Figure 1 shows the review protocol and the data included at each stage. At the end of the initial selection stage 120 articles were selected for content assessment.

Data Extraction Process and Synthesis of Results

The remaining articles were read in full and citations exported. Necessary data was extracted including basic data on the paper (paper titles, author, journal, year of publication, study location, type of work); methodological information (type of target area, sampling methods, data type, analysis method, model type); and results (factors/variables affecting PLOS, classification of factors, the relevant importance for each factor), as shown in Figure 1.

Results

Basic Data on PLOS Research Papers

The final set of research papers is shown in Table 1. Although research on PLOS has been conducted in 18 different countries, a large proportion of the studies were conducted and published in the U.S.A. (18; 31.1%). China produced 19 papers (17.2%) and India seven (12.1%), as shown in Figure 2*a*.

Research on PLOS has been accelerating, as shown in Figure 2*a*, with 25 of the 58 papers published in the few years between 2012 and 2016. Nineteen of the journal articles were published in *Transportation Research Record* and the other 22 were published across a range of journals. Six conference papers were published in the proceedings of the Annual Meeting of the Transportation Research Board (TRB) and the other nine were published in different conference proceedings. There were two key scientific reports (Table 2).

Types of Facilities and Methods of Evaluation

Researchers have been trying to develop PLOS for four key types of facilities, as shown in Table 3. Of the 58 papers: 21 (36%) advanced PLOS for mixed streets; 19 PLOS for footpaths, per se; 14 (24%) PLOS for intersections (14 papers, 24%), and 4 (7%) PLOS for mid-block crossings.

Many papers used quantitative data only (39.7%). We label these researchers the *geometricians* given their focus on capacity, volume, and geometric factors. Only nine papers (15.5%) used qualitative data only—a group we label the *experientialists* given their focus on non-geometric factors that influence the pedestrian experience. A larger proportion used a combination of quantitative and qualitative data (44.8%) with many of these studies bridging the two camps.

Methods included surveys, observations, video, or audits. Twenty-one studies (36.2%) used mixed methods (two or more) for data collection; 14 studies (24%) used surveys (whether questionnaires, field surveys, or intercept surveys), nine (15%) used observations and eight (13.8%) used video. Interestingly, none of the studies used focus groups or Delphi techniques with experts to help identify and rank PLOS factors, which at least in theory would offer a viable approach.

Researchers have also tended to use four main approaches for analyzing data to produce PLOS outputs: a points system (12 papers, 20.6%) or some form of empirical regression (40 papers, 69%), simulation (1 paper), or fuzzy neural network approach (2 papers). Of the latter approaches, capacity-based modeling is particularly common. That said, only four papers (6.9%) used a capacity-based model alone; 40 papers (69%) used a combination of capacity-based modeling with other factors.

Factors Used to Measure PLOS

We themed and grouped the factors used in the studies to measure PLOS, under each facility type, as shown in Table 4. There is considerable variation. Factors included distinct objects in the built environment (i.e., the presence of bollards, water fountains, or pedestrian signals) collected from audits or geographic information systems and more subjective factors that represent an environment's effect on pedestrians (i.e., concepts of comfort, connectivity, or security). Geometrical factors dominate with 45 of the 83 factors used in PLOS for mixed streets, 22 of the 62 factors used in PLOS for footpaths, 15 of 41 of the factors used in PLOS for intersections and 18 of the 30 in PLOS for mid-block crossings. User behavioral factors were generally given the least attention, and often only by the experientialists. Methods used to select factors were usually unreported in these studies. Very few

Table 1. Previous Studies on PLOS Identified via the Review

Author of study	Year	Country of study	Data collection	Target area	Analysis method	Model type	Data type	Types of work	Journal title
Fruin (*30*)	1971	U.S.A.	Time lapse photography	Mixed area	Linear relationship	Capacity based	Quantitative	Journal article	*Mechanical Engineering*
Polus et al. (*9*)	1983	Israel	Videotape recorder and digital clock	Footpath	Linear speed-density regression	Capacity based	Quantitative	Journal article	*Journal of Transportation Engineering: ASCE*
Mori and Tsukaguchi (*31*)	1987	Japan	Video	Footpath	Regression	Combination	Mixed	Journal article	*Transportation Research Part A: Policy and Practice*
Tanaboriboon et al. (*10*)	1989	Thailand	Portable video camera	Footpath	Linear relationship	Capacity based	Quantitative	Journal article	*ITE Journal: Institute of Transportation Engineers*
Sarkar (*13*)	1993	Munich and Rome	Observation	Mixed area	Point system	Street characteristics	Qualitative	Journal article	*Transportation Research Record*
Khisty (*14*)	1994	U.S.A.	Observation	Mixed area	Point system	Street characteristics	Qualitative	Journal article	*Transportation Research Record*
Mozer (*18*)	1994	U.S.A.	Worksheet	Mixed area	Point system	Combination	Mixed	Journal article	International Bicycle Fund
Dixon (*32*)	1996	U.S.A.	Audit	Mixed area	Point system	Street characteristics	Mixed	Journal article	*Transportation Research Record*
Milazzo et al. (*33*)	1999	U.S.A.	Videotape recorder	Signalized and unsignalized intersection crossing	Linear relationship	Capacity based	Quantitative	Journal article	*Transportation Research Record*
Miller et al. (*16*)	2000	U.S.A.	Survey	Intersection	Simulation and point system	Street characteristics	Qualitative	Journal article	*Transportation Research Record*
Jaskiewicz (*7*)	2000	U.S.A.	Observation	Mixed area	Point system	Street characteristics	Qualitative	Journal article	*Transportation Research Circular,* TRB
Gallin (*2*)	2001	Australia	Audit	Mixed area	Point system	Combination	Mixed	Journal article	*Road and Transport Research*
Landis et al. (*5*)	2001	U.S.A.	Observation	Mixed area	Stepwise multi-variable regression analysis	Combination	Mixed	Journal article	*Transportation Research Record*
Baltes and Chu (*34*)	2002	U.S.A.	Archived inventory and video camera	Mid-block	Ordinary least-squares statistical method (regression)	Combination	Mixed	Journal article	*Transportation Research Record*
Sarkar (*17*)	2003	U.S.A.	Audit	Footpath	Point system	Street characteristics	Qualitative	Journal article	*Transportation Quarterly*
Zhang and Prevedouros (*35*)	2004	U.S.A.	Video	Signalized intersection	Regression	Street characteristics	Quantitative	Journal article	*Transportation Research Record*
Steinman and Hines (*36*)	2004	U.S.A.	Survey	Signalized intersection	Point system	Combination	Mixed	Journal article	*Transportation Research Record*
Petritsch et al. (*37*)	2005	U.S.A.	Field observation and a video simulation	Signalized Intersections	Pearson correlation analyses and stepwise regression	Combination	Quantitative	Journal article	*Transportation Research Record*
Muraleetharan et al. (*38*)	2004	Japan	Field survey	Footpath	Linear relationship	Combination	Quantitative	Journal article	*Journal of the Eastern Asia Society for Transportation Studies*

(continued)

Table 1. (continued)

Author of study	Year	Country of study	Data collection	Target area	Analysis method	Model type	Data type	Types of work	Journal title
Muraleetharan et al. (*39*)	2005	Japan	Field survey	Intersection	Stepwise multiple regression model	Combination	Mixed	Journal article	*Infrastructure Planning Review*
Petritsch et al. (*20*)	2006	U.S.A.	Field observation	Mixed area	Pearson correlation analyses and stepwise regression	Combination	Quantitative	Journal article	*Transportation Research Record*
Kim et al. (*40*)	2006	U.S.A.	Field observation	Footpath	Linear relationship	Combination	Quantitative	Journal article	*Transportation Research Record*
Tan et al. (*41*)	2007	China	Intercept survey	Footpath	Stepwise regression	Combination	Quantitative	Journal article	*Journal of Transportation Systems Engineering and Information Technology*
Muraleetharan and Hagiwara (*42*)	2007	Japan	Field survey	Mixed area	Stepwise multiple regression model	Combination	Mixed	Journal article	*Transportation Research Record*
Jensen (*43*)	2007	Denmark	Respondents, video clips, and questionnaire	Mixed area	Cumulative logic regression	Combination	Mixed	Journal article	*Transportation Research Record*
Huang and Chiun (*44*)	2007	Taiwan	Observation	Mixed area	Stepwise regression	Combination	Quantitative	Conference paper	Eastern Asia Society for Transportation Studies
Daniel et al. (*45*)	2007	Malaysia	Field survey	Signalized intersections	Multiple linear regression	Combination	Mixed	Conference paper	Malaysia University of Technology
NCHRP (*46*)	2008	U.S.A.	Survey and video clip	Mixed area	Regression	Combination	Quantitative	Journal article	
Petritsch et al. (*47*)	2008	U.S.A.	Video simulation laboratories	Mixed area	Linear regression modeling	Combination	Quantitative	Journal article	*Transportation Research Record*
FDOT (*48*)	2009	U.S.A.	Observation	Mixed area	Stepwise regression	Combination	Quantitative	Journal article	
Bian et al. (*49*)	2009	China	Intercept survey	Intersection	Stepwise regression	Combination	Mixed	Journal article	*Transportation Research Record*
Asadi-Shekari and Zaly-Shah (*21*)	2011	Malaysia and Singapore	Observation	Mixed area	Point system	Combination	Mixed	Journal article	Penang, Malaysia
Hidayat et al. (*22*)	2011	Thailand and Indonesia	Interview and questionnaire surveys with pedestrians	Footpath	A multiple linear regression	Street characteristics	Qualitative	Journal article	*Journal of the Eastern Asia Society for Transportation Studies*
Vedagiri. and Anithottam (*50*)	2012	India	Videographic and field survey	Footpath	Stepwise regression	Combination	Mixed	Conference Paper	*Transportation Research Board*
Christopoulou and Pitsiava-Latinopoulou (*51*)	2012	Greece	Questionnaire Survey	Footpath	Point system	Combination	Mixed	Conference paper	*Procedia – Social and Behavioral Sciences*
Bunevska and Malenkovska (*52*)	2012	Macedonia	Video Simulation	Footpath	Simulation	Combination	Mixed	Journal article	*Transport Problems*
Nagraj and Vedagiri (*53*)	2013	India	Video graphic technique, questionnaire survey, and field measurement	Signalized intersections	Stepwise regression	Street characteristics	Quantitative	Journal article	*Transportation Research Record*
Ling et al. (*54*)	2013	China	Contingent field survey and extensive video	Signalized intersections	Stepwise regression	Combination	Mixed	Conference paper	*Transportation Research Board*

(continued)

Table 1. (continued)

Author of study	Year	Country of study	Data collection	Target area	Analysis method	Model type	Data type	Types of work	Journal title
Kang et al. (*1*)	2013	China	Video clips and field measurement	Footpath	Ordered probit	Combination	Mixed	Journal article	*Transportation Research Part a-Policy and Practice*
Jensen (*55*)	2013	Denmark	Video clips and questionnaire and field measurement	Signalized and unsignalized intersections and roundabouts	CLM stepwise regression	Combination	Mixed	Conference paper	*Transportation Research Board*
Kim et al. (*56*)	2013	Korea	A pedestrian intercept survey, a field survey	Footpath	Stepwise regression	Combination	Quantitative	Journal article	*International Journal of Urban Sciences*
Bian, Zhao and Lu (*57*)	2013	China	Questionnaire survey and field survey	Unsignalized intersections	Stepwise regression	Street characteristics	Quantitative	Conference paper	*Applied Mechanics and Materials*
Asadi-Shekari et al. (*58*)	2013	Singapore	Results of guidelines	Mixed area	Point system	Street characteristics	Mixed	Journal article	*Journal of Transportation Engineering: ASCE*
Zhao et al. (*59*)	2014	China	Questionnaire survey and field survey	Unsignalized mid-block crossings	Stepwise regression	Combination	Quantitative	Journal article	*Advances in Mechanical Engineering*
Meng et al. (*60*)	2014	China	Questionnaire survey	Mixed area	Stepwise regression	Combination	Mixed	Conference paper	*Applied Mechanics and Materials*
Kim et al. (*61*)	2014	Korea	Pedestrian questionnaire survey and video recordings	Footpath	Multiple linear regressions	Combination	Mixed	Journal article	*Journal of Advanced Transportation*
Kadali and Vedagiri (*62*)	2014	India	Pedestrian questionnaire survey and video	Mid-block crossing	Ordered probit	Street characteristics	Mixed	Conference paper	*Transportation Research Board*
Asadi-Shekari, Moeinaddini and Shah (*63*)	2014	Malaysia	Results of guidelines	Mixed area	Point system	Combination	Mixed	Journal article	*Land Use Policy*
Zhao et al. (*64*)	2015	China	Observation and questionnaire survey	Footpath	Fuzzy mathematics method	Combination	Quantitative	Conference paper	Transportation Research Board
Ye et al. (*65*)	2015	China	Video technique and questionnaire survey	Signalized intersection	Linear regression technique	Combination	Quantitative	Journal article	*Transportation Research Record*
Kadali and Vedagiri (*66*)	2015	India	Questionnaire survey and video graphic survey	Unsignalized mid-block crosswalks	Ordered probit	Combination	Mixed	Journal article	*Transportation Research Part A: Policy and Practice*
Hasan et al. (*67*)	2015	Bangladesh	Questionnaire survey	Footpath	–	Combination	Mixed	Journal article	*Transportation Research Record*
Archana (*68*)	2015	India	Visual surveys and field surveys	Intersection	Multiple linear regressions	Combination	Quantitative	Journal article	*International Journal of Students' Research in Technology and Management*
Lazou et al. (*69*)	2015	Greece	Questionnaire survey	Mixed area	Ordinal regression model (ordered logit)	Street characteristics	Qualitative	Conference paper	7th International Congress on Transportation Research

(continued)

Table 1. (continued)

Author of study	Year	Country of study	Data collection	Target area	Analysis method	Model type	Data type	Types of work	Journal title
Zhao et al. (70)	2016	China	Questionnaire survey	Footpath	Fuzzy neural network method	Combination	Qualitative	Journal article	*Journal of Intelligent and Fuzzy Systems*
Raghuwanshi and Tare (71)	2016	India	Video graphic survey and field survey	Footpath	Multiple linear regression	Combination	Quantitative	Journal article	*Research Journal of Engineering and Technology*
Daniel et al. (72)	2016	Malaysia	On-site measurement, video, and visual walkthrough surveys	Footpath	Multiple linear regression	Combination	Quantitative	Conference paper	EDP Sciences
Chandana et al. (73)	2016	India	Questionnaire survey	Mixed area	Inverse variance method	Street characteristics	Qualitative	Conference paper	ICRTSTM-16

papers used primary research with users to select network, environmental, or behavioral factors, seemingly relying instead on the researchers' own expert judgement.

To answer our final research question, we sought to identify the priority being given to these factors across the many studies for each facility type. This is not easy given the range of factors and their potential overlap. We decided that the simplest measure of priority would be produced by tabulating the number of times the factor was included across the 58 studies in the dataset, as shown in Table 4. Given the reasonable sample size, this frequency measure offers the best objective measure we can provide of how the entire set of researchers valued each factor. Many factors were not repeated across studies: 32 of the factors (39%) were only used once; 23 (28%) were used twice, and seven (8%) were used three times.

The geometry factor of *footpath width* was the most used factor for PLOS measurement for both street and footpath studies, appearing in 12 of the PLOS for streets studies and 10 of the PLOS for footpaths studies. Of the environmental factors, *obstructions* was used in six PLOS for mixed streets studies, and seven PLOS footpath studies. Of the traffic characteristics factors, *pedestrian flow* was used in ten PLOS for footpath studies; *vehicle speed* was used in nine PLOS for streets studies. None of the user behavior factors were commonly used.

The function "Word Clouds" was used to create a compound word cloud image and highlight the more frequently used factors for PLOS measurement for each facility type. Compound factors were developed using a "word-word" formula. The word clouds for each facility type are provided in Figure 3.

Discussion

The growth in the development of different approaches to measure PLOS since 2013 is indicative of increased policy and planning interest in walking provision. The review dataset shows that PLOS studies initially adopted capacity-based modeling approaches. This approach continues in the work of those we call the *geometricians*, however, it has been challenged by the alternative approaches of the *experientialists*. These two competing camps conceive PLOS in fundamentally different ways: the former as a problem of facility capacity, flow, and geometric design; the latter as about the more subjective and experiential factors of the built environment that produce pedestrian amenity. Many researchers have sought to bridge these two conceptions, merging aspects of each in their work. But the litany of different approaches and methods shows that the field remains in significant dispute as to how authorities should proceed. There are increasing numbers of studies but increasing

(a)

Year	Number of publication	Percentage
2016	4	6.9%
2015	6	10.3%
2014	5	8.6%
2013	7	12.1%
2012	3	5.2%
2011	2	3.5%
2009	2	3.5%
2008	2	3.5%
2007	5	8.6%
2006	2	3.5%
2005	3	5.2%
2004	2	3.5%
2003	1	1.7%
2002	1	1.7%
2001	2	3.5%
2000	2	3.5%
1999	1	1.7%
1996	1	1.7%
1994	2	3.5%
1993	1	1.7%
1989	1	1.7%
1987	1	1.7%
1983	1	1.7%
1970	1	1.7%
Total	**58**	**100%**

(b)

Country	Number of PLOS Studies	Percentage
USA	18	31.1%
China	10	17.2%
India	7	12.1%
Malaysia	4	6.9%
Japan	4	6.9%
Korea	2	3.5%
Denmark	2	3.5%
Greece	2	3.5%
Australia	1	1.7%
German & Italy	1	1.7%
Thailand	1	1.7%
Indonesia & Thailand	1	1.7%
Singapore	1	1.7%
Bangladesh	1	1.7%
Israel	1	1.7%
Taiwan	1	1.7%
Macedonia	1	1.7%
Total	58	100%

Figure 2. Annual and geographic distribution of the 58 published research papers: (*a*) number of papers published per year; (*b*) geographic distribution of the papers.

disagreement. Even within each camp the sets of factors proposed differs widely across authors, with little convergence in recent years.

The advantages of the geometricians' approach are relatively straightforward. Their measures of PLOS are relatively objective, based on the fundamentals of well understood volume/capacity ratios and with variations allowing for factors such as turning movements or buffers from moving vehicles. Their approaches are both empirical and relatively replicable. The disadvantages of their approaches are also the reason for the emergence of the competing camp. The geometricians might rate a footpath that offers very good freedom of movement to the pedestrian very highly even if it offers no shade, comfort, or aesthetic amenity, and is located immediately adjacent to a noisy and polluted, ten-lane, high-speed arterial.

The advantages of the experientialists' approach are in its capacity to include almost any factor via use of a points system or another form of multi-criteria evaluation. They can incorporate the most subjective factors, such as comfort and attractiveness. But the experientialists cannot seem to agree on what those factors might be, or how one should measure them. Too many of their studies rely on "best guess" approaches from learned scholars, rather than surveys with pedestrians to inform factor selection, albeit the best guess approach is a beneficial method in qualitative research to open new perspectives for further inquiry (*74*). Worryingly, none of the studies reviewed using audit tool approaches seemed to report any inter-rater reliability testing (the agreement between two or more observers using the same tool) meaning that we were not certain how replicable these approaches might be even within the same urban context, let alone if transferred to other places.

Those who try to bridge the two camps attempt to get the best of both worlds. But the limitations of the research base on subjective factors, in particular, makes such attempts problematic. A reset may not be needed, but it would be helpful to try to articulate a research agenda that can overcome these limitations. We suggest:

1. Improved methods for selecting and including factors in PLOS calculations are needed, using techniques that go well beyond the individual researcher to include the input of a representative

Table 2. Journal and Conference Distribution of 58 Papers on PLOS Models Assessed in This Study

Publication type	Title	Number of papers included	Percentage
Journal	*Transportation Research Record*	19	46.35
	Transportation Research Part A: Policy and Practice	3	7.30
	Journal of Transportation Engineering: ASCE	2	4.8
	Journal of the Eastern Asia Society for Transportation Studies	2	4.8
	Research Journal of Engineering and Technology	1	2.45
	Journal of Intelligent and Fuzzy Systems	1	2.45
	International Journal of Students' Research in Technology and Management	1	2.45
	Land Use Policy	1	2.45
	Journal of Advanced Transportation	1	2.45
	Advances in Mechanical Engineering	1	2.45
	International Journal of Urban Sciences	1	2.45
	Transport Problems	1	2.45
	Journal of Transportation Systems Engineering and Information Technology	1	2.45
	Infrastructure Planning Review	1	2.45
	Transportation Quarterly	1	2.45
	Road and Transport Research	1	2.45
	International Bicycle Fund	1	2.45
	ITE Journal: Institute of Transportation Engineers	1	2.45
	Mechanical Engineering	1	2.45
	Total	41	100
Conference	Transportation Research Board	6	40.0
	Applied Mechanics and Materials	2	13.3
	Procedia: Social and Behavioral Sciences	1	6.67
	Malaysia University of Technology	1	6.67
	International Transport Research Conference	1	6.67
	Eastern Asia Society for Transportation Studies	1	6.67
	7th International Congress on Transportation Research	1	6.67
	ICRTSTM-16	1	6.67
	EDP Sciences	1	6.67
	Total	15	100

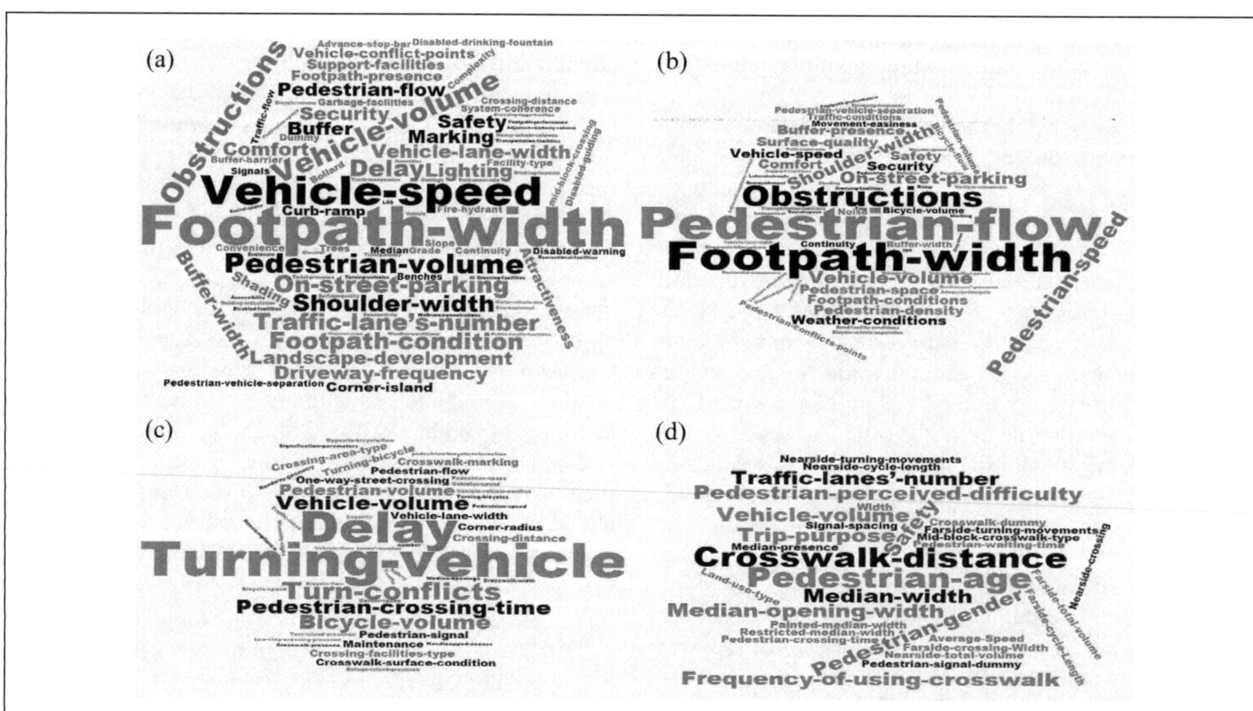

Figure 3. Word clouds demonstrating relative frequency of factors used to determine PLOS by facility type: (*a*) PLOS for streets; (*b*) PLOS for footpaths; (*c*) PLOS for intersections; (*d*) PLOS for mid-block crossings.

Table 3. Methods of Research Papers Assessing PLOS by Facility Type (1971–2017)

Category	No. of PLOS papers for mixed areas	No. of PLOS papers for footpaths	No. of PLOS papers for intersections	No. of PLOS papers for mid-block crossings	Total
Target area	21	19	14	4	58
Sampling methods					
1. Survey (questionnaire, field, intercept)	4	5	5	–	14
2. Observations	8	1	–	–	9
3. Video	2	4	2	–	8
4. Audit	3	1	–	–	4
5. Results of guidelines	2	–	–	–	2
6. Mixed	2	8	7	4	21
Data type					
1. Quantitative	6	9	7	1	23
2. Qualitative	5	3	1	–	9
3. Mixed	10	7	6	3	26
Data analysis					
1. Regression	11	13	12	4	40
• Linear	2	9	5	–	16
• Stepwise	7	3	7	1	18
• Ordinal	2	1	–	3	6
2. Simulation	–	1	–	–	1
3. Fuzzy neural network	–	2	–		2
4. Point system	9	2	1	–	12
5. Other	1	1	1	–	3
Approach type					
1. Capacity based	1	2	1	–	4
2. Street characteristics	7	2	4	1	14
3. Combination	13	15	9	3	40

sample of relevant experts (i.e., via Delphi studies) or a representative sample of pedestrian facility users (i.e., via surveys or interviews).

2. There is a lack of studies identifying the importance of specific factors and especially pedestrian treatments. This issue goes well beyond PLOS, per se, and is needed to inform a range of transport and land use planning policies and programs. One under-researched factor is pedestrian connectivity. While analysis methods such as Space Syntax offer interesting takes on connectivity and are becoming widely used in academia and practice, more real-world analysis of pedestrian networks based on sound theoretical conceptions is needed to identify the key effects of network design.
3. A wider research base from more regions of the world is needed. The majority of the literature is from North America. Walking rates in the U.S.A. are especially low by world standards. PLOS is a critical issue for cities in Saudi Arabia, Iraq, Iran, and India, where the world's largest pilgrimages occur, and where stampedes have resulted in catastrophic loss of life (*75*, *76*). While sunlight may be considered a desirable pedestrian factor in the northern U.S.A. or Canada, shade may be much more desirable in equatorial climates. Developing factors that can be adapted across places within an overall framework would assist.
4. Meaningful inter-rater reliability testing is needed to prove that PLOS methods produce replicable results, at least at the level of the urban context where they were developed and applied, if not beyond.
5. There is little research examining PLOS factors for disabled or special needs users in a wide range, except for the study by Asadi-Shekari et al. (*58*). Understanding how the use of PLOS for able-bodied pedestrians might impact on those with disabilities would be useful.
6. Further disaggregation by facility type, particularly for different types of streets, may make sense. Developing separate PLOS measures suitable for commercial streets, for mixed-use streets, or for suburban streets, may have significant merit.

Table 4. Factors Used to Measure PLOS on Different Facilities

PLOS for streets		PLOS for footpaths		PLOS for intersections		PLOS for mid-block crossings	
Factors detail	No.	Factors detail	No.	Factors detail	No.	Factors detail	No.
1. Factors Related to the Pedestrian Environment							
a. Geometric/physical characteristics							
Footpath width	12	Footpath width	10	Corner radius	2	Crosswalk distance	3
On-street parking	5	On-street parking	4	Crossing distance	2	Number of traffic lanes	2
Shoulder width	5	Shoulder width	4	Crosswalk marking	2	Median opening width	2
Number of traffic lanes	5	Buffer presence	3	Crosswalk surface condition	2	Median width	2
Buffer presence	4	Surface quality	3	Pedestrian signal	2	Crosswalk dummy	1
Buffer width	4	Buffer width	2	Number of traffic lanes	2	Far-side crossing width	1
Driveway frequency	4	Pedestrian–vehicle separation	2	Vehicle lane width	2	Far-side cycle length	1
Lighting	4	Adequate footpath	1	Corner space	1	Median presence	1
Marking	4	Benches	1	Crosswalk width	1	Mid-block crosswalk type	1
Vehicle lane width	4	Bicycle–vehicle separation	1	Disabled access	1	Near-side crossing width	1
Corner island	3	Crossing facilities	1	Median openings	1	Near-side cycle length	1
Curb ramp	3	Disabled guiding	1	Refuge island presence	1	Pedestrian crossing time	1
Shading	3	Driveway frequency	1	Turn island presence	1	Pedestrian waiting time	1
Support facilities	3	Lateral streets	1	Two-step crossing presence	1	Pedestrian signal dummy	1
Advance stop bar	2	Marking	1	Warning signs	1	Signal spacing	1
Benches	2	Ramp	1	–	–	Width of painted median	1
Bollard	2	Shading	1	–	–	Width of restricted median	1
Buffer barrier	2	Support facilities	1	–	–	Land use type	1
Crossing distance	2	Transportation facilities	1	–	–	–	–
Disabled drinking fountain	2	Tree	1	–	–	–	–
Disabled guiding	2	Vehicle lane width	1	–	–	–	–
Disabled warning	2	Road facility conditions	1	–	–	–	–
Facility type	2	–	–	–	–	–	–
Fire hydrant	2	–	–	–	–	–	–
Garbage facilities	2	–	–	–	–	–	–
Grade	2	–	–	–	–	–	–
Median	2	–	–	–	–	–	–
Mid-block crossing	2	–	–	–	–	–	–
Pedestrian–vehicle separation	2	–	–	–	–	–	–
Signals	2	–	–	–	–	–	–
Slope	2	–	–	–	–	–	–
Trees	2	–	–	–	–	–	–
Awnings	1	–	–	–	–	–	–
Crossing facilities	1	–	–	–	–	–	–
Crossing opportunities	1	–	–	–	–	–	–
Disabled facilities	1	–	–	–	–	–	–
Enclosure	1	–	–	–	–	–	–
Elevator	1	–	–	–	–	–	–
Public health facility	1	–	–	–	–	–	–
Recreational facilities	1	–	–	–	–	–	–

(continued)

Table 4. (continued)

PLOS for streets		PLOS for footpaths		PLOS for intersections		PLOS for mid-block crossings	
Factors detail	No.	Factors detail	No.	Factors detail	No.	Factors detail	No.
Surface quality	1	–	–	–	–	–	–
Toilet presence	1	–	–	–	–	–	–
Transportation facilities	1	–	–	–	–	–	–
Trash receptacles	1	–	–	–	–	–	–
b. Network characteristics or (environmental characteristics)							
Obstructions	6	Obstructions	7	Businesses' presence	2	–	–
Footpath condition	5	Comfort	3	Aesthetics	2	–	–
Comfort	4	Footpath condition	3	Footpath performance	2	–	–
Landscape development	4	Safety	3	Pollution	2	–	–
Safety	4	Security	3	Crosswalk presence	1	–	–
Security	4	Weather conditions	3	Safety	1	–	–
Attractiveness	3	Footpath continuity	2	Security	1	–	–
Footpath presence	3	Ease of movement	2	–	–	–	–
Complexity	2	Noise	2	–	–	–	–
Continuity	2	Businesses' presence	1	–	–	–	–
Convenience	2	Aesthetics	1	–	–	–	–
Accessibility	1	Footpath performance	1	–	–	–	–
Amenities	1	Pollution	1	–	–	–	–
Building articulation	1	Social space	1	–	–	–	–
Connectivity	1	–	–	–	–	–	–
Footpath performance	1	–	–	–	–	–	–
Maintenance	1	–	–	–	–	–	–
Transparency	1	–	–	–	–	–	–
Transportation facilities	1	–	–	–	–	–	–
Social space	1	–	–	–	–	–	–
System coherence	1	–	–	–	–	–	–
Weather conditions	1	–	–	–	–	–	–
2. Factors Related to the Traffic System and Users							
a. Traffic characteristics							
Vehicle speed	9	Pedestrian flow	11	Turning vehicle	10	Vehicle volume	2
Pedestrian volume	6	Pedestrian speed	5	Delay	9	Far-side total volume	1
Vehicle volume	6	Vehicle volume	4	Turn conflicts	5	Far-side turning movements	1
Delay	5	Pedestrian density	3	Bicycle volume	4	Near-side total volume	1
Pedestrian flow	4	Pedestrian space	3	Pedestrian crossing time	4	Near-side turning movements	1
Vehicle conflict points	3	Vehicle speed	3	Vehicle volume	4	Pedestrian speed	1
Traffic flow	2	Bicycle flow	2	Pedestrian volume	3	–	–
Adjacent roadway volume	1	Bicycle volume	2	Turning bicycle	3	–	–
Bicycle volume	1	Pedestrian volume	2	Pedestrian flow	2	–	–
Heavy vehicle volumes	1	Traffic conditions	2	Bicycle flow	1	–	–
Motor vehicle mix	1	Adjacent roadway volume	1	Bicycle speed	1	–	–
Pedestrian space	1	Bicycle speed	1	Opposite bicycle flow	1	–	–
Turning vehicles	1	Electric bicycle volume	1	Pedestrian space	1	–	–
Vehicle LOS	1	Mix of path users	1	Pedestrian speed	1	–	–
–	–	Pedestrian crossing time	1	Vehicle flow	1	–	–

(continued)

Table 4. (continued)

PLOS for streets		PLOS for footpaths		PLOS for intersections		PLOS for mid-block crossings	
Factors detail	No.	Factors detail	No.	Factors detail	No.	Factors detail	No.
–	–	Vehicle conflict points	–	Vehicles speed	–	–	–
–	–	Vehicle flow	–	Vehicle-vehicle Conflict	–	–	–
b. Users' characteristics and behaviors							
Encroachment	–	Day time trip	–	Noncompliance	–	Pedestrian age	3
Mix of path users	–	Economical	–	Pedestrian–bicycle interaction	–	Frequency of using crosswalk	2
–	–	Mix of path users	–	–	–	Pedestrian gender	2
–	–	Pedestrian age	–	–	–	Trip purpose	2
–	–	Pedestrian conflicts points	–	–	–	Pedestrian perceived difficulty	2
–	–	Encroachment	–	–	–	Pedestrian perceived Safety	2
–	–	Horizontal maneuvers	–	–	–	–	–
–	–	Vendors attraction	–	–	–	–	–
–	–	Vertical maneuvers	–	–	–	–	–

Note: (–) = no additional factors existent; No. = the number of times the factor was included across the 58 studies in the dataset; PLOS = pedestrian level-of-service.

7. Studies of the use of PLOS by agencies and the outcomes they obtain would be desirable. It is not yet known quite how influential different approaches to measuring PLOS are for real-world transport and built environment outcomes, or how design changes in the transport/land use system influence user behavior. Comparative studies may be possible where local authorities under similar national/state policies may take different approaches to measuring PLOS, creating different outcomes in street, link, or intersection design, and eventually changes in behavior.

Acknowledgments

Nowar Raad's PhD research is supported by the Higher Committee for Education Development in Iraq (HCED) (D-12-1352). Matthew Burke was supported by an Australian Research Council – Discovery Future Fellowship (FT120100976) during the early stages of this research. We wish to thank our colleagues at the Cities Research Institute at Griffith University who helped us, and Dr. Rodney Tolley for his advice.

Author Contributions

The authors confirm contribution to the paper as follows: study conception and design: Nowar Raad, Matthew Burke; data collection: Nowar Raad; analysis and interpretation of results: Nowar Raad, Matthew Burke; draft manuscript preparation: Nowar Raad, Matthew Burke. All authors reviewed the results and approved the final version of the manuscript.

References

1. Kang, L., Y. Xiong, and F. Mannering. Statistical Analysis of Pedestrian Perceptions of Sidewalk Level of Service in the Presence of Bicycles. *Transportation Research Part A: Policy and Practice*, Vol. 53, 2013, pp. 10–21.
2. Gallin, N. Quantifying Pedestrian Friendliness: Guidelines for Assessing Pedestrian Level of Service. *Road & Transport Research*, Vol. 10, No. 1, 2001, p. 47.
3. Asadi-Shekari, Z., M. Moeinaddini, and M. Zaly Shah. Non-Motorised Level of Service: Addressing Challenges in Pedestrian and Bicycle Level of Service. *Transport Reviews*, Vol. 33, No. 2, 2013, pp. 166–194.
4. Ensley, J. O. Application of Highway Capacity Manual 2010 Level-of-Service Methodologies for Planning Deficiency Analysis. Master's Thesis, University of Tennessee, Knoxville, TN, 2012. http://trace.tennessee.edu/utk_gradthes/1373
5. Landis, B., V. Vattikuti, R. Ottenberg, D. McLeod, and M. Guttenplan. Modeling the Roadside Walking Environment: Pedestrian Level of Service. *Transportation Research Record: Journal of the Transportation Research Board*, 2001. 1773: 82–88.
6. Boodlal, L. *Accessible Sidewalks and Street Crossings—An Informational Guide*. Federal Highway Administration,

U.S. Department of Transportation, Washington, D.C., 2004.
7. Jaskiewicz, F. Pedestrian Level of Service Based on Trip Quality. Presented at the Urban Street Symposium of the Transportation Research Board, Dallas, TX, 2000.
8. Pushkarev, B., and J. Zupan. *Urban Space for Pedestrian*. A Report of the Regional Planning Association. The MIT Press, Cambridge, MA, 1975.
9. Polus, A., J. Schofer, and A. Ushpiz. Pedestrian Flow and Level of Service. *Journal of Transportation Engineering: ASCE*, Vol. 109, No. 1, 1983, pp. 46–56.
10. Tanaboriboon, Y., and J. Guyano. Level of Service Standards for Pedestrian Facilities in Bangkok: A Case Study. Institute of Transportation Engineers. *ITE Journal*, Vol. 59, No. 11, 1989, pp. 39–41.
11. Lam, W., J. Morrall, and H. Ho. Pedestrian Flow Characteristics in Hong Kong. *Transportation Research Record: Journal of the Transportation Research Board*, 1995. 1487: 56–62.
12. Davis, D., and J. Braaksma. Level-of-Service Standards for Platooning Pedestrians in Transportation Terminals. Institute of Transportation Engineers. *ITE Journal*, Vol. 57, No. 4, 1987, pp. 31–35.
13. Sarkar, S. Determination of Service Levels for Pedestrians, with European Examples. *Transportation Research Record: Journal of the Transportation Research Board*, 1993. 1405: 35–42.
14. Khisty, C. Evaluation of Pedestrian Facilities: Beyond the Level-of-Service Concept. *Transportation Research Record: Journal of the Transportation Research Board*, 1994. 1438: 45–50.
15. Henson, C. Levels of Service for Pedestrians. *Institute of Transportation Engineers. ITE Journal*, Vol. 70, No. 9, 2000, p. 26.
16. Miller, J., J. Bigelow, and N. Garber. Calibrating Pedestrian Level-of-Service Metrics with 3-D Visualization. *Transportation Research Record: Journal of the Transportation Research Board*, 2000. 1705: 9–15.
17. Sarkar, S. Qualitative Evaluation of Comfort Needs in Urban Walkways in Major Activity Centers. *Transportation Quarterly*, Vol. 57, No. 4, 2003, pp. 39–59.
18. Mozer, D. Calculating Multi-Mode Levels-of-Service. *International Bicycle Fund*, Seattle, WA, 1994.
19. Muraleetharan, T., T. Adachi, K.-E. Uchida, T. Hagiwara, and S. Kagaya. A Study on Evaluation of Pedestrian Level of Service Along Sidewalks and at Crosswalks Using Conjoint Analysis. *Infrastructure Planning Review*, Vol. 21, 2004, pp. 727–735.
20. Petritsch, T., B. Landis, P. McLeod, H. Huang, S. Challa, C. Skaggs, M. Guttenplan, and V. Vattikuti. Pedestrian Level-of-Service Model for Urban Arterial Facilities with Sidewalks. *Transportation Research Record: Journal of the Transportation Research Board*, 2006. 1982: 84–89.
21. Asadi-Shekari, Z., and M. Zaly-Shah. Practical Evaluation Method for Pedestrian Level of Service in Urban Streets. *Presented at International Transport Research Conference*, Penang, Malaysia, 2011.
22. Hidayat, N., K. Choocharukul, and K. Kishi. Pedestrian Level of Service Model Incorporating Pedestrian Perception for Sidewalk with Vendor Activities. *Journal of the Eastern Asia Society for Transportation Studies*, Vol. 9, 2011, pp. 1012–1023.
23. Moher, D., A. Liberati, J. Tetzlaff, D. Altman, and P. The. Preferred Reporting Items for Systematic Reviews and Meta-Analyses: The Prisma Statement. *PLoS Med*, Vol. 6, No. 7, 2009, p. e1000097.
24. Pickering, C., and J. Byrne. The Benefits of Publishing Systematic Quantitative Literature Reviews for PhD Candidates and Other Early-Career Researchers. *Higher Education Research & Development*, Vol. 33, No. 3, 2014, pp. 534–548.
25. Ballantyne, M., and C. Pickering. The Impacts of Trail Infrastructure on Vegetation and Soils: Current Literature and Future Directions. *Journal of Environmental Management*, Vol. 164, 2015, pp. 53–64.
26. Guitart, D., C. Pickering, and J. Byrne. Past Results and Future Directions in Urban Community Gardens Research. *Urban Forestry & Urban Greening*, Vol. 11, No. 4, 2012, pp. 364–373.
27. Roy, S., J. Byrne, and C. Pickering. A Systematic Quantitative Review of Urban Tree Benefits, Costs, and Assessment Methods Across Cities in Different Climatic Zones. *Urban Forestry & Urban Greening*, Vol. 11, No. 4, 2012, pp. 351–363.
28. Steven, R., C. Pickering, and J. Guy Castley. A Review of the Impacts of Nature Based Recreation on Birds. *Journal of Environmental Management*, Vol. 92, No. 10, 2011, pp. 2287–2294.
29. Petticrew, M. Systematic Reviews from Astronomy to Zoology: Myths and Misconceptions. *BMJ: British Medical Journal*, Vol. 322, No. 7278, 2001, pp. 98–101.
30. Fruin, J. *Pedestrian Planning and Design*. Metropolitan Association of Urban Designers and Environmental Planners, New York, NY, 1971.
31. Mori, M., and H. Tsukaguchi. A New Method for Evaluation of Level of Service in Pedestrian Facilities. *Transportation Research Part A: Policy and Practice*, Vol. 21, No. 3, 1987, pp. 223–234.
32. Dixon, L. Bicycle and Pedestrian Level-of-Service Performance Measures and Standards for Congestion Management Systems. *Transportation Research Record: Journal of the Transportation Research Board*, 1996. 1538: 1–9.
33. Milazzo II, J., N. Rouphail, J. Hummer, and D. Allen. Quality of Service for Interrupted-Flow Pedestrian Facilities in Highway Capacity Manual 2000. *Transportation Research Record: Journal of the Transportation Research Board*, 1999. 1678: 25–31.
34. Baltes, M., and X. Chu. Pedestrian Level of Service for Midblock Street Crossings. *Transportation Research Record: Journal of the Transportation Research Board*, 2002. 1818: 125–133.
35. Zhang, L., and P. Prevedouros. Signalized Intersection Level of Service Incorporating Safety Risk. *Transportation Research Record: Journal of the Transportation Research Board*, 2004. 1852: 77–86.
36. Steinma, N., and D. Hines. Methodology to Assess Design Features for Pedestrian and Bicyclist Crossings at Signalized Intersections. *Transportation Research Record:*

Journal of the Transportation Research Board, 2004. 1878: 42–50.

37. Petritsch, T., B. Landis, P. McLeod, H. Huang, S. Challa, and M. Guttenplan. Part 2: Pedestrians: Level-of-Service Model for Pedestrians at Signalized Intersections. *Transportation Research Record: Journal of the Transportation Research Board*, 2005. 1939: 53–62.
38. Muraleetharan, T., T. Adachi, T. Hagiwara, and S. Kagaya. Method to Determine Overall Level-of-Service of Pedestrians on Sidewalks Based on Total Utility Value. Presented at 83rd Annual Meeting of Transportation Research Board, Washington, D.C., 2004.
39. Muraleetharan, T., T. Adachi, T. Hagiwara, and S. Kagaya. Method to Determine Pedestrian Level-of-Service for Crosswalks at Urban Intersections. *Journal of the Eastern Asia Society for Transportation Studies*, Vol. 6, 2005, pp. 127–136.
40. Kim, K., L. Hallonquist, N. Settachai, and E. Yamashita. Walking in Waikiki, Hawaii: Measuring Pedestrian Level of Service in an Urban Resort District. *Transportation Research Record: Journal of the Transportation Research Board*, 2006. 1982: 104–112.
41. Tan, D., W. Wang, J. Lu, and Y. Bian. Research on Methods of Assessing Pedestrian Level of Service for Sidewalk. *Journal of Transportation Systems Engineering and Information Technology*, Vol. 7, No. 5, 2007, pp. 74–79.
42. Muraleetharan, T., and T. Hagiwara. Overall Level of Service of Urban Walking Environment and Its Influence on Pedestrian Route Choice Behavior: Analysis of Pedestrian Travel in Sapporo, Japan. *Transportation Research Record: Journal of the Transportation Research Board*, 2007. 2002: 7–17.
43. Jensen, S. Pedestrian and Bicyclist Level of Service on Roadway Segments. *Transportation Research Record: Journal of the Transportation Research Board*, 2007. 2031: 43–51.
44. Huang, T. H., and C. Chiun, Jr. Modeling Level of Service on Pedestrian Environment. Presented at the 7th International Conference of Eastern Asia Society for Transportation Studies, Saigon, Vietnam, 2007. https://www.doi.org/10.11175/eastpro.2007.0.226.0
45. Daniel, B., S. Choong, K. Ambak, and M. Abdullah. Pedestrian Level of Service Model for Crosswalks at Signalized Intersections. Presented in seminar at Malaysia University of Technology, 2007. http://eprints.uthm.edu.my/1913/
46. Anderson, D. G. *NCHRP Report 611: Seismic Analysis and Design of Retaining Walls, Buried Structures, Slopes, and Embankments*. Transportation Research Board of the National Academies, Washington, D.C., 2008.
47. Petritsch, T., B. Landis, H. Huang, and R. Dowling. Pedestrian Level-of-Service Model for Arterials. *Transportation Research Record: Journal of the Transportation Research Board*, 2008. 2073: 58–68.
48. FDOT. *Quality/Level of Service*. Florida Department of Transportation, Tallahassee, FL, 2009.
49. Bian, Y., J. Ma, J. Rong, W. Wang, and J. Lu. Pedestrians' Level of Service at Signalized Intersections in China. *Transportation Research Record: Journal of the Transportation Research Board*, 2009. 2114: 83–89.
50. Vedagiri, P., and D. Anithottam. Pedestrian Flow Characteristics and Level of Service for Sidewalks in Mixed Traffic Conditions. Presented at 91st Annual Meeting of the Transportation Research Board, Washington, D.C., 2012.
51. Christopoulou, P., and M. Pitsiava-Latinopoulou. Development of a Model for the Estimation of Pedestrian Level of Service in Greek Urban Areas. *Procedia – Social and Behavioral Sciences*, Vol. 48, 2012, pp. 1691–1701.
52. Bunevska Talevska, J., and M. Malenkovska Todorova. Contribution to Sidewalk Pedestrian Level of Service Analysis. *Transport Problems*, Vol. 7, No. 3, 2012, pp. 5–13.
53. Nagraj, R., and R. Vedagiri. Modeling Pedestrian Delay and Level of Service at Signalized Intersection Crosswalks Under Mixed Traffic Conditions. *Transportation Research Record: Journal of the Transportation Research Board*, 2013. 2394: 70–76.
54. Ling, Z., Y. Ni, C. Cherry, and K. Li. Pedestrian Level of Service at Signalized Intersections in China Using Contingent Field Survey and Pedestrian Crossing Video Simulation. *Transportation Research Record: Journal of the Transportation Research Board*, 2015. 2519: 75–84.
55. Jensen, S. Pedestrian and Bicycle Level of Service at Intersections, Roundabouts and Other Crossings. Presented at 92nd Annual Meeting of Transportation Research Board, Washington, D.C., 2013.
56. Kim, I., and H. Kang. *A Study of Delay-Based Level of Service on Pedestrian Facility*. Presented at the 36th Australasian Transport Research Forum, Brisbane, Australia, 2013.
57. Bian, Y., J. Lu, and L. Zhao. Method To Determine Pedestrians Level of Service for Unsignalized Intersections. *Applied Mechanics and Materials*, Vol. 253, 2013, pp. 1936–1943.
58. Asadi-Shekari, Z., M. Moeinaddini, and M. Shah. Disabled Pedestrian Level of Service Method for Evaluating and Promoting Inclusive Walking Facilities on Urban Streets. *Journal of Transportation Engineering: ASCE*, Vol. 139, No. 2, 2013, pp. 181–192.
59. Zhao, L., Y. Bian, J. Lu, and J. Rong. Method To Determine Pedestrian Level of Service for the Overall Unsignalized Midblock Crossings of Road Segments. *Advances in Mechanical Engineering*, Vol. 6, 2014, pp. 1–9.
60. Meng, J., Z. Zhu, and J. Zeng. *The Research on the System of Level of Service of the City Commercial Pedestrian Street*, Applied Mechanics and Materials, Vol. 505, 2014, pp. 656–665.
61. Kim, S., J. Choi, S. Kim, and R. Tay. Personal Space, Evasive Movement and Pedestrian Level of Service. *Journal of Advanced Transportation*, Vol. 48, No. 6, 2014, pp. 673–684.
62. Kadali, B. R., and P. Vedagiri. Modeling Pedestrian Crosswalk Level of Service (LoS) Under Mixed Traffic Condition at Mid-Block Locations. Presented at 93rd Annual Meeting of the Transportation Research Board, Washington, D.C., 2014.
63. Asadi-Shekari, Z., M. Moeinaddini, and M. Shah. A Pedestrian Level of Service Method for Evaluating and

Promoting Walking Facilities on Campus Streets. *Land Use Policy*, Vol. 38, 2014, pp. 175–193.
64. Zhao, L., Y. Bian, J. Rong, X. Liu, and S. Shu. Modeling Pedestrian Level of Service on Sidewalks with Multi-Factors Based on Different Pedestrian Flow Rates. Presented at 94th Annual Meeting of the Transportation Research Board, Washington, D.C., 2015.
65. Ye, X., J. Chen, G. Jiang, and X. Yan. Modeling Pedestrian Level of Service at Signalized Intersection Crosswalks under Mixed Traffic Conditions. *Transportation Research Record: Journal of the Transportation Research Board*, 2015. 2512: 46–55.
66. Kadali, B., and P. Vedagiri. Evaluation of Pedestrian Crosswalk Level of Service (Los) in Perspective of Type of Land-Use. *Transportation Research Part A: Policy and Practice*, Vol. 73, 2015, pp. 113–124.
67. Hasan, T., A. Siddique, M. Hadiuzzaman, and S. Musabbir. Determining the Most Suitable Pedestrian Level of Service Method for Dhaka City, Bangladesh, Through a Synthesis of Measurements. *Transportation Research Record: Journal of the Transportation Research Board*, 2015. 2519: 104–115.
68. Archana, G. Analysis of Pedestrian Level of Service for Crosswalk at Intersections for Urban Condition. *International Journal of Students' Research in Technology & Management*, Vol. 1, No. 6, 2015, pp. 604–609.
69. Lazou, O., A. Sakellariou, S. Basbas, E. Paschalidis, and I. Politis. Assessment of LOS at Pedestrian Streets and Qualitative Factors: A Pedestrians' Perception Approach. Presented at the 7th International Congress on Transportation Research, Athens, Greece, 2015.
70. Zhao, L., Y. Bian, J. Rong, X. M. Liu, and S. Shu. Evaluation Method for Pedestrian Level of Service on Sidewalks Based on Fuzzy Neural Network Model. *Journal of Intelligent & Fuzzy Systems*, Vol. 30, No. 5, 2016, pp. 2905–2913.
71. Raghuwanshi, A., and V. Tare. Assessment of Pedestrian Level of Service for Mixed Lane. *Research Journal of Engineering and Technology*, Vol. 7, No. 1, 2016, p. 11.
72. Daniel, B., S. Nor, M. Rohani, J. Prasetijo, M. Aman, and K. Ambak. Pedestrian Footpath Level of Service (FOOTLOS) Model for Johor Bahru. In *MATEC Web of Conferences*, 2016.
73. Chandana, K., M. Ibrahim, and R. Kumar. *Analysis of Pedestrian Level of Service on Footpath in Hyderabad*. Presented at the 3rd International Conference on Science, Technology, and Management, New Delhi, India, 2016.
74. Baxter, J., and J. Eyles. Evaluating Qualitative Research in Social Geography: Establishing "Rigour" in Interview Analysis. *Transactions of the Institute of British Geographers*, Vol. 22, No. 4, 1997, pp. 505–525.
75. Aljohani, A. *Pilgrim Crowd Dynamics*. University of Birmingham, Birmingham, UK, 2015.
76. Haghighati, R., and A. Hassan. Modeling the Flow of Crowd During Tawaf at Masjid Al-Haram. *Jurnal Mekanikal*, Universiti Teknologi Malaysia, Vol. 36, 2013, pp. 2–18.

The Standing Committee on Pedestrians (ANF10) peer-reviewed this paper (18-03950).

Article

TRR
JOURNAL OF THE TRANSPORTATION RESEARCH BOARD

Transportation Research Record
2018, Vol. 2672(35) 118–128

DOI: 10.1177/0361198118787630
journals.sagepub.com/home/trr
SAGE

Walkability in the Connected and Automated Vehicle Era: A U.S. Perspective on Research Needs

Elizabeth Shay[1], Asad J. Khattak[2], and Behram Wali[2]

Abstract

Walkability and walking activity are of interest to planners, engineers, and health practitioners for their potential to improve safety, promote environmental and public health, and increase social equity. Connected and automated vehicles (CAVs) will reshape the built environment, mobility, and safety in ways we cannot know with certainty—but which we may anticipate will change the meaning of "walkability." The CAV era may provide economic, environmental, and social benefits, while potentially disrupting the *status quo*. This paper considers the concept of walkability in light of the approaching transition to CAVs, considering literature in engineering, information technology, built environment, land use, and public health, to support a discussion on research needs. To add depth, we subject a collection of research papers and technical reports to text analytics.

Walkability and Walking: Past and Future

Walkability is a key element of communities that are active, accessible, livable, efficient, safe, and just. There is both public demand for walkability (good for health, households, and community) and a clear economic case (good for businesses, economic growth, and property values). While connected and automated vehicles (CAVs) will reshape the built environment, mobility, and safety in ways we cannot know with certainty, we can anticipate that the meaning of "walkability" may change in the CAV era.

Walking can reasonably be described as the most sustainable of all transportation modes: affordable and available to most people, health-promoting, free of emissions or fuel consumption, posing little threat to other travelers and—in the right conditions—sociable and pleasant. In the US, a recent surge of interest in walking represents a paradigm shift—a re-examination of a mode once undervalued, now promising to support human and environmental health, promote economic prosperity, and advance social equity.

Forsyth and Southworth (*1*) describe pedestrianism declining through the 20th century, as successive transportation advances (horsecar, streetcar, rail, auto) flourished at the expense of walking—disrupting pedestrian networks, hindering easy movement, and degrading intimate lively streets to service roads. After decades advocating for walkability, planners are buoyed by renewed demand for walkable environments as individuals, governments, and businesses increasingly recognize the potential benefits for physical and mental health, economic development, and efficient use of land (*2*).

Walking is on the rise. The 2009 National Household Travel Survey (*3*) found that walking accounted for over 10% of all trips in the US—a gain of nearly two percentage points over 2001.

Several early 21st-century trends coexist in some degree of tension: walkable urbanism as an explicit goal for planning and economic development; the emergence of a sharing economy that alters travel choice sets and behaviors; and rapid technological advances that promise a transition to driverless vehicles and networked systems—sooner rather than later.

This paper reviews recent literature on walkability within the context of a rapidly changing transportation landscape in which CAVs will mix with—and eventually largely supplant—manually driven vehicles. This supports ongoing inquiry into the most promising technology, designs, and policies to reduce risk to vulnerable road users, to better understand and influence behavior of drivers and walkers, and to identify moral dilemmas and gray areas that may arise in the CAV era.

[1]Department of Geography and Planning, Appalachian State University, Boone, NC
[2]Civil & Environmental Engineering, University of Tennessee, Knoxville, TN

Corresponding Author:
Address correspondence to Elizabeth Shay: shayed@appstate.edu

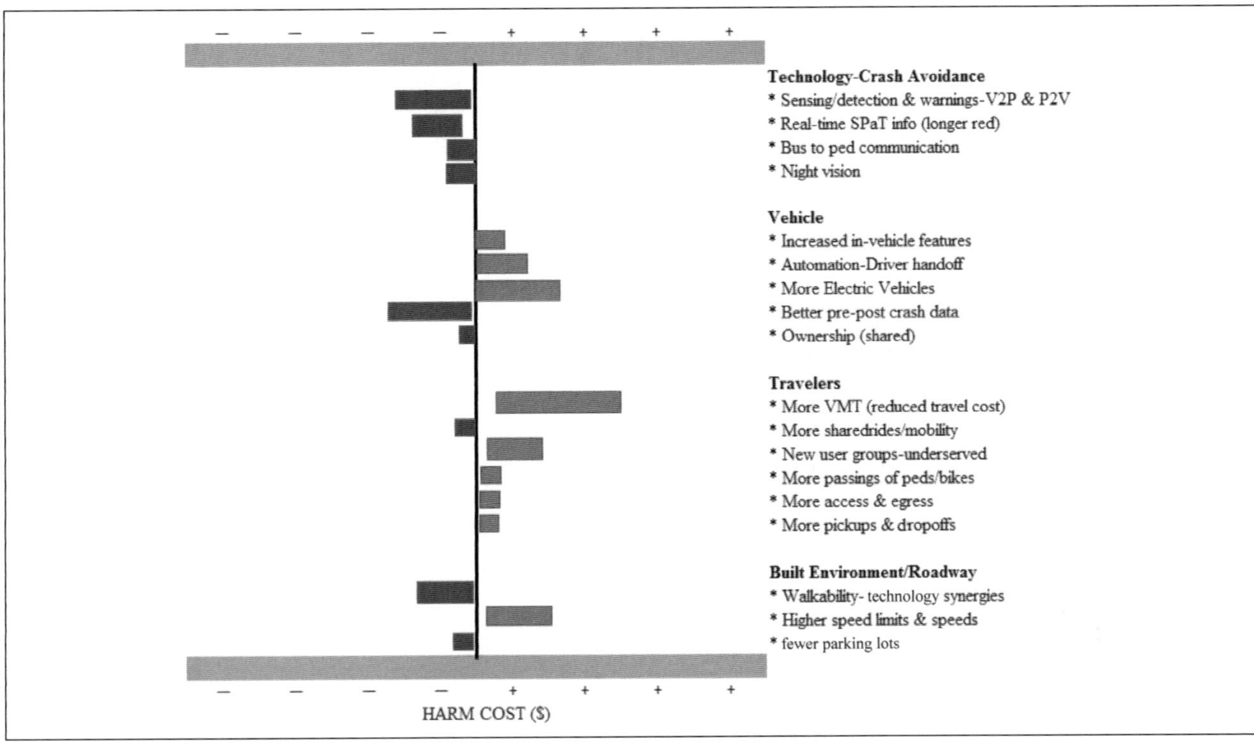

Figure 1. Framework for mechanisms that can affect harm cost.

We reviewed literature on walkability and urban form, travel behavior, traffic safety, public health, information technology, and engineering, searching on a controlled list of 14 terms in four comprehensive databases (TRID, ScienceDirect, Web of Science, Google Scholar). Of several hundred empirical papers, technical reports, book chapters, and reviews scanned, a subset relating to walkability and CAVs were inventoried and subjected to text analysis. We take a primarily US perspective, while acknowledging extensive work under way across the world in engineering, policy, and planning. The rapidly changing transportation landscape, in which emerging CAV technology intertwines with a push for walkable communities and mobility equity, motivated our examination of literature from diverse fields and our effort to illuminate points of intersection and gaps in knowledge.

Framework

Figure 1 provides—for illustrative purposes—a framework, with mechanisms for automation shown on the right. The pedestrian and bicycle safety impacts of CAVs are shown as potential reductions in harm costs (sum of crash costs in a year) for a region. Notably, crash avoidance technology (vehicle–pedestrian communication) may lower crash risks and costs (blue bars). However, potential vehicle miles traveled (VMT) increases in automated vehicles (AVs) may offset these gains and increase the risk and cost of collisions with pedestrians and bicyclists. Estimates of safety costs range from a ~50% decrease to a 30% increase. This paper highlights uncertainties about whether net CAV impacts on pedestrian and bicycle safety will be positive or negative, and the need to identify mechanisms for enhancing safety for these modes.

Text Analysis

To probe views on walkability and CAVs, we performed statistical pattern learning analysis on 70 research and technical papers. This involved extracting key topics and graphically representing them based on proximity/conception distance among them.

Frequency analysis identified common keywords, from which an inclusion dictionary was manually constructed to define categories and group similar words together. To refine that dictionary, a combination of natural language processing and statistical (factor analysis) techniques was applied on the otherwise random textual data in the collection (*4*, *5*). Table 1 presents the results of topic extraction, in broad categories:

- vehicle-related
- energy consumption
- technology and applications
- safety
- built environment

Table 1. Results of Text Mining via Factor Analysis Procedure for Topics Extraction

Broader category	Topic	Keywords	Eigenvalue	% Variance	Frequency	Cases	% Cases
Vehicle-related	Automated vehicles; driving	Driving; automation; driver; vehicle; human; vehicles; task; systems; Society of Automotive Engineers; automotive; control	6.15	1.85	5762	66	94.29
	Shared automated vehicles	AV; shared automated vehicles; Scenarios; adoption; shared; scenario; penetration	2.90	1.59	571	24	34.29
	Liability, security, and privacy	Liability; privacy; legal; security; law; insurance	1.55	1.08	1674	59	84.29
	Morals and ethics	Moral; ethics; ethical; machine; algorithms	1.54	1.01	1628	52	74.29
Energy consumption	Truck platooning and fuel consumption	Platoon; trucks; platooning consumption; energy; emissions; fuel	2.17	1.19	0	0	0.00
	Electric power and oil	Electric; electric vehicles; power; cars; oil; driverless	1.65	0.99	2974	63	90.00
Technology and applications	Wireless communications; mobile applications	Applications; communications; wireless; mobile; connected; dedicated short-range communication; smart	2.01	1.21	878	47	67.14
	Adaptive cruise control; controls	Cruise; adaptive; adaptive cruise control; control; lane	1.89	1.41	1627	59	84.29
	Collision avoidance systems	collision; avoidance; warning; collisions	1.80	1.23	2868	63	90.00
Safety	Pedestrian injuries; bumper	Bumper; injuries; hood; injury; crashes; pedestrian	2.27	1.24	1363	57	81.43
	Safety	Pedestrians; users; traffic; drivers; safety	1.38	1.06	467	26	0.3714
Built environment	Built environment and physical activity	Environment; physical; built; activity; walking; cycling; health; active	2.46	1.25	11,913	65	92.86
	African Americans, submarkets	African; Americans; town; city; suburbs; family submarket; sub-markets; central business district; costar; markets; downtown; office	1.42	0.93	896	48	68.57
	Parking space, shared	Space; parking; shared; sharing; ownership	1.35	0.93	490	47	67.14
Walking	Walkable urban metros	Metros; walkups; metro; rental; drivable; urbanism; walkup; family; walkable	2.65	1.60	1294	56	80.00
	Vibrant suburban	Vibrant; centers; suburban; center; town	4.23	1.40	698	33	47.14
	Walking trips	Trips; trip; mode; transit; travel; vehicle miles traveled; walking; distance; station	2.12	1.36	623	29	41.43
	Walkability Score	Score; walk; properties; retail; property; office neighborhood; walkable; neighborhoods; walkability; social; capital; residents	1.77	1.07	3427	62	0.8857
	Exercise and motivation	Intrinsic; motivation; motives; exercise med; activity; physical	1.52	0.89	423	26	37.14

(continued)

Table 1. (continued)

Broader category	Topic	Keywords	Eigenvalue	% Variance	Frequency	Cases	% Cases
Government assistance	Government programs	Low-income housing tax credit (LIHTC); subsidized; households; voucher	1.73	1.29	1309	56	80.00
	Project-based rental assistance; housing units	Project-based rental assistance (PBRA); units; poverty; tenants; housing; voucher; rental	3.40	1.39	1749	45	64.29
Outreach	International conferences	Conference; IEEE; international; intelligent; systems; journal	1.76	1.03	872	38	0.5429

Note: Percentage variance shows the percentage of variance explained by each topic. Note that the smaller the text phrases in each topic, the lower the percentage of variance explained, at least theoretically; Eigenvalues are calculated and used in deciding how many "text phrases" to extract for a specific topic. In factor analysis, "the topic" with largest Eigenvalue has the most variance, and topics with Eigenvalues greater than 1.00 are traditionally considered worth analyzing.

- walking
- government assistance/subsidy
- outreach.

Appearing 5762 times in 66 articles (94% of the collection), the factor "automated vehicles; driving" has the highest Eigenvalue at 6.15, and explains 1.85% of the variance in the collection. The next most frequent is "vibrant suburban," with an Eigenvalue of 4.23, explaining 1.40% of the variance, and appearing in 47% of the articles. All key topics shown in Table 1 have an Eigenvalue greater than 1, indicating the topics are worth analyzing (*5*). Interestingly, "liability, security, and privacy" and "moral and ethics" also appeared prominently. However, cyber-security, insurance, and attitudes toward automation, encouragement, and enforcement are beyond our scope.

Concept maps were constructed using proximity values computed on all the key topics in Table 1, using constrained multidimensional scaling (*6*). Key topics that plot closer tend to occur together in the text. This technique visualizes the co-presence of keywords across all key topics. The network visualization (Figure 2) shows clustering of words that are significantly connected. Technology-related keywords and topics cluster to the right, and planning-related keywords to the left. Broader concepts written next to the clusters improve readability. Some keywords serve as bridges in the visual network, such as the broader concept of "wireless communications" appearing between "automation" and "collision avoidance systems; adaptive cruise control," suggesting that ideas related to wireless communications are discussed in close proximity to automation or collision avoidance systems. Shared AVs and market penetration, adoption, and scenario development appear near the concept of automation. Finally, planning concepts, that is, built environment, walkability, and government subsidies, cluster together. Concept maps are useful for revealing how key topics are connected through their keyword distribution and relate to walkability and CAVs. Note that the output reflects the selection of documents analyzed; results are illustrative, not determinative.

Walkability and Walking: Environment and Behavior

This discussion builds on an earlier hierarchy (*7*) of walkability (environments) and walking (behavior), considering technology that may impact walkability and walking, and describing how walkability discourse is changing on the cusp of a transition to AVs. Although appreciation for the potential benefits of non-motorized travel has risen along with the number of walkable-by-design communities (neo-traditional, new urbanist, smart growth, and so forth) built or retrofitted across the US, there is no single authoritative definition of walkability. In practice and research, the term applies to a variety of settings displaying a range of features, particularly relating to accessibility, connectivity, aesthetics, comfort, and safety.

The transportation profession has long called for more walkable communities, in part to address traffic congestion and mobile-source pollution. Health professionals and planners increasingly view the physical environment as an intervention opportunity in the US, where over three-quarters of adults are obese or overweight (*8*) and only 20% of adults meet physical activity guidelines (*9*). Physical activity helps manage weight; build strong bones, muscles, and joints; reduce falls; reduce anxiety and depression; and limit medical visits (*10*). Several decades of empirical work have probed how built environments relate to routine physical activity, and strengthened earlier links between planning and public health (*11–15*).

Walkability Described

Much of the walkability literature is concerned with features to promote a pedestrian-friendly environment. Forsyth (*16*) provides a general descriptive framework; Litman (*2*) offers

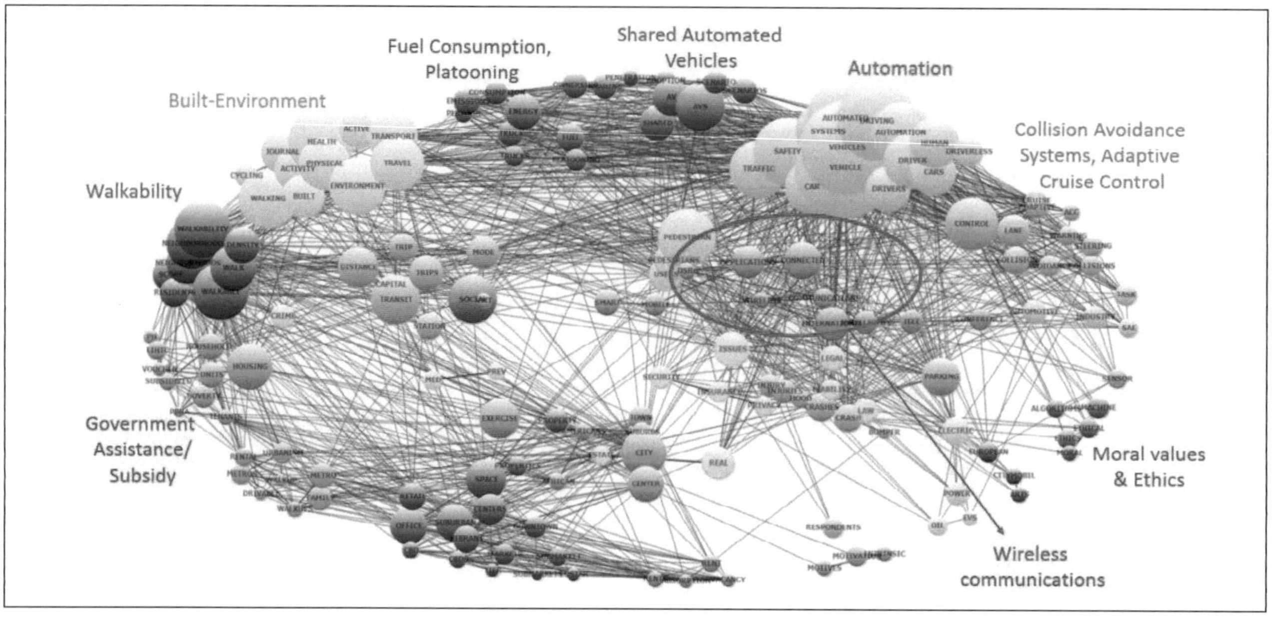

Figure 2. Co-presence structure of words across key topics.

prescriptive guidelines for walkability, including network (paths, sidewalks, crossings), connectivity (connections, directness), security, density, and access. Walkability has been described (*17*, *18*) with a host of inventories, checklists, and audits, used by planners, engineers, advocates, consultants, and citizens to assess whether an environment promotes efficient, safe, comfortable, and pleasant walking.

Measuring active travel support is challenging when "ways to measure active accessibility are as varied as the number of scholars that measure them" (*19*), and when urban form is entangled with socioeconomic patterns such that walkability and walking vary across environments (e.g., urban/rural, heavy/light traffic) and demographic measures such as income or race (*20–25*).

The literature on environment and walking is extensive, with hundreds of entries in the Active Living Research database (Robert Wood Johnson Foundation). Ewing and Cervero's 2010 meta-analysis (*26*) of over 200 empirical papers and more than a dozen reviews concluded that individual measures may be relatively insensitive to environmental features, but the combined effects may influence travel. Talen and Koschinsky (*13*) found claims for social benefits to be relatively weaker than those for physical activity or economic value, but the body of empirical research largely corroborating claims that walkable neighborhoods increase walking.

Numerous studies have tested the relationship of walking with environmental features such as density, proximity, connectivity, street pattern, navigability, traffic, crime, open space, and aesthetics, as well as travel companions (people, pets), social capital, and connectedness (*27–33*). This deep body of work has yielded both discernable relationships and some mixed findings.

Walkability Valued

Increasingly, walkability is viewed as a tool to drive commerce, improve public health and safety, and contribute to general social wellbeing. A 2015 survey (*34*) found half of the public assigning a high priority to a neighborhood's walkability. Frank and Kavage (*35*) call pedestrian-supportive places essential to "averting what is currently a market failure where the supply and demand for walkable environments is misaligned."

Walkability spills beyond engineering and planning, into economics, housing, and design. The economic case and willingness to pay for walkability has been covered in numerous studies (*36–39*). This raises the specter of walkable gentrification, where increasingly popular location-efficient walkable neighborhoods become affordable to fewer people (*40*). Complete Streets are another manifestation of public demand for more widely shared benefits of urban efficiency and walkability (*41*).

Walking and Safety

Worldwide, vehicle crashes take a heavy toll (*42*), with about 270,000 pedestrian deaths annually; children and elderly are particularly vulnerable. In the US, pedestrian fatalities have dropped in raw numbers over four decades, from 7516 reported in 1975 by FARS (Fatality Analysis Reporting System) to 4910 in 2014, rebounding to nearly 6000 in 2016

(*43*). Injuries dropped from 110,000 pedestrians in 1988, when non-fatal injury reporting began through the General Estimates System (GES) of the National Automotive Sampling System, to 70,000 in 2015. With growing population and per capita travel over most of this period, these numbers testify to net declines in death and injury per mile of exposure.

Road features serving pedestrians include sidewalks and paths, medians and crosswalks, intersection and roadway controls and warning devices, over/underpasses, circulation patterns, traffic volumes, and traffic management tools. Travelers with disabilities may benefit from such features as textured pavements, ramps, audible or vibrating signals and strong signage. Which of these features will remain useful and common in the CAV environment is a subject of debate and conjecture. As speeds, spacing, lane-keeping, and pathfinding change with CAVs, the relative value of various pedestrian features may also change. Zegeer and Bushell (*44*) note the particular vulnerability of pedestrians in nations with weak enforcement, as auto use grows globally.

The 2015 FARS and GES data show that while roughly half of US pedestrian *injuries* occur at intersections, over 70% of pedestrian *fatalities* occur at non-intersection locations, underscoring the lethality of pedestrian attempts to negotiate infrastructure without traffic controls. Pedestrian behaviors most frequently cited in fatalities are failure to yield, improper crossing, lack of visibility, and improper activity in streets (*43*). For drivers, hitting non-occupants (pedestrian or bicyclist) accounts for 18% of "first harmful events," compared with hitting another vehicle (38%) or another fixed object (31%). Pedestrians and cyclists are only lightly covered in Highway Safety Manual procedures, and in Intelligent Transportation Systems architecture—where two recent additions deal with sensing and warning systems, and vehicle-to-device communications in transit. A future safe systems approach should draw from engineering, planning, policy, and public health, among other fields, to combine CAV technology and traditional countermeasures to reduce the risk of crashes and severity of injuries.

Walkability and Technology

Layered over the question of how pedestrians, cyclists, and drivers negotiate shared or adjacent space is the rapidly changing technology that suffuses the travel landscape, ranging from personal devices that retrieve information and support decision-making, to IT systems like incident management, to in-vehicle technology—from simple driver assists through full automation. Travel safety is deeply personal—but increasingly impacted by forces beyond the control of the individual traveler.

Emerging technologies by which vehicles detect and respond to objects in the environment may dramatically reduce danger for pedestrians, but need further refinement, especially in complex situations such as school or work zones. The potential for reducing pedestrian injury and fatality under varying automation scenarios is an open question deserving rigorous research—particularly in light of the recent surge in pedestrian fatalities after decades of decline.

That the looming CAV era will challenge pedestrian safety is nearly certain—but murky in the details. Sparrow and Howard (*45*) identify a threshold: When CAVs become safer for third parties than manually driven cars, human drivers become the "moral equivalent of drunk robots" and should be banned. Goodall (*46*) discusses the challenge of coding moral complexity into technology, and calls for ethical crash algorithms involving rationality, artificial intelligence, and natural language.

The rising profile of walkability—in research and practice, as well as in public discourse—exists in apparent tension with growing acceptance and penetration of CAVs. Schwartz (*47*) describes a synergy of technology soon to appear on our roads, and new shared modes that "will force cities to confront a host of uncertainties on issues ranging from safety, ethics, insurance, and regulatory requirements to technological necessity, pricing, and the scale of widespread adoption," as cities wrestle with complex decisions about land use, infrastructure, and mobility. Glancy (*48*) notes that CAV technology promises vastly safer and more efficient infrastructure, while raising legal and policy questions relating to privacy and liability. New technology may feed competing trends: The space-efficiency of CAV operation may *reduce* congestion in the short term, while lower perceived costs may *induce* travel and attendant sprawl and congestion; further, CAVs in a ride-sharing market may *dampen* private vehicle ownership.

The New-er Urbanism?

Leinberger and Rodriguez (*49*) see walkable urbanism as marking the end of sprawl, taking hold even at the cusp of an automated era, as cities promote pedestrianism. They found walkable urban places gaining market share over drivable suburbs, commanding higher prices, and correlating with higher education, wealth, and social equity.

Functional walkability may be impacted by a range of technological advances, from information and messaging that supports travel decision-making (weather, traffic, routes, fares) and devices that reduce pedestrian injury in vehicle crashes (*50*), to an array of technologies that increase automation and reduce crash frequency. Smart travel technology may reduce cost, time, and emissions; it also may introduce additional risks—for both drivers and walkers—of distracted travelers (*51*). While increased safety is a primary claim of CAV advocates, potential negative safety impacts merit scrutiny, yet are understudied (*52*).

Even as technology and systems innovations penetrate engineering, policy, and planning, and—to some extent—capture the attention of the traveling public, other trends are taking different, possibly conflicting, directions. Shared space for multiple modes, well-established in Europe, is gaining ground in some US cities as a low-cost way to reclaim right-of-way for non-motorized travelers and to reorder the hierarchy of road users.

Underwood et al. (*53*) argue that AVs and associated technologies will not halt progress in pedestrian, bicycle, and transit travel, but rather may increase safety, possibly returning urban space to pedestrians when shared vehicles are used more efficiently. Questions remain about whether space liberated by CAVs will return to public uses—or be swamped by demand for driverless services. Interactions between CAVs and other travelers are complex, as modeled by Millard-Ball (*54*): "Because autonomous vehicles will be risk-averse, the model suggests that pedestrians will be able to behave with impunity," such that pedestrian-oriented neighborhoods may flourish as CAV adoption is dampened by slow urban traffic.

CAVs and Pedestrian Safety

A key potential CAV benefit is fewer collisions owing to human error—by drivers and pedestrians. As AVs penetrate the market, they will need to accommodate other road users, who may not follow rules of the road or behave predictably—requiring CAVs to adapt, such as reducing speeds near vulnerable road users.

The CAV safety premium for walkers assumes tight communication between pedestrians and vehicles, making all parties more aware of each other's behavior and better able to predict and negotiate conflict points. The expected benefits of CAVs include safety (fewer crashes), efficiency (as drivers become passengers, relieved of driving duties), and equity of mobility for non-drivers, with rationalized use of urban space and infrastructure and solutions for the first/last mile problem of short and multimodal trips (*55*, *56*). Meeder et al. (*57*) note that driverless cars may change demand for on-street parking and increase careless pedestrian crossings.

The general consensus on safety potential of CAVs also introduces questions of values and ethics with which society must wrestle (*58*). Blyth et al. (*59*) argue that the technology has weighty implications for users and societies: "In particular, values such as privacy, security, and responsibility may be changed for good or bad, in both the short and long-term." The transition to full automation involves stage-wise technological advances (*60*), which raise questions about mixing levels of automation and the CAV transition (*61*, *62*).

Automation promises mobility and safety gains for pedestrians—both general and special populations that are currently underserved, such as young and old, mobility-limited, and low-income travelers. Despite general consensus on net fewer crashes, CAVs do not promise zero risk to travelers, and will sometimes have to choose between sacrificing themselves and passengers or pedestrians (*63*). Writing the code by which CAVs will make what are essentially moral decisions has daunting implications for pacing and pathways for adoption (*63*), forcing decisions with no good choices (*64*). Anderson et al. (*65*) suggest the careful policymaking necessary to both minimize disadvantages of CAVs and maximize social benefits is lagging behind technological progress.

How Will Walkability Change When Vehicles Are Automated?

Is the relationship of people and their cars changing, and will CAVs play a role? A decade ago, the sociologist Urry (*66*) described the car–traveler relationship as demanding intense flexibility that "forces people to juggle tiny fragments of time so as to deal with the temporal and spatial constraints that it itself generates." If auto ownership declines in favor of sharing-based AV mobility (*67*), will the relationship of people, environment, and travel behavior change? Will auto ownership, recently peaking, continue to drop? Moss (*68*) suggests that private vehicle travel will decline, a delivery economy will resurge, and self-contained communities will rule, aided by networks and smart devices.

The CAV era may unfold in stages (*69*): first lower congestion, more miles but less travel time, giving way to second-order reduced infrastructure and auto-ownership, with shifting but uncertain land uses; downstream third-order ripples may bring lower fuel consumption and emissions and enhanced safety, with mixed outcomes for energy, pollution, health, and equity.

Noyman et al. (*70*) sketch out future CAV scenarios: (1) passive—*ad hoc* and patchwork changes, responding to the market; (2) adaptive—bureaucratic and moderately flexible; or (3) active planning—with ground rules, building on the "capability and willingness of certain cities to elect different planning paths." This active stance may see cities choose walkers over drivers and greenspace over infrastructure, in a bid to own and shape their own futures.

General consensus on the inevitability and potential of CAVs does not constitute unanimity; questions and doubts remain. Shladover (*71*) suggests that self-driving cars have been overhyped, while the challenge of managing vehicle–environment interactions in complex and dynamic settings with sensing and visualization technology have been underestimated. Applications that rely on human skills for backup driving are problematic because of the human–technology interface; at the same time, there is an array of narrowly targeted applications and technologies that are realistically part of the near future (*71*). Some observers are bullish; Niles (*72*) sees the goal of reducing impacts of autos—on safety, economy, and environment—as realistic, and key to livable sustainable communities.

The technical challenges of designing, operating, and controlling CAV systems raise human-factor and socioeconomic questions that demand attention if CAVs are to deliver the safety, equity, and efficiency gains they promise (*73*).

The "mechanisms that allow us to accurately model the minds of other people and interpret their communicative signals of attention and intention will be challenged in the context of non-human, autonomous moving entities such as self-driving cars" (*74*) even as humans relinquish direct control of vehicles (*75*). A recent discussion guide (*76*) describes several archetypal problems facing the AV/non-motorized traveler relationship (e.g., detection, networking, communication, right-of-way, speed, pickup/dropoff, driver handoff, mode shift, and data), then sketches out policy implications.

Research Needs

Many planning, engineering, and public health professionals value walkability that promotes safe mobility. Crayton and Meier (*77*) call for a public health agenda to shape the coming CAV-based environment, to "provide a basis to frame autonomous vehicle policies that best support the public's health." Giles-Corti et al. (*78*), summarizing the global challenge of city planning and health in a technology-rich future, focus largely on design for access and equity, and on opportunities for non-motorized travel as part of efficient, safe, affordable, and livable communities.

While evidence suggests that walkability increases walking, it is not a complete response to safety threats, congestion, physical inactivity, and other economic, environmental, and social problems in highly motorized societies. Local context and shared goals should guide planning, with solutions found in technology, markets, and behavior. The literature for walkability and walking is necessarily broad, given the complex interactions between human behavior and richly diverse environments. Future research should continue to build links among engineering, planning, policy, public health, and other fields, and account for the emergence and penetration of CAVs. Social and cognitive psychology may offer useful insights on self-determination theory and exercise motivation, and on conscious and unconscious health-related behavior (*79*, *80*). Open questions include how human-driven vehicles will interact with AVs; how CAV penetration will influence walking behavior; and how CAVs will respond to pedestrians' changing behavior as they test CAV programming in dynamic environments.

With several decades of research and practice in walkable environments and active travel yielding evidence and demonstrations of effective design, the recent paradigm of walkable, livable, efficient, and equitable urban form faces uncertain threats from the CAV transition, with thorny questions about the walking/CAV relationship. Below we offer questions that might constitute a research agenda relating to this relationship.

Vehicles: Owned, Rented, Shared

If interest in owning vehicles fades in favor of access and connectivity, will shared mobility actuated through CAVs make walkability less or more valued? Will stratified ownership models (private, shared, mixed) change propensity to walk?

Built Environment

How will CAV infrastructure impact pedestrian comfort and safety? The implications of promised space-efficiency of CAV fleets and reduced parking demand are murky. More compact infrastructure with less parking may reduce driving *and* walking; excess infrastructure could be returned to the pedestrian or public realm—but the responsibility and champions for such conversions are far from clear. Who would design, site, and control? Will the advent of CAVs dampen, reverse, or change the trend toward compact active walkable cities?

Trip-Making

How might CAVs change the mix and character of trips for utilitarian and recreational purposes? Will general and special (underserved) populations increase walking trips to access CAV services, or might a door-to-door model dampen walking-for-access?

Safety

What safety gains do CAVs offer—to drivers and pedestrians? Safety claims rest on extensively networked IT to support communications among vehicles, infrastructure, and non-drivers. Using data from drivers and other road users to model and predict behavior presupposes ubiquitous use of personal devices. Some walkers may not have financial or knowledge resources—or desire—to be connected to a network. Will walking become criminalized in some environments? What data should be collected—by and from whom, and for whose consumption and action? How have causes of crashes changed over time—and what role do distraction and technology failure play?

Engineering

The CAV transition requires a foundation of sophisticated technology. How will it account for human behavior (sometimes confounding the expectations of rational predictive models), and support planners and service operators in providing safe mobility and accessibility to travelers of multiple modes with an eye to comfort, convenience, and equitable access, while preserving recent livability gains? What are the expectations for communication among vehicles and with the environment and pedestrians, and how might that affect walkability? How will infrastructure and signage change? Will we still need marked crosswalks?

Urban Form, Land Use Planning, Policy

How will CAVs reorder the land use and transportation relationship, and alter urban form and travel patterns? What will more efficient use of facilities by CAVs mean for public space? Who will determine and manage policy changes that honor the core mission of the transportation system—to provide safe mobility, and economic and social opportunity?

Given heady claims that CAVs can approach the Vision Zero goal of eradicating traffic fatalities while also making travel more affordable, accessible, comfortable, and convenient for all travelers, where should pedestrian safety and walkability research focus in the coming years?

While new technologies appear to offer substantial safety gains, mobility options must be protected and walkability improved in communities through better planning and design, training, and education. Research should tap diverse data and advanced analytical methods to generate new knowledge that helps communities achieve safe mobility and build an efficient and equitable multimodal transportation network for all users.

Given the high cost of mobility—in dollars as well as lives and health, the nation should prioritize safety and accessibility for all modes, with strategically selected technology. In particular, research is needed on new connected and automated technologies that can avoid crashes and meet the mobility needs of diverse users and stakeholders. Crosscutting safety and technology-oriented research, development, and deployment efforts should encompass CAVs and infrastructure, and safety technologies and strategies to promote safe and smart communities. Game-changing solutions may lie between and among the many disciplines that touch transportation safety research, and address access to critical destinations. A comprehensive research agenda should use partnerships among academia, public-sector agencies, and leaders in the private sector, to identify innovative life-saving and community-strengthening models for safe and walkable environments.

Acknowledgments

This work was supported by the US Department of Transportation through the Collaborative Sciences Center for Road Safety, a national University Transportation Center consortium led by The University of North Carolina at Chapel Hill in partnership with The University of Tennessee (and other members).

Author Contributions

The authors confirm contribution to the paper as follows: study conception and design: Shay, Khattak; data collection: Shay; analysis and interpretation of results: Shay, Khattak, Wali; draft manuscript preparation: Shay. All authors reviewed the results and approved the final version of the manuscript.

References

1. Forsyth, A., and M. Southworth. Cities Afoot—Pedestrians, Walkability and Urban Design. *Journal of Urban Design*, Vol. 13, No. 1, 2008, pp. 1–3.
2. Litman, T. *Economic Value of Walkability*. Victoria Transport Policy Institute. http://www.vtpi.org/walkability.pdf. Accessed March 7, 2018.
3. National Household Travel Survey. Bureau of Transportation Statistics. https://nhts.ornl.gov/2009/pub/stt.pdf. Accessed June 30, 2018.
4. Jolliffe, I. T. Principal Component Analysis and Factor Analysis. In *Principal Component Analysis*. Springer, New York, 1986, pp. 115–128.
5. Provalis. *WORDSTAT 7 User's Guide*. Available at: https://provalisresearch.com/Documents/WordStat7.pdf. Accessed June 30, 2018.
6. Breiger, R. L., S. A. Boorman, and P. Arabie. An Algorithm for Clustering Relational Data with Applications to Social Network Analysis and Comparison with Multidimensional Scaling. *Journal of Mathematical Psychology*, Vol. 12, No. 3, 1975, pp. 328–383.
7. Shay, E., A. J. Khattak, and S. C. Spoon. *Walkable Environments and Walking Activity*. Southeastern Transportation Center, University of Tennessee, 2006.
8. National Center for Health Statistics. *Prevalence of Overweight and Obesity Among Adults*. https://www.cdc.gov/nchs/data/hestat/obesity_adult_13_14/obesity_adult_13_14.htm. Accessed March 7, 2018.
9. Centers for Disease Control. *Physical Activity Basics*, 2015. https://www.cdc.gov/physicalactivity/basics/index.htm. Accessed March 7, 2018.
10. Centers for Disease Control and Prevention. *Physical Activity and Good Nutrition: Essential Elements to Prevent Chronic Diseases and Obesity*. Centers for Disease Control and Prevention, Atlanta, 2008.
11. Doyle, S., A. Kelly-Schwartz, M. Schlossberg, and J. Stockard. Active Community Environments and Health: The Relationship of Walkable and Safe Communities to Individual Health. *Journal of the American Planning Association*, Vol. 72, No. 1, 2006, pp. 19–31.
12. Sallis, J., and N. Owen. *Physical Activity and Behavioral Medicine*. SAGE Publications, Thousand Oaks, CA, 1999.
13. Talen, E., and J. Koschinsky. The Walkable Neighborhood: A Literature Review. *International Journal of Sustainable Land Use and Urban Planning*, Vol. 1, No. 1, 2013, pp. 42–63.
14. Northridge, M. E., E. D. Sclar, and P. Biswas. Sorting out the Connections between the Built Environment and Health. *Journal of Urban Health*, Vol. 80, No. 4, 2003, pp. 556–568.
15. Buehler, R., T. Gotschi, and M. Winters. *Moving Toward Active Transportation*. Robert Wood Johnson Foundation, Princeton, NJ, 2016.
16. Forsyth, A. What is a Walkable Place? The Walkability Debate in Urban Design. *URBAN DESIGN International*, Vol. 20, No. 4, 2015, pp. 274–292.
17. Maghelal, P. K., and C. J. Capp. Walkability: A Review of Existing Pedestrian Indices. *URISA Journal*, Vol. 12, No. 2, 2011.
18. Weinberger, R., and M. Sweet. Integrating Walkability into Planning Practice. *Transportation Research Record: Journal of the Transportation Research Board*, 2012. 2322: 20–30.
19. Vale, D. S., M. Saraiva, and M. Pereira. Active Accessibility: A Review of Operational Measures of Walking and Cycling Accessibility. *Journal of Transport and Land Use*, Vol. 9, No. 1, 2015.
20. Terzano, K., and A. Gross. Do Wealthier Neighborhoods Have Better Conditions for Walking? *Transportation Research Part F: Traffic Psychology and Behaviour*, Vol. 42, 2016, pp. 227–231.

21. Marshall, J. D., M. Brauer, and L. D. Frank. Health Neighborhoods: Walkability and Air Pollution. *Environmental Health Perspectives*, Vol. 117, 2009, pp. 1752–1759.
22. James, P., J. E. Hart, R. F. Banay, F. Laden, and L. B. Signorello. Built Environment and Depression in Low-Income African Americans and Whites. *American Journal of Preventive Medicine*, Vol. 52, No. 1, 2017, pp. 74–84.
23. Manaugh, K., and A. El-Geneidy. Validating Walkability Indices: How Do Different Households Respond to the Walkability of their Neighborhood? *Transportation Research Part D: Transport and Environment*, Vol. 16, No. 4, 2011, pp. 309–315.
24. Houston, D., V. Basolo, and D. Yang. Walkability, Transit Access, and Traffic Exposure for Low-Income Residents with Subsidized Housing. *American Journal of Public Health*, Vol. 103, No. 4, 2012, pp. 673–678.
25. Kim, Y.-J., and A. Woo. What's the Score? Walkable Environments and Subsidized Households. *Sustainability*, Vol. 8, No. 4, 2016, p. 396.
26. Ewing, R., and R. Cervero. Travel and the Built Environment. *Journal of the American Planning Association*, Vol. 76, No. 3, 2010, pp. 265–294.
27. Koohsari, M. J., H. Badland, T. Sugiyama, S. Mavoa, H. Christian, and B. Giles-Corti. Mismatch between Perceived and Objectively Measured Land Use Mix and Street Connectivity. *Journal of Urban Health-Bulletin, NYAM*, Vol. 92, No. 2, 2015, pp. 242–252.
28. Park, S., E. Deakin, and K. Jang. Can Good Walkability Expand the Size of Transit-Oriented Developments? *Transportation Research Record: Journal of the Transportation Research Board*, 2015. 2519: 157–164.
29. Adkins, A., J. Dill, G. Luhr, and M. Neal. Unpacking Walkability: Testing the Influence of Urban Design Features on Perceptions of Walking Environment Attractiveness. *Journal of Urban Design*, Vol. 17, No. 4, 2012, pp. 499–510.
30. Kaczynski, A. T., and T. D. Glover. Talking the Talk, Walking the Walk: Examining the Effect of Neighbourhood Walkability and Social Connectedness on Physical Activity. *Journal of Public Health*, Vol. 34, No. 3, 2012, pp. 382–389.
31. Khattak, A. J., and D. A. Rodriguez. Travel Behavior in Neo-Traditional Neighborhood Developments. *Transportation Research. Part A: Policy and Practice*, Vol. 39, No. 6, 2005, pp. 481–500.
32. Rogers, S. H., J. M. Halstead, K. H. Gardner, and C. H. Carlson. Examining Walkability and Social Capital as Indicators of Quality of Life at the Municipal and Neighborhood Scales. *Applied Research in Quality of Life*, Vol. 6, No. 2, 2011, pp. 201–213.
33. Van Dyck, D., G. Cardon, B. Deforche, and I. De Bourdeaudhuij. Do Adults like Living in High-Walkable Neighborhoods? *Health & Place*, Vol. 17, No. 4, 2011, pp. 971–977.
34. *America in 2015: A ULI Survey of Views on Housing, Transportation, and Community*. Urban Land Institute, Washington, D.C., 2015.
35. Frank, L., and S. Kavage. A National Plan for Physical Activity: The Enabling Role of the Built Environment. *Journal of Physical Activity & Health*, Vol. 6, 2009, S186–S195.
36. Diao, M., and J. Ferreira, Residential Property Values and the Built Environment. *Transportation Research Record: Journal of the Transportation Research Board*, 2010. 2174: 138–147.
37. Hack, G. *Business Performance in Walkable Shopping Areas*. Active Living Research Technical Report, Robert Wood Johnson Foundation, 2013.
38. Malizia, E. *Preferred Office Locations*, NAIOP Research Foundation, 2014.
39. Pivo, G., and J. D. Fisher. The Walkability Premium in Commercial Real Estate Investments. *Real Estate Economics*, Vol. 39, No. 2, 2011, pp. 185–219.
40. Koschinsky, J., and E. Talen. Location Efficiency and Affordability: A National Analysis of Walkable Access and HUD-Assisted Housing. *Housing Policy Debate*, Vol. 26, Nos. 4–5, 2016, pp. 835–863.
41. Sanders, R., and J. Cooper. Do All Roadway Users Want the Same Things? *Transportation Research Record: Journal of the Transportation Research Board*, 2013. 2393: 155–163.
42. Riaz, F., and M. A. Niazi. Road Collision Avoidance Using Vehicular Cyber-Physical Systems: A Taxonomy and Review. *Complex Adaptive Systems Modeling*, Vol. 4, 2016, p. 15.
43. National Highway Traffic Safety Administration. *Traffic Safety Facts 2016*. https://www.nhtsa.gov/press-releases/usdot-releases-2016-fatal-traffic-crash-data. Accessed March 7, 2018.
44. Zegeer, C. V., and M. Bushell. Pedestrian Crash Trends and Potential Countermeasures from Around the World. *Accident Analysis & Prevention*, Vol. 44, No. 1, 2012, pp. 3–11.
45. Sparrow, R., and M. Howard. When Human Beings are like Drunk Robots: Driverless Vehicles, Ethics, and the Future of Transport. *Transportation Research Part C: Emerging Technologies*, Vol. 80, 2017, pp. 206–215.
46. Goodall, N. J. Ethical Decision-Making During Automated Vehicle Crashes. *Transportation Research Record: Journal of the Transportation Research Board*, 2014. 2424: 58–65.
47. Schwartz, S. *Driverless Future: A Policy Roadmap for City Leaders*, Arcadis/HR&A, 2017.
48. Glancy, D. J. Law for Smart Cities: Regulation, Technology, and the Future of Cities. *Fordham Urban Law Journal*, Vol. 41, 2014, pp. 1617–1664.
49. Leinberger, C. B., and M. Rodriguez. *Foot Traffic Ahead: Ranking Walkable Urbanism in America's Largest Metros*, George Washington University School of Business, Washington, D.C., 2016.
50. Hu, J., and K. D. Klinich. *Designing Pedestrian-Friendly Vehicles*. UMTRI-2012-19. University of Michigan Transportation Research Institute, Ann Arbor, MI, 2012.
51. Siuhi, S., and J. Mwakalonge. Opportunities and Challenges of Smart Mobile Applications in Transportation. *Journal of Traffic and Transportation Engineering (English Edition)*, Vol. 3, No. 6, 2016, pp. 582–592.
52. Cavoli, C., B. Phillips, T. Cohen, and P. Jones. *Social and Behavioural Questions Associated with Automated Vehicles: A Literature Review*. UCL Transport Institute, London, 2017.
53. Underwood, S. E., S. Marshall, and J. Niles. *Automated, Connected, and Electric Vehicles*, 2014. http://graham.umich.edu/media/files/LC-IA-Final-Underwood.pdf. Accessed March 7, 2017.
54. Millard-Ball, A. Pedestrians, Autonomous Vehicles, and Cities. *Journal of Planning Education and Research*, Vol. 38, No. 1, 2016, pp. 6–12.
55. Ohnemus, M., and A. Perl. Shared Autonomous Vehicles: Catalyst of New Mobility for the Last Mile? *Built Environment*, Vol. 42, No. 4, 2016, pp. 589–602.

56. Zellner, M., D. Massey, Y. Shiftan, J. Levine, and M. J. Arquero. Overcoming the Last-Mile Problem with Transportation and Land-Use Improvements. *International Journal of Transportation*, Vol. 4, No. 1, 2016, pp. 1–25.
57. Meeder, M., E. Bosina, and U. Weidmann. Autonomous Vehicles: Pedestrian Heaven or Pedestrian Hell? In *17th Swiss Transport Research Conference*, Ascona, May 17–19, 2017.
58. Fleetwood, J. Public Health, Ethics, and Autonomous Vehicles. *American Journal of Public Health*, Vol. 107, No. 4, 2017, pp. 532–537.
59. Blyth, P. -L., M. N. Mladenovic, B. A. Nardi, N. M. Su, and H. R. Ekbia. Driving the Self-Driving Vehicle. *2015 IEEE International Symposium on Technology and Society*, IEEE, New York, 2015.
60. Becker, J., M.-B. A. Colas, S. Nordbruch, and M. Fausten. Bosch's Vision and Roadmap Toward Fully Autonomous Driving. In *Road Vehicle Automation*, Springer, New York, 2014, pp. 49–59.
61. Turnbull, K. F., L. Cherrington, Z. Elgart, J. Zmud, T. Baker, J. Hudson, and J. Wagner. *Automated and Connected Vehicle (AC/CV) Test Bed to Improve Transit, Bicycle, and Pedestrian Safety*. Technical Report No. 0-6875. Texas A&M Transportation Institute, College Station, TX, 2017.
62. Zohdy, I., R. Kamalanathsharma, S. Sundararajan, and R. Kandarpa. Automated Vehicles from Modeling to Real World. *Road Vehicle Automation 2* (G. Meyer and S. Beiker, eds.), Springer-Verlag, Berlin, 2015, pp. 187–191.
63. Bonnefon, J. -F., A. Shariff, and I. Rahwan. The Social Dilemma of Autonomous Vehicles. *Science*, Vol. 352, No. 6293, 2016, pp. 1573–1576.
64. Kirkpatrick, K. The Moral Challenges of Driverless Cars. *Communications of the ACM*, Vol. 58, No. 8, 2015, pp. 19–20.
65. Anderson, J. M., K. Nidhi, K. D. Stanley, P. Sorensen, C. Samaras, and O. A. Oluwatola. Oluwatola. *Autonomous Vehicle Technology: A Guide for Policymakers*, RAND Corporation, Santa Monica, CA, 2016.
66. Urry, J. Inhabiting the Car. *Sociological Review*, Vol. 54, 2006, pp. 17–31.
67. Lavieri, P. S., V. M. Garikapati, C. R. Bhat, R. M. Pendyala, S. Astroza, and F. F. Dias. Modeling Individual Preferences for Ownership and Sharing of Autonomous Vehicle Technologies. *Transportation Research Record: Journal of the Transportation Research Board*, 2017. 2665: 1–10.
68. Moss, S. End of the Car Age: How Cities are Outgrowing the Automobile. *Guardian*, 2015. https://www.theguardian.com/cities/2015/apr/28/end-of-the-car-age-how-cities-outgrew-the-automobile. Accessed March 7, 2018
69. Milakis, D., B. van Arem, and B. van Wee. Policy and Society Related Implications of Automated Driving. *Journal of Intelligent Transportation Systems*, 2017, pp. 1–25.
70. Noyman, A., A. Stibe, and K. Larson. *Roadmap for Autonomous Cities: Sustainable Transformation of Urban Spaces*. https://www.media.mit.edu/publications/roadmap-for-autonomous-cities-sustainable-transformation-of-urban-spaces/. Accessed March 7, 2018.
71. Shladover, S. E. The Truth About "Self-Driving" Cars. *Scientific American*, Vol. 314, No. 6, 2016, pp. 53–57.
72. Niles, J. S. *Connected, Automated, Zero-Emission Cars are Essential for Improving Livable, Sustainable Communities*. Center for Advanced Transportation and Energy Solutions, Seattle, 2015.
73. Schwartz, C., G. Thomas, K. Nelson, M. McCrary, N. Sclarmann, and M. Powell. *Towards Autonomous Vehicles*. Mid-America Transportation Center, Lincoln NE, 2013.
74. Surden, H., and M. -A. Williams. Technological Opacity, Predictability, and Self-Driving Cars. *Cardozo Law Review*, Vol. 8, 2016, 121.
75. Banks, V. A., and N. A. Stanton. The Unknown Paradox of "Stop the Crash" Systems: Are We Really Improving Driver Safety? In *Advances in Human Aspects of Transportation* (N. A. Stanton, S. Landry, G. Di Bucchianico, and A. Vallicelli, eds.), Vol. 484, Springer, New York, 2017, pp. 525–533.
76. Sandt, L., and J. M. Owens. *Discussion Guide for Automated and Connected Vehicles, Pedestrians, and Bicyclists*. Pedestrian and Bicycle Information Center, Chapel Hill, NC, 2017.
77. Crayton, T. J., and B. M. Meier. Autonomous Vehicles: Developing a Public Health Research Agenda to Frame the Future of Transportation Policy. *Journal of Transport & Health*, Vo. 6, 2017, pp. 245–252.
78. Giles-Corti, B., A. Vernez-Moudon, R. Reis, G. Turrell, A. L. Dannenberg, H. Badland, and N. Owen. City Planning and Population Health: A Global Challenge. *The Lancet*, Vol. 388, No. 10062, 2016, pp. 2912–2924.
79. Teixeria, P. J., E. V. Carraca, D. Markland, M. N. Silva, and R. M. Ryan. Exercise, Physical Activity, and Self-Determination Theory: A Systematic Review. *International Journal of Nutrition and Physical Activity*, Vol. 9, No. 78, 2012.
80. Sheeran, P., P. M. Gollwitzer, and J. A. Bargh. Nonconscious Processes and Health. *Health Psychology*, Vol. 32, No. 5, 2013, pp. 460–473.

The Standing Committee on Pedestrians (ANF10) peer-reviewed this paper (18-00972).

TRR
JOURNAL OF THE TRANSPORTATION RESEARCH BOARD

Article

Transportation Research Record
2018, Vol. 2672(35) 129–138

DOI: 10.1177/0361198118782036
journals.sagepub.com/home/trr
SAGE

The Role of Accessibility to Public Transport and Quality of Walking Environment on Urban Equity: The Case of Santiago de Chile

Ignacio Tiznado-Aitken[1], Juan Carlos Muñoz[1], and Ricardo Hurtubia[1,2]

Abstract

Most studies of public transport accessibility have focused on proximity to stops, walking distances or time to reach them. This approach ignores other accessibility barriers, such as the quality of the urban environment of these walks. The aim of this research is to analyze both accessibility to public transport stops and the quality of the urban walking environment, exploring fairness issues across the city in terms of these two dimensions. The proposed methodology considers the definition of two indicators: walking accessibility to public transport stops and quality of the walking environment, considering different attributes and dimensions. These indicators are later used to develop a fairness analysis at the local and metropolitan level, using Lorenz curves, Gini coefficient, and Foster-Greer-Thorbecke (FGT) poverty measures. A diagnosis based on these indicators for Santiago de Chile allows us to suggest public policy priorities to improve accessibility to public transport and promote modal shift. The results show that 12 out of 34 municipalities in Santiago are deprived of one or both dimensions, not managing to achieve minimum fairness standards. Moreover, since a correlation between low income and poor access and urban space quality is detected, urban fairness across the city would benefit from centralized urban policies not depending on the budget of each administrative division.

Urban transport planning has traditionally focused on improving the performance of its system. The main goal has been to increase the efficiency (i.e., high speed and capacity) and target congestion, mostly leaving aside equity considerations. This has created a privileged scenario for those who have transport alternatives such as the car because very often new infrastructure projects are car-oriented. This has generated a significant deterioration for those who are excluded from the use of cars and usually reach their daily activities in other transport modes (*1*).

Poor accessibility to transport has relevant social consequences since it reduces opportunities for work, study, recreation, health services and social interaction, which impact people's social and economic development and may cause social exclusion (*2*, *3*). Given this context, multimodal transport planning based on accessibility aiming at how easily people can reach opportunities and basic services becomes paramount.

To promote this alternative transport planning paradigm, public transport plays a key role in terms of urban fairness and sustainability. From a fairness perspective, the most vulnerable groups are often those lacking a car and forced to use both active and public transport. From a sustainability perspective, this paradigm provides alternatives to car use, which has grown significantly in Latin America during the last decade. In Santiago de Chile between 2001 and 2012, the use of the car increased by 5.1% while the number of trips walking and by public transport decreased by 10.1% (*4*). It can be argued that despite the authorities' stated purpose of fostering sustainable transport modes and hindering car use, the current planning model is not achieving its goals and, therefore, should be questioned.

Most study approaches to public transport accessibility focus on proximity to stops in terms of walking distance or time to reach them (e.g., *5*, *6*). However, other important types of accessibility barriers have received less attention, such as the quality of the walking

[1]Department of Transport Engineering and Logistics, Pontificia Universidad Católica de Chile, Macul, Santiago, Chile
[2]School of Architecture & Department of Transport Engineering and Logistics, Pontificia Universidad Católica de Chile, Macul, Santiago, Chile

Corresponding Author:
Address correspondence to Ignacio Tiznado-Aitken: iatiznad@uc.cl

environment. Urban attributes such as the quality of stops, sidewalks and roads, and the safety level perceived by users, affect public transport accessibility. By incorporating these attributes into accessibility analysis, we would give importance to elements that directly affect the experience of pedestrians and bicycle users, groups commonly ignored by traditional transportation planning (*7*).

Therefore, the aim of this research is to analyze both accessibility to public transport stops and the quality of the urban walking environment under a common framework, exploring fairness issues through sufficientarian and egalitarian approaches (see *8* for a review). By applying this methodology to the case of Santiago de Chile, we determine which areas should be prioritized in terms of public policies to improve the accessibility experience to public transport and promote modal shift.

The paper is structured as follows: the following section presents a brief literature review of accessibility, built environment and walkability, highlighting the research gap this work fills. The next section presents the methodological approach to measure and analyze accessibility to public transport stops and quality of the urban walking environment. Following that, we apply the methodology to the city of Santiago, Chile. Finally, the main results of this research and a brief discussion are presented. The paper closes with policy recommendations for authorities in Santiago and future work.

Quality of Walking Spaces and Access to Public Transport

Accessibility has been the focus of extensive research over the years. The concept has been defined and addressed in several disciplines, like transportation, geography, sociology and economics, among others (*9–12*). In this work, we adopt a traditional definition from the transport field: the ability to reach different activities or opportunities through the transport system from each particular location (*9*).

Murray and Wu point out that accessibility through public transport has two main factors: access to public transport stations or stops given an influence area or buffer (by walking, bicycle or car) and the geographic coverage of public transport given a certain time budget (*13*). However, this approach is limited since it does not account for the level of service. In the case of the first factor, the quality of the walking environment is key, while in the second one attributes like waiting time, comfort, transfers and reliability should be considered.

In this work, we tackle the first gap related to the quality of the walking environment. Most of the studies in public transport accessibility focus on proximity to public transport stops, analyzing walking distances to reach them (*14*, *15*). Nevertheless, considering only proximity is a quite limited approach. Taylor & Fink classify the factors that could improve the public transport experience and promote ridership into internal factors, such as fare or level of service, and external factors, such as convenient and safe access to public transport stops (*16*).

There is extensive research about walkability and the influence of the built environment into modal choice. The literature reports the influence of several built environment elements such as density, diversity, design, accessibility to destinations and distance to public transport in this choice (*17*, *18*). For example, the absence of a pedestrian-friendly environment (i.e., connected walking paths, fewer car lanes, and wide sidewalks, among others) and lack of mixed land use influences people to drive instead of walking and using public transport (*19*).

Regarding the walkability, the literature analyzes the characteristics of the physical environment contributing to a neighborhood being walkable. Some of the most important factors influencing walkability according to these studies are facilities to connect destinations, mixed land use and urban design and (*20*), sidewalk capacity, quality of the environment, perceptions of safety and comfort (*21*) and the attractiveness of pedestrian network (*22*, *23*).

Additionally, pedestrians do not always choose the shortest routes, but instead one that delivers a good level of service in terms of infrastructure (*24*). For example, Tilahun & Li found that crime perception associated with access to a stop or station increases the perceived disutility in a magnitude equivalent to 6 minutes in access time, while the absence of paths does it in the equivalent of 5.9 minutes (*19*).

Therefore, this paper jointly analyzes the access to public transport stops and the quality of users' walking environment, setting priorities in terms of public policies after a fairness analysis that will be detailed in the methodology section.

Methodology

The methodology considers the definition of two indicators: walking accessibility to public transport stops, here referred to as physical accessibility, and the quality of the walking environment, considering different attributes and dimensions. These indicators are later used to develop a fairness analysis at the local and metropolitan level, using Lorenz curves, Gini coefficient and Foster-Greer-Thorbecke (FGT) poverty measures. Below we describe each of them in detail.

Physical Accessibility

To measure accessibility to stops, we propose to use location-based measures, specifically potential measures (i.e., considering all possible destinations within a threshold or buffer). To determine how much a station contributes to the accessibility of the location, we use a decay function based on a resistance parameter (see Equation 1). We prefer this indicator over isochronous access measures that use arbitrary thresholds and equally count the opportunities regardless of proximity. Also, location-based measures have advantages on operationalization, interpretability, and communicability, which are key to influence planning and public policies (*12*).

$$\mathrm{PA}_i = \sum_{j \in \varphi_i} F\left(\frac{d_{ij}}{v}\right) \forall i \in \omega \qquad (1)$$

In Equation 1 i is the sub index for locations (origins) within the city ω, j corresponds to the sub index of all public transport stops that are within a threshold or buffer around location $i\,(\varphi_i)$ and $F\left(\frac{d_{ij}}{v}\right)$ corresponds to decay function, which depends on the distance (d_{ij}) and walking speed (v). The indicator PA_i is later normalized by dividing it by its maximum value across all zones in the city, so all PA_i keep a value between 0 and 1. The results obtained will depend on the decay function chosen, the use of Euclidean or network distance, and the value of the walking speed. This will be discussed later in the case study section.

Urban Walking Environment

Like any transport mode, walking provides a certain 'level of service' determined by its intrinsic characteristics and the environment in which it occurs. We propose to include the presence of urban furniture and the quality of the urban environment and road network (e.g., including amenities while travelers wait at stops) into an indicator that would yield the experience faced from the moment a user starts walking from home towards a stop, until boarding a vehicle.

Since the construction of this indicator strongly depends on data quality and availability, we propose an indicator that can be adapted to different contexts, following the basic structure shown in Equation 2.

$$U_i = \sum_{k \in \vartheta} P_{ki}(s) \forall i \in \omega \qquad (2)$$

In Equation 2, i is the location (origin) and k corresponds to different categories of the urban environment (safety, comfort, infrastructure, furniture, cleanliness, among others). Meanwhile, ϑ is the set of categories of the urban environment and $P_{ki}(s)$ is the quantifiable level of each category in each location i based on scale s (for example, binary to define the presence of some attribute or Likert to evaluate attribute quality). Therefore, every U_i corresponds to the overall sum of urban environment categories from a certain origin.

Fairness Analysis

Distributive fairness analysis is usually addressed through an egalitarian or a sufficientarian approach (*8*). In simple words, the first approach addresses how evenly distributed a certain attribute within population is, and the second one defines a minimum or sufficient standard for that attribute that must be achieved by all people.

In the first approach, we propose the use of Gini coefficient (G), probably the most widely used measure of inequality. While the most common application for this indicator is income distribution, several works have proposed using the Lorenz curves and Gini coefficient as a measure to analyze equity in the public transport supply, the accessibility provided by the system or as an approach for wider transport project assessment (e.g., *25–27*). Typically, we can calculate the Gini coefficient using Equation 3, where X_s represents the cumulative proportion of the households or population variable and Y_s typically represents the cumulative proportion of income variable, with $s = 0 \ldots n$, $X_o = Y_o = 0$ and $X_n = Y_n = 1$.

$$G = 1 - \sum_{s=1}^{n} (X_s - X_{s-1})(Y_s + Y_{s-1}) \qquad (3)$$

In the second approach, we propose the use of a poverty indicator related to private income (y_p). In Equation 4 z is a consumption or income poverty threshold and q corresponds to all subgroups p that are below the minimum threshold ($y_p < z$) with n_p being the population of subgroup p. This formula, well-known as FGT measures, has different interpretations based on the poverty aversion parameter ($\alpha >= 0$, chosen by the analyst), since "a larger α gives greater emphasis to the poorest poor" (*28*). This formulation has been used in several topics, including accessibility fairness index to analyze severity of the accessibility deficiency in a region (*1*).

$$\mathrm{FGT}_\alpha = \frac{1}{N} \sum_{p=1}^{q} n_p \left(\frac{z - y_p}{z}\right)^\alpha \qquad (4)$$

We propose to adapt both indicators (Gini coefficient and FGT measures) to measure fairness by replacing income by accessibility levels and scores of urban walking environment. We describe this process in the next section.

Application to Santiago, Chile

Santiago, the capital of Chile and its largest Metropolitan region, has over 6.5 million inhabitants within an area of approximately 640 km^2. This city shows an important socio-spatial segregation, as typical in Latin America, due to a historical planning process.

Until 1980 Santiago had several slums located in central zones of the city, where the main activities were developed. Due to a new urban policy and the deregulation of the land market, between 1980 and 2002 social housing was mostly relocated in the southern, south and west suburbs of the city (*29*). This development has harmed the low-income population, especially those living in the periphery. As a consequence of the aforementioned housing policies, the distance to activities for those inhabitants has systematically increased over the last few decades. This scenario has deepened dramatically since the northeastern area, where the majority of high-income elite lives, has grown much faster than the rest of the city in the last decades, attracting most of the productive activities, commerce, and services (*30*).

If we look at the transport system, since the late 1970s the collective transport of Santiago was handled as a service to be subject to the market forces rather than being understood as a basic social service needing to be regulated, given the neoliberal policies that prevailed in that period within the country. Several measures led the system to a gradual process of deregulation, which meant for example that the operators were allowed to enter the system freely in 1979 and to modify the lines or operate the service of their choice since 1988 (*31*). Unexpectedly for the local authorities, these policies increased the fares, while developing a supply deficit in peripheral sectors and an oversupply in downtown.

Given this context, in 1990 a process of regulation began as a result of the negative results of the previous policies. In 1992, the "Micros Amarillas (yellow buses)" system achieved a great coverage in the city but kept generating a high percentage of accidents and environmental externalities. Thus, in 2007 a new city-wide public transport system, Transantiago, was implemented formalizing bus operators, integrating the underground and the private bus networks based on a structure of trunk and feeder services and a fare payment by touchless smart card (*32*).

The system has improved after a very rough start, increasing the number of bus routes by 37%, the number of public transport stops by 20% and the length of Metro network by 22% between 2007 and 2015 (*33*). However, Transantiago still does not work as expected. The aforementioned social segregation, and the lack of coordinated and integrated transport and land use planning produces high inequalities regarding access to transport (*34*).

The 34 municipalities covered by Transantiago are analyzed here (Figure 1). Below we describe each indicator and data needed to carry out the previously proposed methodology.

Physical Accessibility Indicator (PAI)

In this work, we approach accessibility through its positive instead of its normative aspects (*35*). This means that we focus our analysis on actual user travel experiences, not on what is reasonable or desirable for people to experience. This means that the impedance functions *F*() are calibrated based on actual travel times.

To compute walking accessibility to public transport stops, we use OpenTripPlanner (OTP), an open source

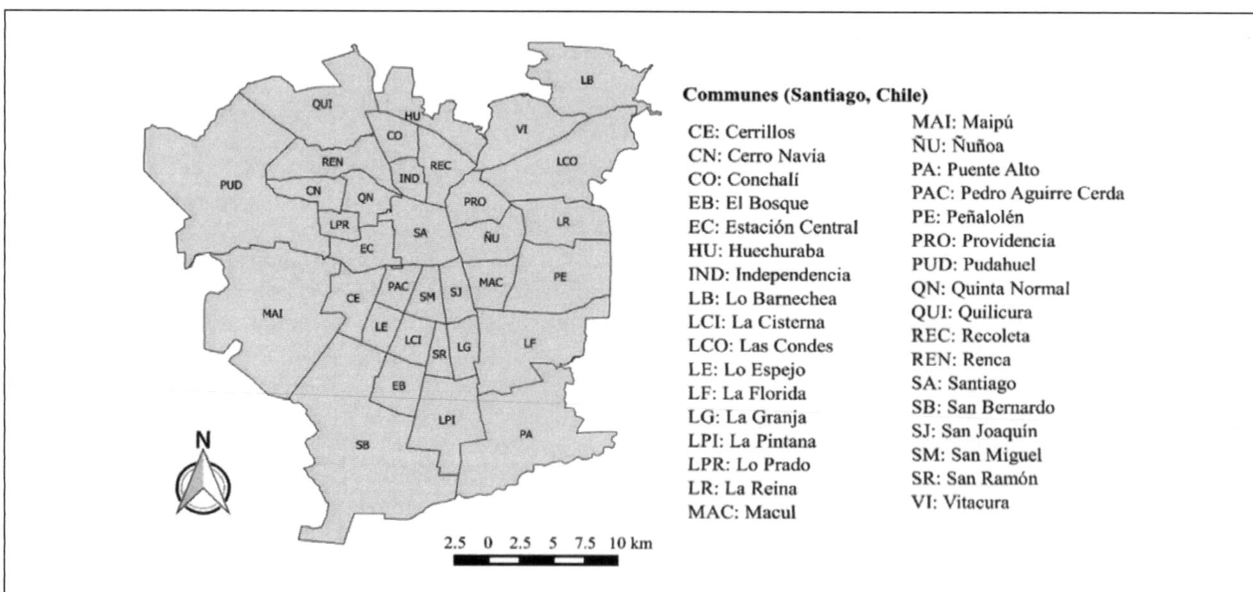

Figure 1. Municipalities of Santiago to be considered for analysis *(Source: Own elaboration)*.

Table 1. EUQI Components and Values (Source: Own Elaboration, Based on INE (*40*))

Index	Components	Range of values
Security/safety	Luminary, road signs and roofed bus stops, and road safety	0–5
Environment	Gardens, seats, sport fields, and playground	0–4
Cleanliness	Garbage bins and rubble	0–2
Infrastructure	Sidewalks and street quality	0–10

platform for multi-modal journey planning (*36*). For the spatial analysis, we subdivide the territory using a grid, with cells of 200 × 200 meters (city blocks) as trip origins. We divided these cells further to make them compatible with the Transantiago Zoning, consisting of 804 zones defined by The Metropolitan Public Transportation Agency, which manages and regulates public transport in Santiago. If a cell shared 2 or more zones, it was divided into smaller cells, resulting in a total of 35,784 cells.

We define the set φ_i of public transport stops accessible from origin i (the centroid of a cell), as those located within 1,000 meters (Euclidian distance) around it. This distance threshold was selected because 95% of bus users and 100% of Metro users in Santiago report a walking distance below that value (*37*). In some locations, the number of stops satisfying this condition was large, thus we limited the calculation to the 10 closest to each location$i$. While this may bias our indicator, all these locations do not have accessibility problems, so it does not affect the analysis regarding the most vulnerable low-accessibility zones, which is the focus of this work.

We obtained walking times from each location centroid to public transport stops based on network distance (d_{ij}) and the OTP average walking speed (v = 4.8 km/hr). Using the formulation shown in Equation 1, we calibrate the parameters through information from the origin destination survey in Santiago (*4*) as explained next.

Using the Mamun et al. method (*13*), we observe that 95% of public transport users walk less than 15 minutes to their initial stop, so we define an access value of 0.05 for a 15-minute walk. Using the same procedure, we define an access value of 0.2 and 0.7 for 10-minute and 5-minute walk respectively, allowing us to calibrate the functions. Despite the Negative Exponential being the most used accessibility impedance function, we use Richards's (*38*) arguing that this function represents better people's perceptions by softening accessibility decay for low travel times, unlike the Exponential function (*39*).

Urban Walking Environment

For a pre-census activity in 2011, groups of trained surveyors evaluated each city block in Santiago in terms of urban quality (considering the quality, presence, and condition of urban furniture) and road network (*40*). The aspects evaluated are mostly binary (1 meaning the presence of the element and 0 meaning the lack of the element), while sidewalk and street quality are rated with a scale of 0 to 5, with 0 meaning very poor conditions and 5 meaning excellent conditions.

We used this georeferenced information (*40*) to obtain a proxy measure of the "walking level of service". Four indexes were created by grouping elements according to their similarity (Table 1). Also, we added a new road safety index using (*41*, *42*). In Equation 5, we show the index considering severity of those involved in all road accidents in the zone i in 2011 (d_a for people who died, s_a for people with serious consequences, ls_a for people with moderate consequences and n_a for people with minor consequences) and the average traffic flow tf_i in every zone i. We grouped the values into 3 main categories: 0 for an unsafe zone, 1 for moderate safety and 2 for good road safety. Together, all these elements and components form the Environment and Urban Quality Index (EUQI) with values ranging between 0 and 21 (Table 1).

$$SV_i = \sum_{a=1}^{q} \frac{d_a + s_a + ls_a + n_a}{tf_i} \quad (5)$$

Fairness Analysis

For the egalitarian approach, we calculate the Lorenz curves and Gini indexes replacing income by PAI and EUQI. In both cases we use the most disaggregated analysis unit: 35,784 zones in the first one and 154,488 zones in the second one.

For the sufficiency approach, we use the framework proposed by Martens (*1*) to calculate Accessibility Fairness Index (AFI) and Environment Fairness Index (EFI). We adapt Equation 4 using a_p as the PAI / EUQI of that municipality, q as the number of groups with PAI / EUQI below the PAI / EUQI minimum threshold($a_p < z$) and n_p represents the total population of the p-th municipality for 2016 (*43*).

We decide to use α = 2 to reflect both poverty and inequality in terms of accessibility to public transport stops and the quality of walking environment. Finally, to determine z we use two different percentiles of the distribution of a_p across all municipalities: 10th and 20th.

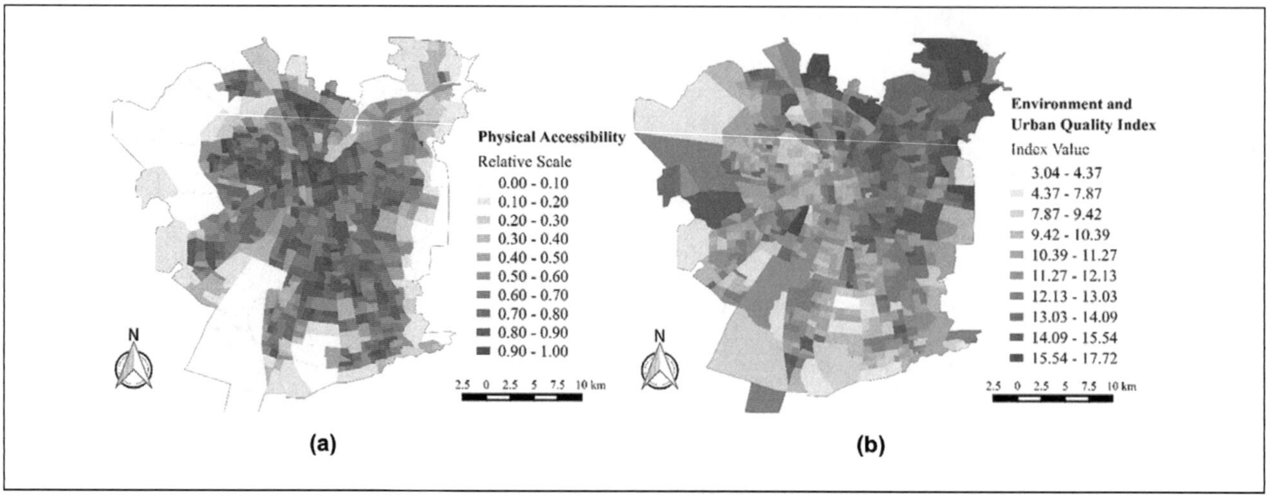

Figure 2. Showing (*a*) physical accessibility, and (*b*) EUQI for Santiago *(Source: Own elaboration).*

Results

In general terms, Santiago has good accessibility to public transport stops, with 0.67 out of 1 as the average value. These areas are mostly located in central and pericentral zones (Figure 2*a*). About 11,000 public transport stops contribute to an extensive network that has achieved wide coverage and has grown significantly in recent years. This is consistent with the proximity perception to public transport that more than 85% of the people from the 34 analyzed municipalities have (*44*).

However, as we expected, the periphery has lower accessibility levels compared to central and greater productive areas of the city (Figure 2*a*). In northwest and southwest sectors of Santiago, there is little accessibility to public transport. Municipalities such as Pudahuel and Lo Barnechea have areas with significantly lower access than more central municipalities such as Santiago and Lo Prado. In practical terms, whereas the areas with greater accessibility have 10 public transport stops reachable in an average of 4 minutes, other less-benefited areas can access fewer stops in an average of 20 minutes or over.

Thus, it is recommended that the implementation of new stops should be focused on deprived zones whose urban growth precedes the presence of public transport. In some cases, these zones also match with lower income municipalities like San Bernardo, Pudahuel or La Pintana. These zones show a very high public transport modal share, so improving proximity is a key factor. Additionally, richer municipalities like Vitacura or Lo Barnechea have the highest modal split of car use. If authorities want to promote more sustainable transport modes, providing these municipalities with good accessibility could be the first step.

As we said before, choosing public transport for a trip depends not only on proximity but the quality of the experience, like pedestrian-friendly environments and sidewalk availability. Figure 2*b* shows that high quality urban environment is concentrated mainly in the northeast sector of the city, composed by 6 municipalities: Vitacura, Las Condes, Lo Barnechea, Ñuñoa, Providencia and La Reina (see Figure 1). Those municipalities have 17% of the population concentrating 62% of the high-income population, and earning more than 6,000 dollars per household-month on average (*45*). These municipalities have an average EUQI indicator of 12.7, while the average of the 28 remaining municipalities has only 10.5.

Also, central zones, with better accessibility as we shown in Figure 2*a*, do not have such high EUQI index values as high-income municipalities such as Vitacura and Lo Barnechea. This creates a clear mismatch between these key factors to promote ridership and a better experience using public transport. A likely explanation for this situation is that each of the 34 municipalities analyzed has its own mayor, budget, and regulations. Therefore, high-income municipalities can provide better infrastructure and public spaces given their budget, producing important differences in terms of urban environment within the city.

Thus, improving infrastructure and services for pedestrians needs planning and policy support through urban design guides (*21*, *22*). Moreover, to deal with the dissatisfaction that exists in the population, where more than 40% evaluate the streets, sidewalks, and parks poorly (*40*), it is necessary to ensure a minimum standard. The definition of this standard or sufficiency threshold may be useful also for physical accessibility, closing the gap

Table 2. AFI and EFI for Santiago's Case

	Name	AFI10	AFI20	EFI10	EFI20
	Standard (z)	0.34 (10th percentile)	0.46 (20th percentile)	9.4 (10th percentile)	9.7 (20th percentile)
Indicators	Value	0.01	0.03	4.30E-04	7.58E-04
Contribution of each municipality to the indicator value (%)	San Bernardo	78.75%	55.20%	0%	4.29%
	Pudahuel	15.28%	21.89%	0%	0%
	Lo Barnechea	5.98%	9.33%	0%	0%
	Quilicura	0%	7.10%	0%	0%
	Maipú	0%	3.38%	0%	0%
	Vitacura	0%	2.96%	0%	0%
	Peñalolén	0%	0.13%	0%	0%
	La Pintana	0%	0%	87.50%	74.05%
	P.A. Cerda	0%	0%	9.58%	12.89%
	Quinta Normal	0%	0%	2.92%	6.43%
	Renca	0%	0%	0%	2.15%
	Conchalí	0%	0%	0%	0.18%
	Total	100%	100%	100%	100%

Source: Own elaboration.

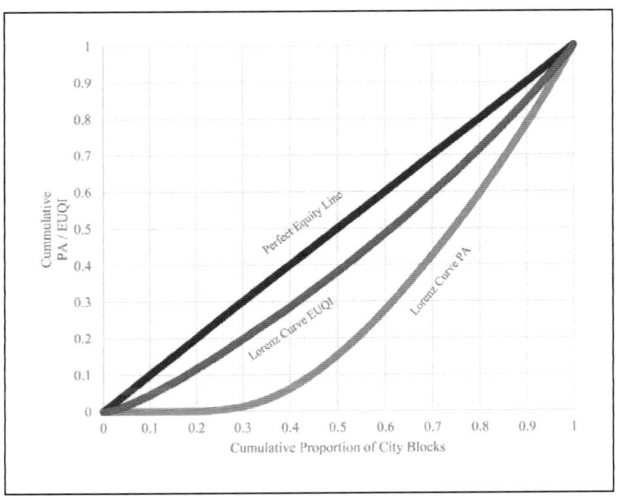

Figure 3. Lorenz curves for PAI and EUQI in Santiago *(Source: Own elaboration).*

between both EUQI and physical accessibility levels to achieve a more equitable scenario.

Adapting the Gini index for PAI and EUQI, we obtain revealing results (Figure 3). The Gini index value for EUQI is equal to 0.168, revealing a relatively equitable scenario. However, for physical accessibility the value is significantly higher: 0.442, which is similar to the Gini coefficient for income distribution in Chile.

The high value presented by the Gini of PAI is affected by many blocks of the city being deprived of access (50% of localities have less than 20% of total access, as can be seen in Figure 3). The value of the Gini index for EUQI may be counterintuitive when looking at Figure 2*b*, where a clear quality of the environment concentration is observed. However, a possible explanation for these results is related to the magnitude of the indicators, where the differences between PA and EUQI are usually lower than those found in terms of income. In addition, the spatial aggregation unit used in the calculation also affects the results obtained for the Gini indicator, so this is an issue that should be analyzed in greater depth in future work.

On the other hand, Table 2 presents the AFI and EFI values for Santiago. In this analysis, we are interested mainly in disadvantaged municipalities, so we calculate these indicators using 10th (AFI10 and EFI10) and 20th percentile (AFI20 and EFI20) as threshold. Since the indicator represents the severity of the deficiency of accessibility to public transport stops and the quality of the walking environment, the greater the percentage contribution to the indicator, the worse the situation of the analyzed municipality is.

We can see that San Bernardo is probably the most deprived municipality of Santiago in these terms, since in the AFI20 contributes more than 50% and 4% in EFI20. This situation is detrimental for its population since San Bernardo made 9% of total trips by public transport and 47% by walking. A similar situation can be observed in Pudahuel, despite its EUQI being above the threshold. This municipality contributes with more than 20% in AFI20 and the modal share is similar to San Bernardo, with 76% of trips made by public transport and walking.

A different situation happens in Lo Barnechea and Vitacura, having the second and sixth best EUQI respectively, but they have low accessibility to public transport stops with a contribution of about 10% and 3%

respectively in AFI20. This may not be a problem for most of their population since more than 60% of their trips are made by car. However, these are high-income municipalities where a strong urban development has been observed in recent years. Thus, citizens from all around the city must travel to Lo Barnechea and Vitacura for their commute, using public transport for their return trip as well. Also, to promote sustainability and a modal shift from car to both public and active transport, walking must be an attractive experience, therefore its conditions should be improved.

The largest contributors in terms of EFI10 and EFI20 are La Pintana and Pedro Aguirre Cerda municipalities. These are low-income municipalities (approximately the 25th percentile of income) (*46*) and therefore they have a low budget to improve their urban and walking environment. They made between 67% and 71% of trips by walking and public transport since less than 40% of households own a car. As captive users of public transport, improving the walking stage of their journey would improve their travel experience.

Unfortunately, Santiago does not have a metropolitan authority nor a system that allows effective fund redistribution between municipalities to ensure a coherent and equitable development of the city, generating discontinuities of the urban environment, basic transport structure for pedestrians, cyclists and public transport users.

In Chile, municipalities frame their budget based on two main funds: permanent own funds and municipal common funds. The former comes from tax collection, traffic licenses, commercial patents and cleanliness rights from each municipality. The latter comes from a redistributive mechanism, where each municipality contributes to the fund proportionally to the permanent own funds generated (*47*). Despite this distributive mechanism, disadvantaged municipalities cannot generate enough revenues to improve their urban environment, and differences remain or increase between municipalities, as shown in Figure 2b.

Policy Recommendations and Future Work

Santiago has grown without integrated land use and transport planning during the last four decades, generating urban segregation that has brought clear inequalities in terms of access to opportunities and urban quality. High-income municipalities have greater proximity to sub-centers that concentrate activities and services, which provide better infrastructure and public spaces. This, in turn, makes these municipalities even more attractive for the location of new activities, services, and high-quality real estate developments, creating a negative loop of segregation and concentration of opportunities around high-income areas. This urban development pattern stresses accessibility conditions for lower income people, who are usually located in the periphery and are more likely to be captive public transport users.

This work highlights the importance of analyzing both proximity to public transport stops and urban walking environment together since both elements are essential to understanding the accessibility experience and use of public transport. In Santiago, the majority of the municipalities have at least a sufficient level of access and urban walking environment. However, 12 out of 34 municipalities have problems that need to be addressed. On the one hand, we have municipalities like San Bernardo that are deprived of both dimensions. On the other hand, there is a mismatch in some municipalities where only one of these two elements achieves a minimum standard. Due to the observed correlation between income level and access or quality of urban space, the city would benefit from a metropolitan authority ensuring a coherent development in the different municipalities, regardless of spatial distribution of income We suggest that priority should be given to those municipalities that actually do not achieve a sufficient standard of access and urban quality, with special emphasis on those that have greater use of public transport and walking.

If we follow what has been proposed by Martens (*1*), the priority in public investments and policies should be placed in improving urban fairness, based on sufficiency and egalitarianism considerations. For this purpose, our framework could be used to identify where the gaps between different zones of the city are located (and, hence, between socioeconomic groups) and how to reduce them.

The analysis proposed here is the first step to improve access for public transport users, but this is still limited. Future work should also include public transport level of service and how people perceive different attributes such as comfort, waiting times, and transfers. This kind of analysis will help to bridge the gap between perceived and measured accessibility, which is a key element if the goal is to retain current users of public transport (or increase them) and reduce the use of cars.

Acknowledgments

The authors would like to thank for the support provided by the Center for Sustainable Urban Development (CONICYT/FONDAP 15110020), the Bus Rapid Transit Centre of Excellence funded by the Volvo Research and Educational Foundation and "AccesoBarrio" project (Fondecyt 1150239).

Author Contributions

The authors confirm contribution to the paper as follows: study conception and design: Ignacio Tiznado-Aitken, Juan Carlos Muñoz, Ricardo Hurtubia; data collection: Ignacio Tiznado-Aitken; analysis and interpretation of results: Ignacio

Tiznado-Aitken, Juan Carlos Muñoz, Ricardo Hurtubia; draft manuscript preparation: Ignacio Tiznado-Aitken, Juan Carlos Muñoz, Ricardo Hurtubia. All authors reviewed the results and approved the final version of the manuscript.

References

1. Martens, K. *Transport Justice: Designing Fair Transportation Systems*. Routledge, New York, 2016.
2. Sanchez, T. W., R. Stolz, and J. S. Ma. *Moving to Equity: Addressing Inequitable Effects of Transportation Policies on Minorities*. The Civil Rights Project at Harvard University, Cambridge, Mass., 2003.
3. Lucas, K. Providing Transport for Social Inclusion within a Framework for Environmental Justice in the UK. *Transportation Research Part A: Policy and Practice*, Vol. 40, No. 10, 2006, pp. 801–809.
4. SECTRA. *O-D Survey from 2012*. Planning and Development Coordination, Transport and Telecommunications Ministry, Government of Chile, 2015.
5. Zhao, F., L. Chow, M. Li, I. Ubaka, and A. Gan. Forecasting Transit Walk Accessibility: Regression Model Alternative to the Buffer Method. *Transportation Research Record: Journal of the Transportation Research Board*, 2003. 1835: 34–41.
6. Olszewski, P., and S. Wibowo. Using Equivalent Walking Distance to Assess Pedestrian Accessibility to Transit Stations in Singapore. *Transportation Research Record: Journal of the Transportation Research Board*, 2005. 1927: 38–45.
7. Koglin, T., and T. Rye. The Marginalisation of Bicycling in Modernist Urban Transport Planning. *Journal of Transport & Health*, Vol. 1, No. 4, 2014, pp. 214–222.
8. Pereira, R. H., T. Schwanen, and D. Banister. Distributive Justice and Equity in Transportation. *Transport Reviews*, Vol. 37, No. 2, 2017, pp. 170–191.
9. Dalvi, M., and K. Martin. The Measurement of Accessibility: Some Preliminary Results. *Transportation*, Vol. 5, No. 1, 1976, pp. 17–42.
10. Ben-Akiva, M., and S. R. Lerman. Disaggregate Travel and Mobility Choice Models and Measures of Accessibility. In *Behavioural Travel Modelling* (D. A. Hensher, and P. R. Stopher, eds.), Croom Helm, Andover, 1979, pp. 654–679.
11. Cass, N., E. Shove, and J. Urry. Social Exclusion, Mobility and Access. *The Sociological Review*, Vol. 53, No. 3, 2005, pp. 539–555.
12. Geurs, K. T., and B. van Wee. Accessibility Evaluation of Land-Use and Transport Strategies: Review and Research Directions. *Journal of Transport Geography*, Vol. 12, No. 2, 2004, pp. 127–140.
13. Murray, A. T., and X. Wu. Accessibility Tradeoffs in Public Transit Planning. *Journal of Geographical Systems*, Vol. 5. No. 1, 2003, pp. 93–107.
14. Lei, T. L., and R. L. Church. Mapping Transit-Based Access: Integrating GIS, Routes and Schedules. *International Journal of Geographical Information Science*, Vol. 24. No. 2, 2010. pp. 283–304.
15. Mamun, S. A., N. E. Lownes, J. P. Osleeb, and K. Bertolaccini. A Method to Define Public Transit Opportunity Space. *Journal of Transport Geography*, Vol. 28, 2013, pp. 144–154.
16. Taylor, B. D., and C. N. Fink. *The Factors Influencing Transit Ridership: A Review and Analysis of the Ridership Literature*. University of California Transportation Center, 2003.
17. Cervero, R., and K. Kockelman. Travel Demand and the 3Ds: Density, Diversity, and Design. *Transportation Research Part D: Transport and Environment*, Vol. 2, No. 3, 1997, pp. 199–219.
18. Ewing, R., and R. Cervero. Travel and the Built Environment: A Synthesis. *Transportation Research Record: Journal of the Transportation Research Board*, 2001. 1780: 87–114.
19. Tilahun, N., and M. Li. Walking Access to Transit Stations: Evaluating Barriers with Stated Preference. *Transportation Research Record: Journal of the Transportation Research Board*, 2015. 2534: 16–23.
20. Humpel, N., N. Owen, and E. Leslie. Environmental Factors Associated with Adults' Participation in Physical Activity: A Review. *American Journal of Preventive Medicine*, Vol. 22, No. 3, 2002, pp. 188–199.
21. Landis, B., V. Vattikuti, R. Ottenburg, D. McLeod, and M. Guttenplan. Modeling the Roadside Environment: A Pedestrian Level of Service. *Transportation Research Record: Journal of the Transportation Research Board*, 2001. 1773: 82–88.
22. Khisty, C. Evaluation of Pedestrian Facilities: Beyond the Level-of-Service Concept. *Transportation Research Record: Journal of the Transportation Research Board*, 1994. 1438: 45–50.
23. Krambeck, H., and J. Shah. The Global Walkability Index: Talk the Walk and Walk the Talk. Presented at Better Air Quality Conference (BAQ), Yogyakarta, Indonesia, 2005.
24. Muraleetharan, T., and T. Hagiwara. Overall Level of Service of Urban Walking Environment and Its Influence on Pedestrian Route Choice Behavior: Analysis of Pedestrian Travel in Sapporo, Japan. *Transportation Research Record: Journal of the Transportation Research Board*, 2007. 2002: 7–17.
25. Niehaus, M., P. Galilea, and R. Hurtubia. Accessibility and Equity: An Approach for Wider Transport Project Assessment in Chile. *Research in Transportation Economics*, Vol. 59, 2016, pp. 412–422.
26. Delbosc, A., and G. Currie. Using Lorenz Curves to Assess Public Transport Equity. *Journal of Transport Geography*, Vol. 19, No. 6, 2011, pp. 1252–1259.
27. Welch, T. F., and S. Mishra. A Measure of Equity for Public Transit Connectivity. *Journal of Transport Geography*, Vol. 33, 2013, pp. 29–41.
28. Foster, J., J. Greer, and E. Thorbecke. A Class of Decomposable Poverty Measures. *Econometrica: Journal of the Econometric Society*, Vol. 52, No. 3, 1984, pp. 761–766.
29. Tapia, R. Vivienda social en Santiago de Chile: Análisis de su comportamiento locacional, período 1980-2002. *Revista INVI*, Vol. 26, No. 73, 2011, pp. 105–131.
30. Suazo, G., and J. C. Muñoz. *CBD Movement in Santiago de Chile in 1990-2015: Impact on Travel Times and Slums.*

Master of Science thesis. Department of Transport Engineering and Logistics, PUC, Chile, 2017.

31. Figueroa, O. La desregulación del transporte colectivo en Santiago: Balance de diez años. *Revista EURE*, Vol. 16, No. 49, 1990, pp. 23–32. http://repositorio.uc.cl/xmlui/bitstream/handle/123456789/3716/000170243.pdf?sequence=1.
32. Muñoz, J. C., and A. Gschwender. Transantiago: A Tale of Two Cities. *Research in Transportation Economics*, Vol. 22, No. 1, 2008, pp. 45–53.
33. DTPM. *Management Reports*. Transport and Telecommunications Ministry, Government of Chile, 2016. http://www.dtpm.cl/descargas/Datos%20Sistema%20Transporte%20P%C3%BAblico%20Metropolitano%20al%202015%20-%20DTPM.xlsx. Accessed July 27, 2017.
34. Shirahige, M., and J. Correa. La desigualdad en el acceso al transporte público en el área metropolitana de Santiago: Análisis mediante la aplicación del modelo PTAL en campamentos y blocks. *Revista del Centro de Investigación Social*, Vol. 18, Vol. 18, 2015, pp. 55–89.
35. Páez, A., D. M. Scott, and C. Morency. Measuring Accessibility: Positive and Normative Implementations of Various Accessibility Indicators. *Journal of Transport Geography*, Vol. 25, 2012, pp. 141–153.
36. OpenTripPlanner. *An Open Source Multi-Modal Trip Planner*. https://github.com/opentripplanner/OpenTripPlanner. Accessed March 20, 2016.
37. BRT. *Benchmark Report that Compares Six Latin American Public Transport Systems*. Database from Walking Distances, 2013. http://brt.cl/wp-content/uploads/2013/02/Estudio-comparativo-ciudades-latinoamericanas-Informe-Final-Anexo-2013-02-06.pdf. Accessed 20 June, 2017.
38. Richards, F. J. A Flexible Growth Function for Empirical Use. *Journal of Experimental Botany*, Vol. 10, No. 2, 1959, pp. 290–301.
39. Martínez, L. M., and J. M. Viegas. A New Approach to Modelling Distance-Decay Functions for Accessibility Assessment in Transport Studies. *Journal of Transport Geography*, Vol. 26, 2013, pp. 87–96.
40. INE. *National Pre-Census*. Government of Chile, 2011.
41. CONASET. *Road Accidents in Chile for 2011*. National Traffic Safety Commission, Ministry of Transport and Telecommunications, Government of Chile, 2011.
42. IDE-OCUC. *Traffic Flow per Hour Based on Origin-Destination Zones in Santiago*. 2016. http://ide.ocuc.cl/layers/geonode:Calculo_de_flujo_de_vehiculos_por_vias_RM. Accessed November 1, 2017.
43. INE. *Population of the Municipalities of Chile*. Government of Chile, 2016.
44. MINVU. *Quality Perception of Urban Life*. Gfk Adimark and Advisory Committee on Housing and Urban Studies, Government of Chile, 2010.
45. Gfk Adimark. Segregation Study: The two faces of Santiago, Chile, 2015. http://www.adimark.cl/es/estudios/documentos/micro%20estudio%20clasismo%20en%20chile_final.pdf. Accessed July 24, 2017.
46. MDS. *National Socioeconomic Characterization Survey (CASEN)*. Ministry of Social Development, Government of Chile, 2013.
47. ICAL. La desigual distribución de recursos municipales. 2015. http://www.ical.cl/2015/11/la-desigual-distribucion-de-recursos-municipales/. Accessed October 19, 2017.

The Standing Committee on Pedestrians (ANF10) peer-reviewed this paper (18-03785).

Article

TRR
JOURNAL OF THE TRANSPORTATION RESEARCH BOARD

Transportation Research Record
2018, Vol. 2672(35) 139–150
© National Academy of Sciences:
Transportation Research Board 2018
Article reuse guidelines:
sagepub.com/journals-permissions
DOI: 10.1177/0361198118787095
journals.sagepub.com/home/trr
SAGE

Walkability: Which Measure to Choose, Where to Measure It, and How?

Gabriel Lefebvre-Ropars[1] and Catherine Morency[1]

Abstract

The urban structure of neighborhoods has a decisive impact on active mobility, but this impact is hard to evaluate in a mode choice model because of the high collinearity between urban form variables and the uncertainty surrounding adequate spatial measurement parameters. Several composite scores, or walkability measures, have emerged from the literature, each using its own method and including different variables. No consensus has been reached on the size of the catchment area that should be used to measure walkability and most studies only measure walkability at the origin of the trip without considering other spatial units. In this paper, a series of four walkability measures: the Pedestrian Index of the Environment (PIE), the Walkability Index (WI), the Pedestrian Potential Index (PPI) and the Neighborhood Destination Accessibility Index (NDAI), are applied to the Greater Montréal Area to examine their correlation with the choice of walking for short trips. Several definitions of a neighborhood are tested for each measure, as well as several spatial units of measurement. A series of binary logistic regressions are then estimated using observed trip data from the 2013 Montreal Origin–Destination Survey to identify which measure, using which spatial parameters, offers the best performance in a mode choice modeling context. Measures using a diversity of urban form variables, the WI and the PIE, are found to offer the best performance, especially when measured at a scale smaller than a 1.6 km radius, while the spatial unit offering the best model fit varies between trip purposes.

More and more cities strive to build walkable neighborhoods and encourage the use of active modes for daily travel (*1*). These policies are based on the knowledge that urban form and transportation are interrelated and that dense, diverse, and walkable neighborhoods reduce automobile dependence (*2*). However, the link between urban form and active behavior is still a pressing research question and many authors have put forward models to describe the urban form as it relates to walking.

Ewing and Cervero (*3*) define the five dimensions of the urban form as the 5Ds: density, diversity, design, destination accessibility, and distance to transit. This categorization has enabled researchers to distinguish the effect of different aspects of urban environments on transportation choices, but the inclusion of those dimensions in a mode choice modeling process is complex because of the strong collinearity between the 5Ds (*4*). To circumvent this problem, many composite measures of the urban form as it relates to walking, or walkability measures, have been created by researchers in the last two decades. A recent review of the literature (*5*) found over 80 measures of pedestrian accessibility, each with its own set of variables and methodology. While the most popular measure, the Walkability Index, has been used in many different spatial contexts, many walkability measures have not been transferred to another urban region or even validated with observed trip data (*4*, *6*).

Some uncertainty remains as to the parameters to use when measuring the walkability of a neighborhood. The definition of the neighborhood itself, in terms of size and physical limits, varies across studies on walkability (*4*, *7*). This can skew model results because the value of the variable representing walkability is altered by the spatial unit used to measure it, a problem called the Modifiable Areal Unit Problem or MAUP (*8–10*). Moreover, existing studies concentrate on measuring walkability at the origin of the trip, often ignoring the other spatial units of the trip as described by Moudon and Lee in their Behavioral Model of Environment (*11*): the destination and the route to destination (or itinerary).

Furthermore, the few studies that have studied the contribution of walkability measures to pedestrian mode choice modeling in the Greater Montréal Area (GMA) have only done so at scales much smaller than the metropolitan region (*7*, *12*) or using audit-based measures on targeted city blocks (*13–15*).

The goals of this paper are twofold. First, this research examines the effect of the variation of the spatial unit and the

[1]Department of Civil, Geological and Mining Engineering, Polytechnique Montréal, Montreal, QC, Canada

Corresponding Author:
Address correspondence to Gabriel Lefebvre-Ropars: gabriel-lee.lefebvre-ropars@polymtl.ca

Table 1. Datasets Used in This Paper

Database	Source	Year	Geometry type
Road network centerlines	OpenStreetMap (OSM)	2016	Lines
Land use	*Communauté métropolitaine de Montréal* (CMM)	2014	Polygons
Origin–Destination Survey, version 13.2b	*Agence métropolitaine de transport* (AMT)	2013	Points
Archived GTFS feeds for all GMA transit operators	GTFS Exchange	2013	Points
Business locations	Environics Analytics Group (EAG)	2012	Points
National Household Survey (NHS) and Census of the Population	Statistics Canada	2011	Polygons

definition of the neighborhood when measuring walkability. Then, four different walkability measures are computed at a metropolitan scale in the GMA. The measures are compared in several mode choice models to understand the effect of the underlying variables and methodology of different measures on the modeled effect of the urban form on walking behavior.

Methods

Datasets

The Origin–Destination survey. The Origin–Destination (OD) survey is a quinquennial travel survey that collects information on daily mobility in the GMA. The most recent survey took place in the fall of 2013 and sampled around 5% of the region's households. Respondents were randomly selected using land-line and mobile phone listings and the sample was adjusted based on data from the 2011 Census of Population (*16*).

The survey collects information on every trip conducted by all household members (aged 5 or older) on the last business day before the interview. For each trip, the geographical coordinates of its origin and destination are collected, along with the trip's purpose, the departure time, and the sequence of modes used to carry it out. Information about the household and household members is also collected, namely the size of the household, the number of cars owned, and the home location, as well as the age, gender, possession of a driver's license, and occupation of its members.

In the OD database, the only trips recorded as walking trips are those where walking is the sole mode used between the origin and the destination. Trip segments carried out by foot to access another mode of transportation (e.g. a parked car or a transit stop) are not recorded as walking trips and are thus excluded from this analysis.

Urban Form Variables. To compute the four walkability measures used in the current paper, several databases (see Table 1) are integrated into a geographic information system (GIS). The land use and census databases are issued and validated by government agencies. GTFS feeds extracted from the (now defunct) GTFS Exchange archive were compiled by transit operators. The Business Locations database is a private database supplied by Environics Analytics Group (EAG). Conversely, the OpenStreetMap (OSM) roadway network is a crowdsourced dataset maintained by volunteers, whose reliability in the context of the representation of pedestrian networks has nevertheless been confirmed in a North American context (*17*).

For the purposes of this paper, the GMA is defined as the common territory between the administrative border of the *Communauté métropolitaine de Montréal* (CMM) and the Montréal Census metropolitan area (CMA).

Building the Trip Database

Preparing the Trip Database. The complete OD database for the 2013 survey contains 452,978 records. Of these, 346,335 are complete records whose origin and destination are within the study area, representing around 8 million trips. For each household in the database, a car access rate is computed by dividing the number of cars in the household by the number of household members with a driver's license. Trips categorized as "business meetings" or "on the road," which represent less than 1% of trips each, are reclassified as work trips.

A shortest path algorithm is used to create a pedestrian itinerary and calculate a trip distance. Hence, for each complete OD pair in the database, a path is created using a local implementation of the OSRM calculator, an open source tool (*18*). The final database contains 327,991 records representing 7,567,717 trips.

Identifying Transferable Trips. Walking can only be competitive with other transportation modes when certain criteria are met. Too great a distance or too complex a trip chain, for example, can make a trip impossible to carry out by foot. For the purposes of this study, trips considered not transferable to walking are excluded from the trip database to train the binary choice models on a plausible, "walkable" set of trips.

Using a methodology developed previously by Morency et al. (*19*), all transferable trips are identified in the OD database using the following criteria:

1. Trips should not be longer than the threshold distance for the person's demographic segment (age and

gender), where the threshold distance corresponds to the 80th percentile of observed distances in the OD Survey for a given segment.

2. Every trip in the parent trip chain should meet condition 1. as well.
3. The total length of the parent trip chain should not exceed four times the threshold distance for the person's demographic segment.

The resulting database contains 24,803 records representing 565,553 trips. The modal share of walking for these trips (observed walking trips + transferable trips) is 60.33%.

Computation of the Walkability Measures

A selection of four walkability measures is computed for the GMA: the Pedestrian Index of the Environment (PIE), the Walkability Index (WI), the *Indice de potentiel piéton* (Pedestrian Potential Index, or PPI), and the Neighborhood Destination Accessibility Index (NDAI). The measures are computed for each of the 80 m by 80 m cells contained in a grid spanning the entire GMA, with the exception of water bodies. This size corresponds to the distance that can be covered on foot in one minute (*20*). The use of a grid eliminates the border effects that occur when measuring urban form variables within statistical or administrative borders, while allowing the computation of a walkability score for the whole metropolitan territory. The following section presents a summary of the steps followed to compute the selected walkability measures. A detailed methodology can be found in the original papers (*20–23*).

The Pedestrian Index of the Environment. The PIE was developed by Clifton et al. (*20*) as a tool for representing the urban form in pedestrian demand modeling. The measure was originally based on a proprietary dataset, Oregon Metro's Context Tool, but was successfully transferred to Montréal using local datasets in a previous research effort (*6*). The PIE includes variables relating to four of the 5Ds: density, design, distance to transit, and destination accessibility. The variables used are the following:

1. Infrastructure comfort
2. Block size
3. Activity density
4. Transit frequency
5. Local amenities

It is worth noting that this measure is the only one, out of the four selected measures, to include a notion of pedestrian experience. This is done through the inclusion of *Infrastructure comfort*, which relates to the perceived safety of a street as represented by the street's hierarchical classification. Each variable is computed using a kernel density algorithm to weigh each element according to its distance from the target grid cell. The value produced by the kernel density algorithm is then reclassified in a score of 1 to 5 using a natural breaks (Jenks) algorithm applied to the whole grid. The 1-to-5 scores for each of the five variables are then weighted using weights derived from a single-variable mode choice model. Finally, variables are scaled so that the sum of the five dimensions of the PIE spans a range of 20 to 100, which is the measure's scale. The innovative use of a decay function and of a modeling-based internal weighing makes the PIE an interesting candidate to include in a comparative walkability modeling approach to evaluate the added value of those innovative techniques.

The Walkability Index. The WI was one of the first composite walkability measures to emerge from the literature (*21*, *24*) and remains the most widely used (*5*). It includes three of the 5Ds: density, diversity, and design. The variables used by the measure are the following:

1. Residential density
2. Intersection density
3. Land use mix
4. Floor area ratio for commercial uses

Each variable is computed, for each grid cell, by calculating the average value of all objects within the search radius. A Z-score is then computed for each of the four variables and those scores are weighted and summed to produce the final measure. A weight of 2 is applied to the *Intersection density* variable, while the other variables retain a weight of 1.

The Pedestrian Potential Index. The PPI was developed by the *Institut national de santé publique du Québec* (INSPQ), a public research institute that provides expertise to the provincial government in the field of public health (*22*). The PPI is a variation of the WI made to be easily computed using data available to civil employees in the province of Québec, which could make it a very useful tool for practitioners and decision makers in the study area if it performs well in a mode choice modeling context. The measure has never been validated in a mode choice modeling context until now. The PPI includes the following variables:

1. Residential density
2. Intersection density
3. Land use mix
4. Destination density

The unweighted Z-score of each variable is then summed to produce a final score.

The Neighborhood Destination Accessibility Index. The NDAI was developed in New Zealand to study the relation between the presence of amenities in a neighborhood and walking behavior (*23*). The measure has not been transferred to

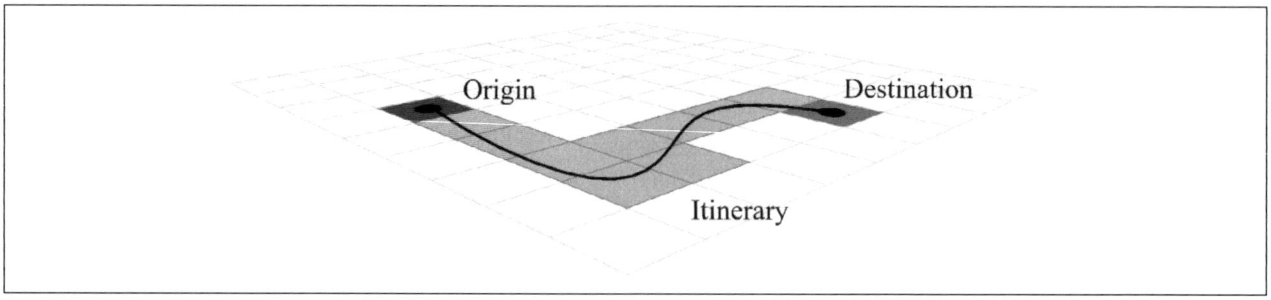

Figure 1. Spatial units used in this paper.

another country until now. This measure was selected because it differs in nature from the three other indices, with only two of the 5Ds being represented in the NDAI: access to transit and destination accessibility. For the purposes of this paper, amenities (or subdomains) are grouped into the following categories (or domains):

1. Education
2. Public transit frequency
3. Leisure
4. Culture and entertainment
5. Restaurants and food retail
6. Financial services
7. Healthcare

The number of amenities in each subdomain is reclassified in binary variables or terciles, as detailed in the original paper (*23*), and then summed into a domain score. The score of each domain is then weighted and summed to produce a final score.

Defining the Neighborhood

Measuring the walkability of a neighborhood implies defining the size of that same neighborhood. Different definitions of the neighborhood are used in this paper to identify the scale at which walkability measures offer the best performance in a mode choice modeling context. The neighborhood is defined using two parameters: the search radius and the spatial unit of measurement.

Search Radius. The search radius is the straight-line distance around the centroid of the cell within which urban form variables are measured. This paper compares six different search radii:

- 200 m
- 400 m
- 800 m
- 1,200 m
- 1,600 m
- 2,000 m

Every cell of the grid receives a score for each of the four selected walkability measures, for a total of 24 different scores.

Spatial Unit. In order to assign a walkability score to a trip for mode choice modeling purposes, the walkability of the urban environment producing the trip must be measured at a defined point, or spatial unit. A total of four different spatial units are compared in this paper:

- The cell containing the origin of the trip
- The cell containing the destination of the trip
- The average value of the cells containing the trip's origin and destination
- The average value of all cells crossed by the calculated itinerary of the trip

For each of these spatial units (shown in Figure 1), trips are assigned the value of a walkability measure for every search radius. Each trip is thus assigned a total of 96 different walkability scores: four measures computed within six search radii for four distinct spatial units.

Modeling the Choice of Walking for Short Trips

To examine the explanatory power of different walkability measures and spatial units, a series of binomial logistic regressions are estimated on trip data from the OD survey segmented according to the four most frequent trip purposes: work, school, shopping, and leisure. Each dataset is then split into a training set (70%) and a validation set (30%). Every record of the validation set is replicated n times, where n represents the record's rounded survey weight.

A total of 24 models are trained for each of the four walkability measures described above using R, version 3.3.1 (*25*). Every model corresponds to a distinct combination of spatial unit and search radius. The walkability measure is inputted as a continuous variable along with a series of household-, individual-, and trip-related binary variables. Walkability measures are not standardized as their underlying methodologies are deemed too different to

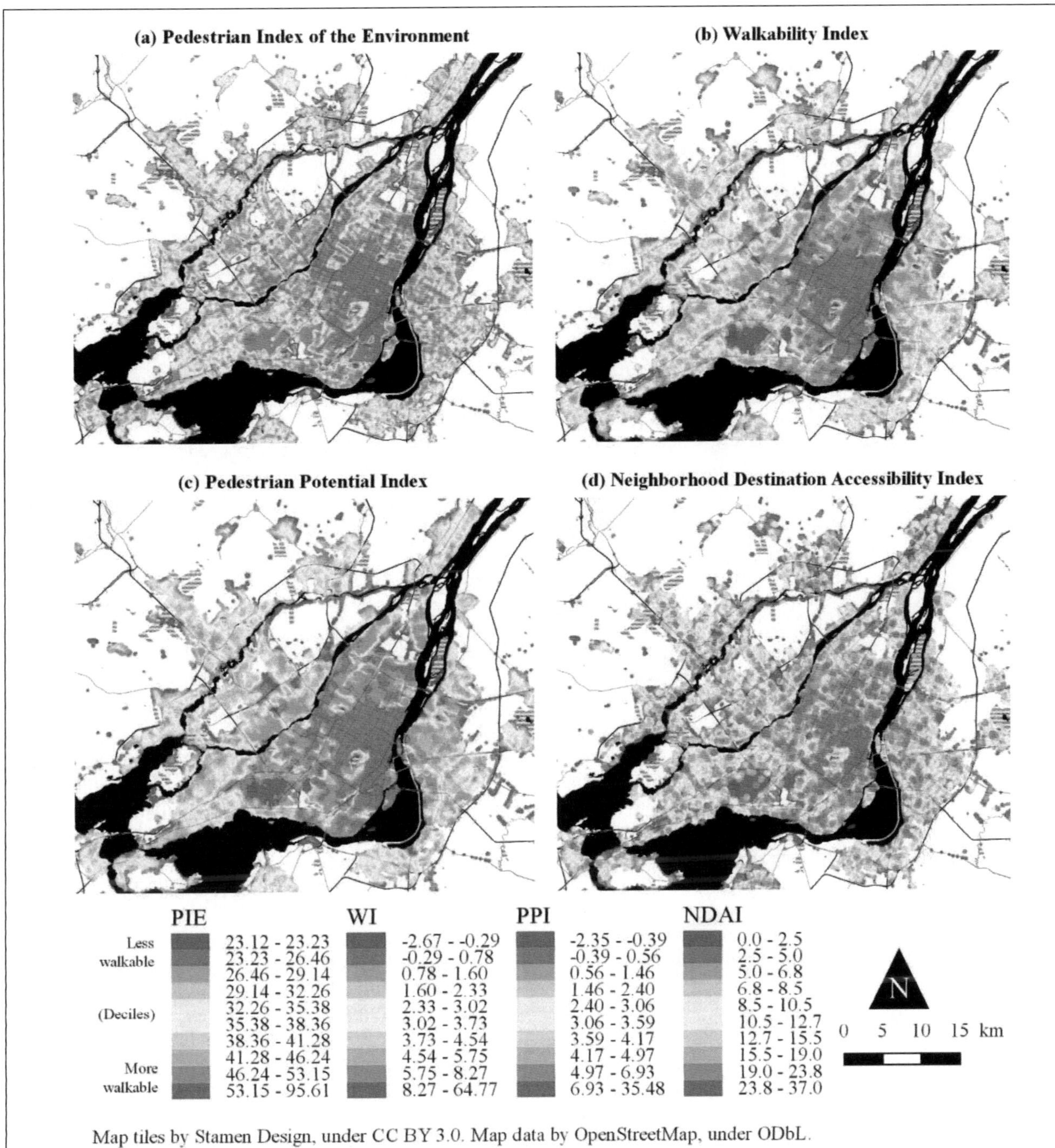

Figure 2. Spatial distribution of walkability scores in the GMA with a 400-m search radius, ranked by decile: (*a*) PIE; (*b*) WI; (*c*) PPI; (*d*) NDAI.

compare their direct effects. Each model controls for the household's size and car access, the person's age and gender, as well as the trip length and the complexity of the trip chain. The models are then applied to the validation dataset to test their ability to predict the choice of walking for each individual trip.

Results

Spatial Distribution of Walkability in the GMA

As shown in Figure 2, which uses a 400-m search radius, the selected walkability measures offer a consistent large-scale portrait of walkability. Downtown Montréal, central neighborhoods, and densely populated areas rank in the highest walkability deciles, while low-density suburbs and exurbs, industrial areas and large transportation infrastructures (ports, airports) intuitively rank in the lowest walkability deciles.

While similar at the metropolitan scale, each measure has a particular way of capturing walkability at a smaller scale. The PIE (Figure 2*a*) seems to follow certain streets—often commercial streets—very closely and leave out specific areas like large parks and industrial spaces. This can be explained by the presence of the *Infrastructure comfort* variable, which weighs calm residential streets higher than arteries, as well as its inclusion of proximity amenities. The PIE does not measure green spaces or parks, which explains the low walkability score of these areas when using a 400-m search radius (*20*).

The WI (Figure 2*b*) behaves differently in that it decreases less sharply, spatially speaking, than the PIE. This translates into larger, smoother zones that do not follow transportation infrastructure as closely. Large parks rank in the highest walkability deciles, a corollary of the inclusion of green spaces in the *Land use mix* variable (*21*). Low-density neighborhoods in peripheral towns or leap-frog suburban developments rank in lower deciles with the WI than with the PIE, while many central urban areas rank higher. The PPI (Figure 2*c*) offers a very similar representation of walkability to that of the WI, with no apparent difference in the walkability ranking of areas in the GMA. This is to be expected, as the PPI is based on the WI and differs only in the use of the *Destination density* variable instead of the *Floor area ratio for commercial uses* variable and the absence of any weights applied to the variables.

The NDAI (Figure 2*d*) produces a spatial distribution that is highly focused on commercial zones, with a very clear radial decline in walkability score as the distance from commercial concentrations increases. Some suburban boulevards and malls rank in the highest deciles, while many central neighborhoods and dense suburbs rank lower than with other measures.

Effect of Spatial Unit and Search Radius on Walkability

Varying the search radius and the spatial unit used to assign walkability scores to trips has a major effect on the distribution of values to be used in the mode choice modeling process. Figure 3 presents the distribution of walkability scores for each of the 48 search radius/spatial unit combinations. In comparison with the search radius, the variation of the spatial unit seems to have a minor effect of walkability scores. Indeed, there are few discernable differences within versions of the measures using the same search radius.

The PIE (Figure 3*a*) appears to be the most stable measure throughout the changes in its search radius, although the median score still increases by almost 50% between the lowest and highest radius. This can be explained by the reclassification of variables in 1-to-5 scores prior to the computation of the final score, which helps to even out the distribution of values. Higher scores are more frequent in versions of the PIE with a higher search radius, while lower search radii are characterized by a clustering of PIE values toward the lowest possible values of the score.

The lower three quartiles of the WI (Figure 3*b*) and PPI (Figure 3*c*) remain stable throughout the different search radii and spatial units. Each increase in the search radius seems to reduce the number of high walkability scores assigned to trips. This is to be expected since the WI and the PPI are computed by summing Z-scores. When the search radius is larger, more grid cells will be awarded a high absolute value, which in turn increases the standard deviation of the variables and thus diminishes their Z-score. The highest search radius, 2,000 m, produces values that come closest to a normal distribution.

The NDAI behaves in a totally different manner than the other measures. Since it is simply an additive score, a higher search radius means that more amenities will be captured by the catchment area. Thus, as the radius increases, so does the number of higher scores. Over 75% of trips are assigned a score higher than 30 when the search radius is set at 1,600 m or higher, with a maximum possible value of 37.

Link between Measured Walkability and the Choice to Walk

Table 2 presents a summary of the 384 choice models fitted in this paper by showing the odds ratio of the walkability measures. Every version of the walkability measures was found to be statistically significant at the $p < 0.001$ level, while controlling for the household-, individual-, and trip-related variables mentioned above. The odds ratios represent the increase in odds of walking following an increment of 1 in the explanatory variable. Since the walkability measures are not scaled prior to inclusion in the models, and given the variation in their distribution when modifying their parameters, particularly the search radius (see Figure 3), these odds ratios should only be directly compared between models using the same walkability measure and search radius.

For all measures and combinations of spatial unit and search radius, the difference in correlation between walkability and the choice of walking seems to vary according to the trip purpose. For example, an increase of 1 in the PIE score, when using a 200-m search radius around the origin of the trip, increases the odds of walking for a work trip by 6.3%, but only by 2.9% for a school trip. School trips consistently get lower odds ratios than every other trip purpose, hinting at the fact that people—children, in this case—are likely to walk to school in a greater diversity of urban environments than for other purposes. The choice of walking for leisure trips, on the other hand, appears to be the most correlated to the walkability of the urban environment.

When comparing between the different spatial units, holding the search radius constant, the destination of the trip is

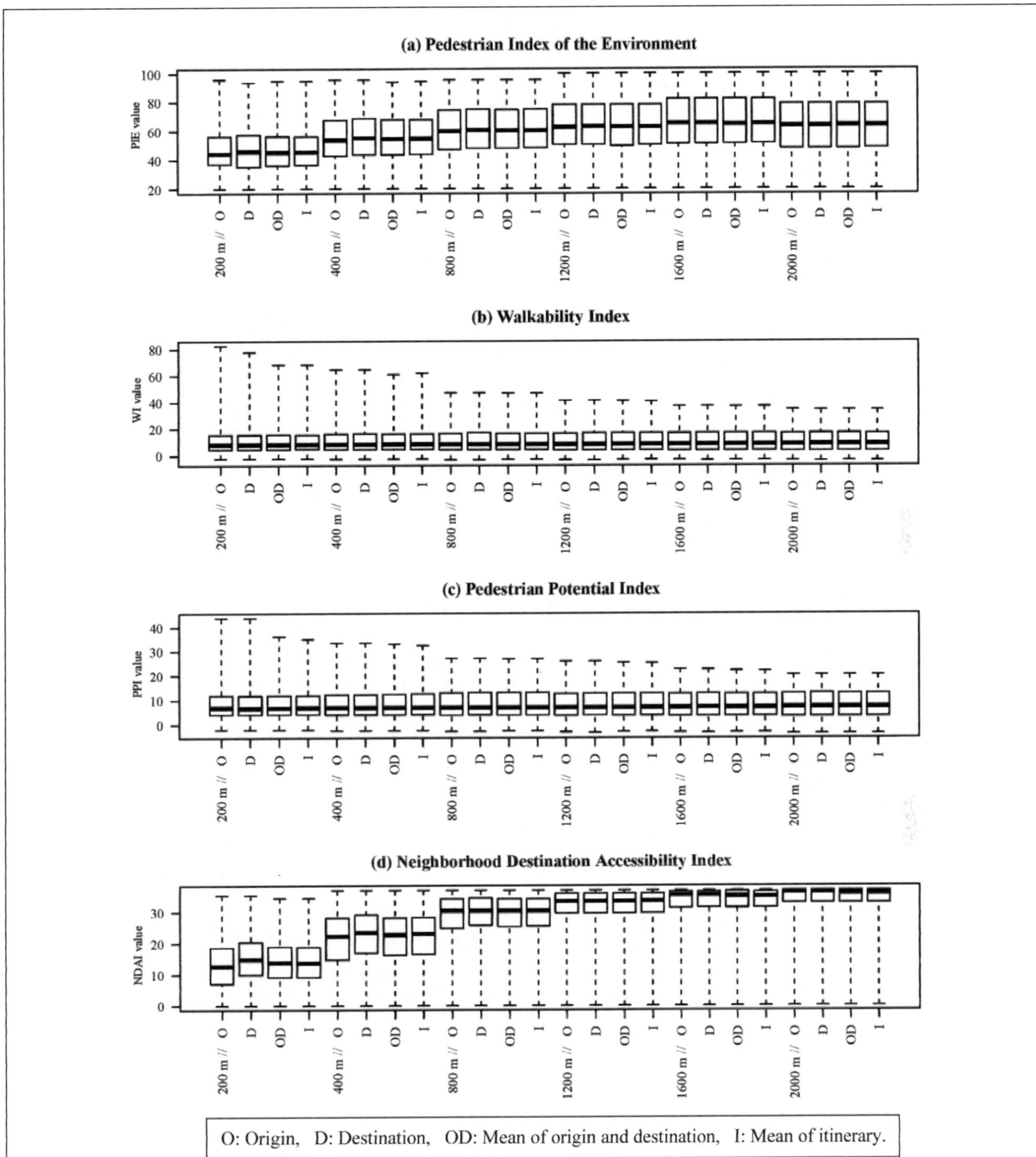

Figure 3. Distribution of walkability measures by search radius and spatial unit: (*a*) PIE; (*b*) WI; (*c*) PPI; (*d*) NDAI.

the spatial unit that is characterized by the lowest odds ratios for all measures. The PIE and the NDAI seem to be more correlated to the choice of walking when they are measured along the itinerary or at both the origin and the destination.

Model Fit

The Akaike information criterion (AIC) is a measure of model fit that decreases when the model increases. Table 3

Table 2. Summary of Mode Choice Models—Odds Ratios of Walkability Measures by Trip Purpose, Spatial Unit, and Search Radius

		PIE				WI				PPI				NDAI			
Spatial Unit	Search Radius	Work	School	Shopping	Leisure	Work	School	Shopping	Leisure	Work	School	Shopping	Leisure	Work	School	Shopping	Leisure
Origin	200 m	1.063	1.029	1.068	1.081	1.099	1.057	1.110	1.127	1.152	1.080	1.165	1.191	1.093	1.037	1.109	1.100
	400 m	1.053	1.025	1.064	1.069	1.103	1.062	1.116	1.139	1.159	1.084	1.180	1.210	1.084	1.038	1.112	1.109
	800 m	1.048	1.023	1.060	1.066	1.105	1.065	1.121	1.142	1.163	1.088	1.187	1.214	1.096	1.038	1.140	1.148
	1200 m	1.048	1.022	1.058	1.063	1.103	1.064	1.120	1.137	1.158	1.084	1.184	1.209	1.117	1.039	1.174	1.187
	1600 m	1.047	1.022	1.057	1.062	1.102	1.061	1.118	1.133	1.155	1.081	1.179	1.202	1.127	1.043	1.211	1.248
	2000 m	1.045	1.020	1.055	1.059	1.101	1.059	1.117	1.132	1.149	1.076	1.175	1.199	1.127	1.038	1.230	1.273
Destination	200 m	1.053	1.027	1.061	1.063	1.083	1.060	1.105	1.104	1.132	1.088	1.159	1.173	1.071	1.033	1.097	1.098
	400 m	1.050	1.024	1.062	1.062	1.095	1.063	1.114	1.124	1.155	1.088	1.174	1.197	1.077	1.032	1.113	1.105
	800 m	1.049	1.020	1.059	1.063	1.101	1.062	1.120	1.127	1.164	1.081	1.187	1.198	1.093	1.034	1.135	1.135
	1200 m	1.048	1.020	1.057	1.062	1.102	1.061	1.120	1.127	1.158	1.081	1.184	1.201	1.113	1.039	1.175	1.177
	1600 m	1.047	1.020	1.056	1.060	1.101	1.059	1.118	1.128	1.155	1.079	1.178	1.199	1.127	1.042	1.211	1.239
	2000 m	1.045	1.020	1.054	1.058	1.102	1.060	1.116	1.128	1.152	1.078	1.174	1.196	1.133	1.042	1.238	1.260
Mean (O+D)	200 m	1.066	1.032	1.076	1.084	1.099	1.067	1.119	1.128	1.161	1.094	1.180	1.206	1.113	1.048	1.148	1.139
	400 m	1.057	1.027	1.070	1.073	1.104	1.066	1.121	1.138	1.167	1.091	1.188	1.215	1.098	1.042	1.141	1.132
	800 m	1.051	1.023	1.063	1.069	1.105	1.064	1.123	1.137	1.167	1.086	1.191	1.211	1.105	1.039	1.153	1.161
	1200 m	1.050	1.021	1.059	1.064	1.104	1.063	1.121	1.133	1.161	1.084	1.186	1.207	1.122	1.041	1.187	1.198
	1600 m	1.048	1.022	1.057	1.063	1.102	1.061	1.119	1.132	1.156	1.081	1.180	1.202	1.133	1.044	1.224	1.261
	2000 m	1.046	1.021	1.055	1.060	1.103	1.060	1.117	1.131	1.152	1.077	1.175	1.199	1.134	1.041	1.244	1.280
Mean (itinerary)	200 m	1.067	1.032	1.078	1.086	1.101	1.068	1.118	1.129	1.165	1.100	1.179	1.210	1.116	1.051	1.164	1.154
	400 m	1.057	1.026	1.071	1.074	1.102	1.066	1.119	1.135	1.164	1.092	1.183	1.211	1.099	1.042	1.146	1.133
	800 m	1.051	1.023	1.063	1.068	1.103	1.064	1.121	1.136	1.165	1.087	1.189	1.211	1.105	1.040	1.151	1.161
	1200 m	1.049	1.022	1.060	1.065	1.103	1.063	1.120	1.132	1.160	1.084	1.185	1.206	1.121	1.042	1.184	1.199
	1600 m	1.048	1.022	1.057	1.063	1.102	1.061	1.118	1.130	1.156	1.081	1.180	1.202	1.133	1.044	1.228	1.254
	2000 m	1.046	1.021	1.055	1.060	1.102	1.059	1.116	1.130	1.152	1.077	1.175	1.197	1.139	1.042	1.243	1.275

Table 3. Summary of Model Fit—Delta AIC of Purpose-Based Segments by Walkability Measure, Spatial Unit, and Search Radius

		Work				School				Shopping				Leisure			
Spatial Unit	Search Radius	PIE	WI	PPI	NDAI	PIE	WI	PPI	NDAI	PIE	WI	PPI	NDAI	PIE	WI	PPI	NDAI
	200 m	63	44	61	119	37	23	29	59	123	111	127	187	89	66	87	212
	400 m	47	24	35	108	28	12	20	37	52	81	70	113	76	26	43	163
	800 m	38	7	22	127	19	2	7	47	28	38	36	163	56	0*	27	165
	1200 m	23	9	29	133	20	0*	8	56	25	27	29	192	64	4	37	195
	1600 m	26	11	32	151	12	4	7	57	34	25	32	199	61	11	47	174
Origin	2000 m	26	23	48	176	21	9	15	72	30	27	42	227	59	14	54	189
	200 m	71	64	98	188	40	9	14	72	90	120	112	256	127	119	131	218
	400 m	46	23	24	151	31	6	11	58	52	85	69	183	109	55	58	199
	800 m	24	6	1	143	39	7	17	60	43	41	28	212	70	31	47	199
	1200 m	20	4	12	139	34	6	13	55	38	33	30	215	68	29	49	203
	1600 m	18	8	24	148	24	7	12	59	44	30	35	221	72	20	48	187
Destination	2000 m	15	11	35	169	23	4	10	67	38	32	42	225	65	24	56	204
	200 m	35	33	53	110	29	5	12	52	43	78	80	118	67	64	75	156
	400 m	21	8	13	97	22	3	11	38	7	60	45	70	61	23	33	139
	800 m	17	0*	4	119	25	2	10	50	12	29	23	155	46	9	30	158
	1200 m	13	2	17	127	25	2	9	53	17	23	24	184	55	12	39	185
	1600 m	15	6	25	144	16	4	9	57	29	22	30	196	59	12	45	169
Mean (O+D)	2000 m	15	14	39	169	20	6	12	69	26	24	39	217	56	16	54	190
	200 m	33	31	43	105	32	4	5	47	43	84	82	65	59	57	69	132
	400 m	26	16	17	96	25	5	7	40	0*	72	54	53	55	30	37	140
	800 m	24	9	11	121	22	4	9	46	14	42	30	167	50	10	27	166
	1200 m	23	7	18	129	22	3	8	51	17	29	25	194	55	15	38	186
	1600 m	21	10	26	143	14	3	8	57	32	29	31	195	59	16	44	175
Mean (Itinerary)	2000 m	18	16	39	161	16	7	12	66	33	31	40	220	56	18	56	193

*Model with lowest AIC for each trip purpose.

shows the Delta AIC of each model, i.e., the difference between the AIC of a given model and the lowest AIC of all compared models. Since AIC scores should only be compared between models estimated on the same dataset, distinct Delta AIC values were calculated for each trip purpose.

The walkability measures offering the best statistical fit according to their AIC are the WI (for work, school and leisure trips) and the PIE (for shopping trips), for most spatial units and search radii. The PPI offers a performance comparable to that of the WI for work and school trips, with a Delta AIC lower than 10 in several cases (*26*).

The formulations offering the best model fit for each walkability measure seem to vary by trip purpose. For example, the WI offers a better model fit for work trips when measured at both the origin and the destination of the trip, but performs better for leisure trips when measured at the origin of the trip. A medium-sized search radius (between 400 m and 1,200 m) seems to offer a better fit. The NDAI, which does not include the density, diversity, and design aspects of the urban form, seems to offer the lowest fit in all cases.

Validation of the models shows that the change in the spatial unit and search radius has a marginal impact on the precision of mode choice prediction. Within a trip purpose, the largest improvement in model accuracy barely exceeds 2%. For example, the best performing model for shopping trips achieves 75.01% accuracy, while the least accurate model for this purpose still achieves 73.17% accuracy. The inclusion of the urban form, as represented by walkability measures, in binary mode choice models thus does not seem to have a major effect on the prediction of short walking trips. Model accuracy varies proportionally to Delta AIC, with the best fitting models achieving the highest accuracies.

Discussion

A look at the spatial distribution of the measures shows that the nature of the data used to compute them has a great impact on their representation of walkability. Measures that include more point-based data (amenities, transit service) like the PIE and the NDAI give higher scores to precise locations, while measures using more zonal data (land use, density) like the WI and the PPI tend to offer a more homogeneous picture of walkable and non-walkable zones. The inclusion of the first three Ds: density, diversity, and design, helps the measures discriminate between walkable and non-walkable amenities, for instance, suburban strip malls and urban commercial streets, which the NDAI was less able to do.

The parameters defining the neighborhood appear important to the structure of a walkability measure. Increasing the search radius has a noticeable effect on the distribution of values in a single measure, which renders any comparison between different versions of the same measure, or between measures, quite complex. While the PIE keeps a somewhat constant distribution due to the standardization of its variables during the computation, the NDAI, for example, drifts from something that resembles a Gaussian distribution to a skewed distribution where almost all trips have high walkability scores. Z-score-based measures like the WI and the PPI stay mostly constant, except for the highest values which become rarer as the search radius increases. This also presents challenges in a modeling context, as the increment of 1 on which odds ratios are based does not represent the same increase in walkability between different versions of a given measure. In light of the modeling results, a moderately sized neighborhood definition (between 400 and 1,200 m) seems to offer the best performance.

The variation in spatial unit appears to have a smaller effect on the distribution of walkability scores than the search radius. Measuring walkability at the destination of the trip or along the itinerary does not impact the distribution of walkability values in a noticeable way, although it does have an impact on the fit of the mode choice model. This could be explained by the short length of modeled trips, since these trips (observed walking trips and transferable short trips) are usually short enough so that both their origin and destination are located inside the larger search radii. Considering the modeling results, it is difficult to establish which spatial unit offers the best performance for capturing the effect of walkability on the choice of walking for short trips. General guidelines can however be derived from these results: for work and school trips, the slight difference in model performance (with Delta AIC lower than 10 when using a search radius between 400 and 1,200 m) does not seem to warrant the added complexity of taking trip destination into account. Measuring walkability at the trip origin for leisure trips offers the best results, while one should definitely consider trip destination when modeling shopping trips because the difference in model performance is notable (Delta AIC > 10).

When modeling the interaction between walkability and the choice of walking for different trip purposes, some conclusions can still be drawn. For instance, the choice of walking for school trips appears to respond weakly to an increase in walkability. This can be explained by the reduced mobility choices for children, but also by the increasing tendency observed in the GMA for parents to drive their children to school, even in dense urban environments (*27*). Unconstrained trip purposes appear to be more sensitive to changes in walkability, stressing the importance of walkable commercial districts and mixed urban environments.

Conclusion

A series of four walkability measures were applied to the Greater Montréal Area (GMA) to compare their performance in a pedestrian mode choice model for short trips. Several variations in neighborhood definition and spatial unit of measurement were computed for all measures to study their sensitivity to spatial parameters. These estimations highlighted the crucial importance of the definition of the neighborhood, as determined by the search radius, in the stability of a walkability measure.

A total of 384 binary logistic regressions, segmented by trip purpose, were then carried out to study the performance of the different walkability measures. The Walkability Index (WI), a pioneering measure in the field of walkability research, stands out as the best performing measure for work, school, and leisure trips. The PIE, which uses a quite different methodology, offers a better model fit for shopping trips as well as the most stable distribution across spatial units and neighborhood definitions. The PPI, tailor-made for local data, appears as a satisfactory measure of walkability for modeling constrained trips, less so for leisure and shopping trips.

This research has some limitations. It must first be noted that the cross-sectional dataset used in this paper cannot reveal any causation between walkability and the choice of walking. The modeling framework does not account for residential self-selection. This framework also assumes that the trip destination and the decision to make said trip are not dependent on the mode used. Moreover, the itineraries were developed with a shortest path algorithm and do not necessarily represent the actual path chosen by the individuals when carrying out their trip.

Further research should investigate standardization methods applied to walkability measures to enable a significant and meaningful comparison between versions of a measure using different neighborhood definitions and spatial units.

Acknowledgments

The authors acknowledge the support and funding provided by the Natural Sciences and Engineering Research Council of Canada (NSERC)'s Discovery Grant Program, the partners of the Mobilité Chair – the City of Montréal, the *Société de transport de Montréal* (STM), the *Ministère des Transports, de la Mobilité durable et de l'Électrification des transports* (MTMDET), and the *Autorité régionale de transport métropolitain* (ARTM), and *Exo*, as well as the *Fonds de recherche du Québec–Nature et technologies* (FRQNT). They also thank Professor Patrick A. Singleton, from Utah State University, and Professor Kelly J. Clifton, from Portland State University, for their assistance.

Author Contributions

The authors confirm contribution to the paper as follows: Study conception and design: Gabriel Lefebvre-Ropars, Catherine Morency; data collection: Gabriel Lefebvre-Ropars; analysis and interpretation of results: Gabriel Lefebvre-Ropars, Catherine Morency; draft manuscript preparation: Gabriel Lefebvre-Ropars, Catherine Morency.

References

1. Vojnovic, I. Urban Sustainability: Research, Politics, Policy and Practice. *Cities*, Vol. 41, 2014, pp. S30–S44.
2. McIntosh, J., R. Trubka, J. Kenworthy, & P. Newman. The Role of Urban Form and Transit in City Car Dependence: Analysis of 26 Global Cities from 1960 to 2000. *Transportation Research Part D: Transport and Environment*, Vol. 33, 2014, pp. 95–110.
3. Ewing, R., and R. Cervero. Travel and the Built Environment: A Meta-Analysis. *Journal of the American Planning Association*, Vol. 76, No. 3, 2010, pp. 265–294.
4. Wong, B. Y.-M., G. Faulkner, and R. Buliung. GIS Measured Environmental Correlates of Active School Transport: A Systematic Review of 14 Studies. *International Journal of Behavioral Nutrition and Physical Activity*, Vol. 8, No. 1, 2011, pp. 1–22.
5. Vale, D. S., M. Saraiva, and M. Pereira. Active Accessibility: A Review of Operational Measures of Walking and Cycling Accessibility. *Journal of Transportation and Land Use*, Vol. 9, No. 1, 2015. https://doi.org/10.5198/jtlu.2015.593
6. Lefebvre-Ropars, G., C. Morency, P. A. Singleton, and K. J. Clifton. Spatial Transferability Assessment of a Composite Walkability Index: The Pedestrian Index of the Environment (PIE). *Transportation Research Part D: Transport and Environment*, Vol. 57, 2017, pp. 378–391.
7. Manaugh, K., and A. El-Geneidy. Validating Walkability Indices: How do Different Households Respond to the Walkability of their Neighborhood? *Transportation Research Part D: Transport and Environment*, Vol. 16, No. 4, 2011, pp. 309–315.
8. Lloyd, C. D. The Modifiable Areal Unit Problem. *Exploring Spatial Scale in Geography*. John Wiley & Sons, Chichester, 2014, pp. 29–44.
9. Madsen, T., J. Schipperijn, L. B. Christiansen, T. S. Nielsen, and J. Troelsen. Developing Suitable Buffers to Capture Transport Cycling Behavior. *Frontiers in Public Health*, Vol. 2, 2014, p. 61.
10. Dong, E., H. F. Liao, and H. Kang. Grocery Shopping: Geographic Scale Matters in Analyzing Effects of the Built Environment on Choice of Travel Mode. Presented at 95th Annual Meeting of the Transportation Research Board, Washington, D.C., 2016.
11. Moudon, A. V., and C. Lee. Walking and Bicycling: An Evaluation of Environmental Audit Instruments. *American Journal of Health Promotion*, Vol. 18, No. 1, 2003, pp. 21–37.
12. Herrmann, T., G. Boisjoly, N. A. Ross, and A. M. El-Geneidy. The Missing Middle: Filling the Gap Between Walkability and Observed Walking Behavior. *Transportation Research Record: Journal of the Transportation Research Board*, 2017. 2261: 103–110.
13. St-Louis, A. *Mesurer le potentiel piétonnier pour les aînés: les cas de Lille et Montréal*. Université du Québec, Institut national de la recherche scientifique, 2015.
14. Gauvin, L., L. Richard, C. L. Craig, M. Spivock, M. Riva, M. Forster, S. Laforest, S. Laberge, M. C. Fournel, H. Gagnon, and S. Gagné. From Walkability to Active Living Potential. *American Journal of Preventive Medicine*, Vol. 28, No. 2, 2005, pp. 126–133.
15. Raulin, F., S. Lord, and P. Negron-Poblete. Évaluation de la marchabilité de trois environnements urbains de la région métropolitaine montréalaise à partir de l'outil MAPPA. *[VertigO] La revue électronique en sciences de l'environnement*, Vol. 16, No. 2, 2016. https://doi.org/10.4000/vertigo.17774
16. Secrétariat à l'enquête Origine-Destination, *Enquête Origine-Destination 2013: La mobilité des personnes dans la région de Montréal*. 2013, Agence métropolitaine de transport: Montréal.
17. Zielstra, D., and H. Hochmair. Using Free and Proprietary Data to Compare Shortest-Path Lengths for Effective Pedestrian Routing in Street Networks. *Transportation Research Record: Journal of the Transportation Research Board*, 2012. 2299: 41–47.
18. Luxen, D., and C. Vetter. Real-Time Routing with OpenStreetMap Data. *Proc., 19th ACM SIGSPATIAL International Conference on Advances in Geographic Information Systems*, ACM, Chicago, Illinois, 2011, pp. 513–516.
19. Morency, C., M. Demers, and E. Poliquin. Shifting Short Motorized Trips to Walking: The Potential of Active Transportation for Physical Activity in Montreal. *Journal of Transport & Health*, Vol. 1, No. 2, 2014, pp. 100–107.
20. Clifton, K. J., P. A. Singleton, C. D. Muhs, and R. J. Schneider. Representing Pedestrian Activity in Travel Demand Models: Framework and Application. *Journal of Transport Geography*, Vol. 52, 2016, pp. 111–122.
21. Frank, L. D., J. F. Sallis, B. E. Saelens, L. Leary, K. Cain, T. L. Conway, and P. M. Hess. The Development of a Walkability Index: Application to the Neighborhood Quality of Life Study. *British Journal of Sports Medicine*, Vol. 44, No. 13, 2010, pp. 924–933.
22. Institut national de santé publique du Québec. *Indicateurs géographiques de l'environnement bâti et de l'environnement des services influant sur l'activité physique, l'alimentation et le poids corporel*. Institut national de santé publique du Québec, Montréal, Québec, Canada, 2009, p. 98.
23. Witten, K., J. Pearce, and P. Day. Neighbourhood Destination Accessibility Index: A GIS Tool for Measuring Infrastructure Support for Neighbourhood Physical Activity. *Environment and Planning A*, Vol. 43, No. 1, 2011, pp. 205–223.

24. Frank, L. D., J. F. Sallis, T. L. Conway, J. E. Chapman, B. E., Saelens, and W. Bachman. Many Pathways from Land Use to Health: Associations between Neighborhood Walkability and Active Transportation, Body Mass Index, and Air Quality. *American Planning Association. Journal of the American Planning Association*, Vol. 72, No. 1, 2006, pp. 75–87.
25. R Core Team. *R: A Language and Environment for Statistical Computing*. R Foundation for Statistical Computing,Vienna, Austria, 2016.
26. Burnham, K. P., D. R. Anderson, and K. P. Burnham. *Model Selection and Multi-Model Inference: A Practical Information-Theoretic Approach*. Springer, New York, 2002, p. 488.
27. Thomas-Maret, I., P. Lewis, A. Laforest, and D. L. Métivier. Localisation des activités métropolitaines: quels impacts sur le navettage à Montréal? *Environnement Urbain*, Vol. 5, 2011, pp. 38–51.

The Standing Committee on Pedestrians (ANF10) peer-reviewed this paper (18-03473).

TRR
JOURNAL OF THE TRANSPORTATION RESEARCH BOARD

Article

Transportation Research Record
2018, Vol. 2672(35) 151–160

DOI: 10.1177/0361198118781652
journals.sagepub.com/home/trr
SAGE

Simulation Modeling of Pedestrian Performance in the New Synchronized and Milwaukee B Interchanges versus Existing Designs

Amirarsalan Mehrara Molan[1] and Joseph E. Hummer[2]

Abstract

Most U.S. highway agencies recognize the importance of designing safe, short, and comfortable paths for pedestrians. Providing such an acceptable condition for pedestrians requires specific attention at intersections and service interchanges due to their interaction with other modes of transportation. The main objective of this research was to analyze pedestrian operation at two new service interchanges—the synchronized and Milwaukee B interchanges—in comparison with four existing designs. The analysis consists of three main parts: (1) a general comparison among the designs for pedestrian performances in terms of safety, travel time, and the level of service; (2) investigation of the effects of vehicles on pedestrian performance; and (3) modeling the impact of pedestrians on the traffic operation of vehicles. A comprehensive series of simulation tests were run using VISSIM and Synchro to study the pedestrian performance of interchanges in various situations of traffic volume, turning traffic ratio, traffic distribution, and percentage of heavy vehicles. The results indicated that a relatively safe condition is expected for pedestrians in the proposed new designs in comparison with the existing interchanges; however, the pedestrians' travel time was lower in all the existing designs than in the new designs except the diverging diamond interchange (DDI). The DDI, one of the most popular alternative interchanges, showed the worst performance in all the aspects of the pedestrian analysis. Despite the very good performance of conventional diamond interchange regarding pedestrian travel time and safety, pedestrians could have a significant negative impact on vehicle travel time through a diamond.

Approximately, 11–15% of highway fatalities in the U.S. and Canada are pedestrians but this rate reaches 25%, and 30%, in China and Poland, respectively (*1–4*). The statistics become very substantial in less developed countries, where the rate increases to 57% in Mumbai, India (*5*). The threat seems to be more critical in intersections and service interchanges (where a freeway intercepts an arterial) due to the interaction of pedestrians and vehicles, especially in urban and suburban areas. In Montreal, almost 60% of pedestrian crashes occurred at intersections (*6*), while the rate in some places in the U.S. is as high as 76% (*7*).

The design and construction of most of the current service interchanges in the U.S. goes back more than 50 years, when there was much less notice taken of pedestrians. The considerable traffic growth during the last two decades caused transportation agencies to improve some old interchanges using alternative designs such as the diverging diamond interchange (DDI). The DDI offers good traffic operations and superior safety (*8*) but questions remain on its friendliness to pedestrians. This point might be one of the reasons for special attention to the topic these days, as a new project entitled NCHRP 07-25 has been recently been funded regarding the pedestrian performance at alternative intersections and interchanges (*9*).

The primary objective of this research was to study the performance of pedestrians in two new service interchanges in comparison with four existing designs. The influence of various traffic variables on pedestrian performance was also analyzed at each type of interchange. In addition, the research identified the impact of pedestrians on the vehicular travel time. There were only a few studies up to this point (*10–12*) which focused on the impact of pedestrians on vehicle opera-

[1]Wyoming Technology Transfer Center, University of Wyoming, Laramie, WY
[2]Mobility and Safety Division, North Carolina Department of Transportation, Raleigh, NC

Corresponding Author:
Address correspondence to Amirarsalan Mehrara Molan: amehrara@uwyo.edu

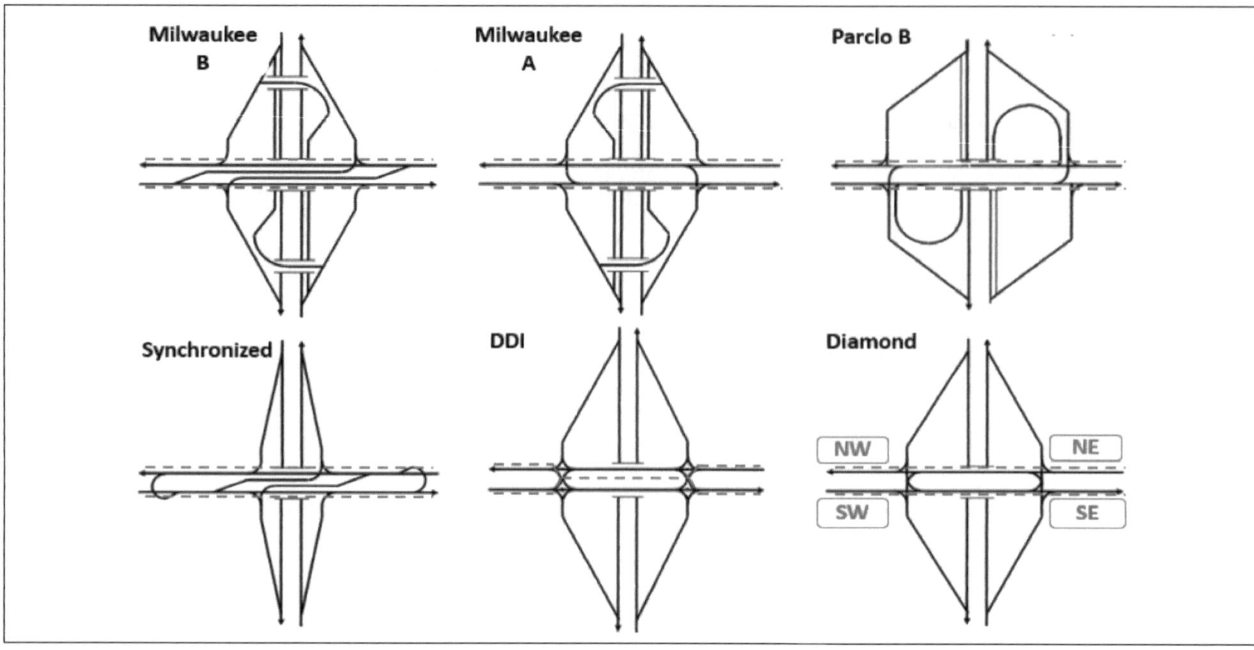

Figure 1. The geometry and the pedestrian path (the dashed red line) in the interchanges tested.

tion at intersections, so this effort hoped to build that knowledge base.

The new interchanges studied here—the synchronized and Milwaukee B interchanges—already showed potential in terms of traffic operation to be viable substitutes for failed conventional interchanges. In a previous work (*13*), the authors compared the new interchanges to the DDI, to the parclo B, to the diamond, and to an existing interchange in Milwaukee, WI that the authors called a Milwaukee A. The authors found that the new interchanges performed very well in terms of vehicle travel time and delay compared to those existing interchanges. Figure 1 provides diagrams of the interchanges considered in this study and in that previous work. In Figure 1 northbound (NB) and southbound (SB) represent the direction of the freeway while eastbound (EB) and westbound (WB) represent the arterial.

The synchronized interchange is essentially the interchange version of a superstreet intersection (also known as synchronized, restricted crossing u-turn, j-turn, or reduced conflict intersection). It has the same features as a superstreet design in prioritizing the traffic flow on the arterial while its geometry limits the access of the traffic on the side street. In this way, the traffic flow of the network is improved when there is not a large volume of left turns from the freeway. Because of the contraflow lanes for the left turns from arterial, and U-turns for the left turns from the freeway, all crossovers on the arterial have just "half-signals" (two-phase signals which only stop one direction of the arterial). Therefore, a strong progression system at any speed, any cycle length, and any crossover spacing is possible with the synchronized interchange.

In addition to the contraflow design for expediting the flow on arterial, two loops are provided in the Milwaukee B interchange for the left turns from the freeway. The Milwaukee B interchange thus mimics the current parclo B interchange. Like the parclo B and the synchronized interchange, the Milwaukee B also just uses half-signals on the arterial and should provide superb signal progression.

The authors chose the comparison interchanges based on popularity, efficiency, and similarity to the new interchanges. The conventional diamond is the most common interchange form in the U.S., and the parclo B design is acknowledged as one of the most efficient designs. The DDI was included because it has become very popular after the research by Chlewicki (*14*), and after the first DDI in the U.S. opened in 2009. The Milwaukee A was also included in this research since its geometry has similarities to the new Milwaukee B design and since no publication has reviewed its performance before to our knowledge.

Background

Alternative Intersections and Interchanges

Pedestrians have been a focus of research attention at alternative intersections and interchanges during the past 12 years or so. The Federal Highway Administration (FHWA) information guides on the restricted crossing u-turn (RCUT), median u-turn, and continuous flow intersections (CFI) all included chapters on how pedestrians should be accommodated in those designs and tried to make the case that the designs were not pedestrian barriers (*15–17*). The CFI has been a particular focus in this regard, with papers such as by Jagannathan and Bared in 2005 (*18*) attempting to improve the pedestrian-friendliness of the design. A research effort on pedestrians at the RCUT sponsored by the North Carolina Department of Transportation (NCDOT) showed that signalized midblock crosswalks would be helpful (*19*).

Since the 1990s, a need to improve the old interchange designs was found in the U.S. due to the considerable increase in the number of vehicles and the lack of budget for building new highways. To respond to the need, single-point interchanges (SPIs) became very popular during the early 1990s. SPIs gained the trust of designers because of their acceptable traffic operation and capacity using only a single three-phase signal. This single signal allowed the through traffic to clear the interchange faster than a conventional diamond, and the opposing left turns could move at the same time (*20*). As for disadvantages, the SPI needs a wide bridge that substantially increases the cost of construction. Besides, the design could be complicated for the users who were not familiar with the design and provide them trouble in finding their way safely. Pedestrians had also four free-flow and four signalized crossings to contend with. Despite the drawbacks, hundreds of SPIs were built across the U.S. until the mid-2000s when its popularity suddenly dropped after the DDI was introduced.

The DDI provides almost the same traffic operations as the SPI but is usually much cheaper and easier to construct. Thus, more than 80 DDIs have been opened in the U.S. since 2009 (*21*). According to the numerous studies on DDI (*14, 22–24*), the DDI has improved the vehicular traffic performance of failed diamond interchanges very well. Yeom et al. (*24*) conducted a before–after study to compare the performance of the interchanges when they had a conventional diamond before construction. Based on the results, queue lengths were decreased between 19 and 83 percent, and the DDI was able to improve the level of service (LOS) from F to C or even B in some cases. On the other hand, the previous studies mentioned a lack of capacity in the DDI, especially in high through traffic conditions. As an example, a DDI was not a proper treatment for improving a conventional diamond interchange for a case in Alabama (*25*). Regarding the pedestrian performance, two different paths have been used at a DDI: (1) a center crossing, and (2) an outside crossing. However, both the paths are like the SPI's path having four free-flowing and four controlled crossings and this added another point to the list of DDI disadvantages.

Pedestrian Studies

Many previous studies had been conducted to analyze and recognize the parameters involved in pedestrian crashes at intersections. Long waiting time for the walk interval, short walk interval, and the high turning volume of vehicles at conflict points with pedestrians on permissive green controls were identified as the most important variables for pedestrian crashes at intersections (*4*). Brosseau et al. (*6*) studied the effect of pedestrians' waiting time on their safety at intersections. The research defined the waiting time as a factor of signal phasing and arrival time and concluded that minimizing waiting time can considerably decrease the dangerous behavior of, and violations by, pedestrians. Clearance time (or the flashing "DO NOT WALK" interval) was introduced as another important factor for pedestrian safety. Pedestrians tend to commit a violation either when the clearance time is longer or shorter than needed.

Regarding pedestrian operation analysis, many important gaps are still observed in the literature despite the recent efforts. Milazzo et al. (*10*) and Hubbard et al. (*12*) claimed that the method in the version of the Highway Capacity Manual (HCM) in force when they did their research was not accurate and the effect of traffic volume was not reflected well. The HCM (*26*) method relates LOS for pedestrians to pedestrian space and delay, while no other variables such as the effect of right-turning traffic on pedestrians are considered. Other possible parameters which are not taken in account in the HCM analysis include the direction of pedestrian movement, pedestrian volume, the time of arrival (whether the pedestrian arrives on time or late to the crossing point), and the crosswalk location (*12*). Milazzo et al. (*10*) examined the capacity of intersections considering the effect of pedestrians. Their results showed that pedestrian volume impacts the vehicular saturation flow, especially when the rate is more than 500 pedestrians per hour in the U.S. Milazzo et al. (*10*) recommended to add new saturation flow adjustment factors to include the effect of pedestrians on the affected lane groups (right turn and left turn).

Simulation Modeling Procedure

Simulation models are in widespread use in various aspects of transportation engineering from the studies on user behavior (*27*) or vehicle dynamics (*28*), to studies related to operation and safety of transportation systems (14, *29, 30*). Based on recent and relevant studies (*4, 31, 32*), it can be concluded that VISSIM is one of the best available tools in this field for modeling vehicles and pedestrians, especially in the U.S.

There is no doubt that signal timing and phasing plays a notable role in the performance of pedestrians at an interchange. To make sure of the accuracy of signal timing in the simulation, all the signals were designed using Synchro. Both VISSIM and Synchro have been validated by many studies (*31–35*) and have provided high accuracy in their outcomes. Molan (*34*), which is a comprehensive research relevant to this paper, conducted a validation using real probe data and found that there is no significant difference between the simulation results and the real data. Ishaque and Noland (*31*) conducted extensive research on the calibration and validation of VISSIM and found that the VISSIM car-following algorithm is strongly adapted to the reality of

Table 1. Division of Traffic Conditions in the Simulation Modeling

			Traffic distribution	
Pedestrians volume	Truck percentage	Turning volume ratios	EB/WB	NB/SB
360 pedestrians per hour	4 % (moderate truck traffic)	Left turn = through = right turn (high turning condition)	EB = WB (equal traffic on EB and WB)	NB = SB (equal traffic on NB and SB)
No pedestrians	8 % (high truck traffic)	Left turn = 0.66 through = right turn (moderate turning condition)	EB = 0.5 WB (higher traffic on WB)	NB = 0.5 SB (higher traffic on SB)
		Left turn = 0.25 through = right turn (low turning condition)		0.5 NB = SB (higher traffic on NB)

pedestrian and vehicle flow even in the most complicated modeling environments. After the Synchro and VISSIM modeling, ANOVA was performed on the results using IBM SPSS to compare the mean values of the measure of effectiveness (MOEs) as well as to investigate the effect of various variables on the performance of each interchange. The following sections describe different aspects of the simulation modeling.

Simulation Scenarios

A fair comparison between interchanges would not be possible without including a wide range of traffic scenarios in the simulation. Vehicle volumes were detemined using the critical lane volume (CLV) procedure, which is a traditional and popular method to obtain an initial ratio of volume over capacity (V/C) at intersections and interchanges. The entry traffic volume of the network was found to be 5330, 5250, and 5140 vehicles per hour considering V/C equal to 1 in a conventional diamond interchange in high turning, moderate turning, and low turning cases, repectively. The authors collected data from 37 random service interchanges in North Carolina and found an average of 4.3% for the truck volume. Also, about 90% of the service interchanges had a truck traffic ratio less than 8%. Therefore, 4% and 8% were selected to simulate moderate and high truck traffic situations in the simulation. To cover balanced and unbalanced traffic situations, six different traffic distributions were considered in the study. As the last factor involved in the simulation, the network was analyzed for two cases of pedestrian volume: (1) when the pedestrian volume was a relatively high 360 per hour total on all sidewalks, and (2) when the network is located in a place with no effective presence of pedestrians. The researchers chose to use a single relatively high pedestrian volume in the experiment for three reasons. First, there is almost nowhere in the U.S. where there are higher pedestrian demands crossing interchanges than those simulated in this study outside of a few downtown areas where these interchanges would not be considered. Second, lower pedestrian demands would start to mean that the results would be based on low sample sizes leading to instability in the results. Third, since pedestrians up to fairly high flow levels travel in packs or bunches rather than lanes, causing each other little extra delay, there should be little variation in the results for different levels of pedestrian demand up to a high level. As shown in Figure 1, there were four origins and destinations (southwest, southeast, northwest, and northeast) for the pedestrians, so, each route handled 90 pedestrians per hour in the model runs with pedestrians. No pedestrians crossed the arterial; all pedestrians crossed the bridge. Table 1 presents a summary of scenarios in the research.

A total of 432 VISSIM simulation scenarios were tested. Each test was repeated two times (to include different seeds during the simulation), and the average (of two runs) was taken as the representative outcome of each test.

Interchange Design

Based on data collected on the geometry of 30 service interchanges in the U.S. (mostly from Michigan), curb radii of right turns, loops radii, and ramp lengths of 40 ft, 250 ft, and 2000 ft, respectively, were used in this research. A 600-ft distance between ramps was defined for all the interchanges except the parclo B which needed a longer distance of 1200 ft. As the base of the simulation model, a typical four-lane arterial (two lanes in each direction) was selected for all the interchanges, and exclusive right turn or left turn lanes were added to each design as needed. All the auxiliary lanes begin at the same place with a storage length of 400 ft and a taper length of 100 ft.

The lengths of the network on each leg were 5280 ft (1 mile) and 1600 ft for vehicles and pedestrians, respectively. Regarding the crossing pattern, as indicated by Figure 1, the outside crossing was chosen for all the interchanges except the DDI, where a center crossing was simulated due to its popularity in comparison with the outside crossing.

Table 2. Mean Values of Pedestrians' Travel Time and Stop in Each Interchange

	Overall		High turning		Moderate turning		Low turning	
Interchange type	Travel time (sec)	Stops (no)	Travel time (sec)	Stops (no)	Travel time (sec)	Stops (no)	Travel time (sec)	Stops (no)
DDI	386	2.05	380	2.13	386	1.96	391	2.07
Diamond	346	0.68	348	0.71	346	0.66	344	0.65
Milwaukee A	342	0.93	343	1.03	342	0.89	342	0.87
Milwaukee B	355	1.25	357	1.29	356	1.32	353	1.13
Parclo B	348	1.34	352	1.20	348	1.10	345	1.09
Synchronized	364	1.34	371	1.27	364	1.15	360	1.02

Traffic Signal Design

Right turn on red (RTOR) was allowed but turning vehicles had to yield to pedestrians in permissive (shared) green intervals. There was no jaywalking allowed in the models. Since pedestrian clearance time is an important component of safety, and there are different practices in setting that time around the U.S., the current research chose clearance time based on field data collection. The authors collected clearance time on all approaches at 25 intersections with pedestrian signals in urban and suburban areas of Detroit, Michigan. Reviewing the data, a wide range of clearance times were observed even in different locations with the same crosswalk length. For example, the range of clearance time in three-lane crossings varied from 9 sec to 16 sec. The data confirmed that no single method is dominant and clearance time mostly depends on designer philosophy or factors not related to intersection geometrics. Mean clearance times of 10.3, 13.7, 17.2, and 18.7 sec were found at two-lane, three-lane, four-lane, and five-lane crosswalks, respectively. Therefore, in these simulations the authors applied a 7-sec clearance time at one-lane crossings and added 3.5 sec for any additional lane.

The maximum of signal cycle length was considered 120 secs in the Synchro. We did not use higher cycle lengths due to increased pedestrian waiting time and chance of committing a violation (like jaywalking) for pedestrians as well as raising the threat of spillback on the traffic flow on main roadways. All the minimum green times of pedestrians were considered during the signal design. Pedestrians at the on-ramp crossings at the diamond interchange had an extra protected green interval simultaneously with the green of off-ramps since there was no conflict between vehicles and pedestrians during that phase.

User Behavior

Pedestrian crossing speed depends on many traffic and non-traffic parameters (such as age, gender, type of crossing, conflicting traffic volume, time of day, day of the week, etc.) and pedestrians usually tend to adjust their speed based on these variables at the particular moment and location. Based on information from previous studies, pedestrian speed generally follows a normal distribution with the majority of speed observations (about 70–80%) near the average. Marisamynathan and Vedagiri (*5*) found most pedestrian speeds in their sample between 4 feet per second (fps) to 4.5 fps, while a range from 4.6 fps to 5.8 fps was noted by Ishaque and Noland (*31*). A study in Poland found the range of pedestrian speeds from 3.6 fps to 4.6 fps (*4*). The current research relied on the field data collected during a previous study on pedestrian performance at superstreet intersections (*19*) and the same speed graphs were used. In that study, pedestrians were categorized into two groups as "walking pedestrians" with 91% of the observations and the average speed of 5 fps, and "running pedestrians" with 9% of the observations and an average speed of 9.6 fps. Regarding the priority rule of vehicles and pedestrians on the free-flow crossings in DDI and Milwaukee A, vehicle drivers had to stop for the pedestrians when pedestrians could find a minimum gap of 3 sec or longer to initiate a crossing.

The speed of cars was 70 mph on the freeway while it was defined as 60 mph for trucks. Both cars and trucks had a speed of 35 mph on arterial. The selected speeds are typical in the U.S. for these classes of roads. Fitzpatrick et al. (*36*) estimated the 85th percentile free-flow speed of vehicles between 13 to 21 mph and 17 to 29 mph in the center and approach of exclusive right-turn lanes, respectively. Therefore, the vehicle speed was simulated as 20 mph on the approach to a turn and 15 mph in the center of a right or left turn.

Data Analysis and Results

Overall Pedestrian Performance

Travel Time and the Number of Stops. The mean travel times and the number of stops in each interchange are provided in Table 2. Regarding pedestrian travel time, the Milwaukee A was the best interchange which provided faster routes for the pedestrians by a slim margin over the diamond and the parclo B. The reason for the strong travel time performance of the Milwaukee A is the existence of only one signalized

Table 3. Cycle Length, Red Light Interval, and the Estimated Waiting Time of Pedestrians

Parameters	Overall			High turning			Moderate turning			Low turning		
	CL[a] (sec)	R[b] (sec)	Waiting time (sec)	CL (sec)	R (sec)	Waiting time (sec)	CL (sec)	R (sec)	Waiting time (sec)	CL (sec)	R (sec)	Waiting time (sec)
DDI	75	43	44	61	36	38	76	42	41	89	49	51
Diamond	120	36	12	120	39	14	120	36	12	120	33	11
Milwaukee A	67	31	14	62	32	16	66	31	14	73	30	13
Milwaukee B	57	28	18	53	30	19	57	28	18	62	27	15
Parclo B	70	32	21	67	34	20	70	32	18	74	31	17
Synchronized	68	28	19	65	30	19	67	27	16	71	26	13

[a] Average cycle length of scenarios.
[b] Average red interval of pedestrians (clearance time of pedestrians is included).

crossing for each route in the geometry; its other crossing is a free-flow one with the right-of-way for pedestrians. If pedestrians had to wait for vehicles at the free-flowing crossing the result would be much different. The Milwaukee A performance was significantly better than the other designs in the high turning conditions, where its margin was wider over the diamond and parclo B. The new Milwaukee B and synchronized interchanges had travel times that were 13 to 22 seconds higher than the Milwaukee A on average. The DDI had the worst performance in terms of travel time and the number of stops. The diamond was the best interchange in terms of the number of stops, with an average of 0.68 per pedestrian, due to its long green interval in each cycle. After the diamond, the Milwaukee A and Milwaukee B had the lowest number of stops with the values of 0.93, and 1.25, respectively. On average, the parclo B and synchronized interchanges had the same number of stops per pedestrian but the parclo B did better with higher turning volumes and the synchronized did better in low turning conditions.

Safety Parameters. The number of stops should be one of the most effective variables looking at pedestrian safety. Pedestrians likely commit more violations as the number of stops increase. To elaborate on this issue, Table 3 examined the waiting time of pedestrians multiplying half of the red interval by the number of stops. Note that the number of stops is used as a factor for the probability of facing a red light in this research since pedestrians had the right-of-way for crossing at any other point of conflict with vehicles so all the stops were because of red lights. The reason for using half the red interval was to consider an average stop length for the pedestrians assuming random arrivals. Table 3 emphasized the poor performance of the DDI, with 3.6 times higher waiting times than the diamond which was the interchange with the lowest waiting times. There were three main reasons for predicting higher waiting time in the DDI in comparison with the other designs: (1) more number of stops based on Table 2, (2) the higher clearance time due to crossing longer crosswalks, especially in crossing the through traffic in the crossovers, and (3) lower ratios of green/cycle length (G/CL) since the pedestrians had to conflict with the main flow (through traffic) in the signalized crossings in the DDI, while the pedestrians of other designs had to be stopped only against the turning traffic and were receiving green time simultaneously with the through traffic of the arterial. The diamond had the lowest waiting times overall and the Milwaukee A had the second best waiting times. The new interchange designs were generally superior to the parclo B for waiting times due to shorter red intervals. The synchronized interchange performed particularly well in low turning scenarios.

Based on the literature review (*4*, *6*, *12*), the type, frequency, and size (length) of conflict points with vehicles are important parameters for pedestrian safety. The volume of conflicting traffic is also important to pedestrian safety. Table 4 shows the details regarding vehicle–pedestrian conflict points for each interchange considered. In Table 4, a "free-flowing conflict" happens where there is no signalized control and a "permissive crossing" is when both vehicles and pedestrians receive a green light simultaneously at the conflict point. The results that stand out in Table 4 are for the DDI and the Milwaukee A. All the other designs provided the same performance with four crossings of six total lanes and a total conflicting volume of 2270 vehicles per hour. On the other hand, two of the conflict points in Milwaukee A are free-flowing crossings and there are eight total lanes to cross. Most notably, because they cross and re-cross the through arterial lanes, pedestrians in a DDI face more and longer conflicting points with a significantly higher total conflicting volume.

Effective Variables on the Pedestrian Performance

Table 5 presents the effect of traffic variables on pedestrian travel time and stops, including turning volume ratio, traffic distribution, percentage of truck volume, and the interactions between them. Based on consistently high F-values and consistently low significance levels, the turning condition had the

Table 4. Geometric and Traffic Features of Each Interchange Network in the Conflict Points with Vehicles

	Free-flow crossing			Permissive crossing			Protected crossing			Total		
Parameters	N[a]	L[b]	V[c]	N	L	V	N	L	V	N	L	V
DDI	4	4	1514	0	0	0	4	20	2752	8	24	4266
Milwaukee A	2	4	1516	0	0	0	2	4	1512	4	8	3028
Other designs	0	0	0	2	2	758	2	4	1512	4	6	2270

[a]Number of crossings.
[b]Total Length (number of lanes).
[c]Total Conflicting Volume (veh/hr).

most influence among all the parameters on pedestrian travel time in all the interchanges. The DDI and the synchronized interchange were the designs most sensitive to the turning case. The DDI performed better as the turning ratio raises while the synchronized showed an opposite reaction. The parclo B had the same behavior as the synchronized, with better performance in lower turning ratio cases, while the rest of the interchanges did not change significantly with different turning ratios. Traffic distribution was found to be important to travel time for all designs. None of the interchanges showed a significant relationship at the 0.05 level between truck percentage and travel time. Among the two-way interactions, only the interaction between turning case and traffic distribution was statistically significant in most cases examined.

Impact of Pedestrians on Traffic Operation

As the last part of the evaluation in this research, the impact of pedestrians on vehicle travel time was analyzed. Table 6 presents the travel times extracted from VISSIM for runs with 360 pedestrians per hour and for runs without pedestrians. Generally, the interchanges with better vehicle travel time performance showed less vulnerability to the presence of pedestrians. According to an ANOVA conducted on the mean differences, the diamond was the only design with a significant difference at the 0.05 level between its results with and without pedestrians. The synchronized interchange and the DDI had mean differences of 6.6 sec, and 5.4 sec between pedestrian and no pedestrian cases, respectively, which were higher impacts than the Milwaukee A, Milwaukee B, and parclo B.

Table 6 also shows the Milwaukee B as the best interchange in terms of vehicle travel time either in the presence or absence of pedestrians. This confirms findings from our previous research (*13*). Table 6 shows that the synchronized interchange can be a good substitute for a DDI in low turning cases while the DDI was more promising during higher turning scenarios.

Conclusions

This paper compared the pedestrian performance of the new synchronized and Milwaukee B designs to four existing interchanges. Pedestrian service quality is an increasingly important part of interchange alternative analyses being conducted by agencies all over the U.S., so we hoped to contribute to that portion of those comprehensive analyses. The primary research method was a series of simulation models using VISSIM and Synchro, and evaluation with ANOVA.

Overall, these results should be helpful to highway agencies making choices on interchange design and trying to serve all roadway users well. The conventional diamond showed the best pedestrian operation among all the interchanges considered in the research. The Milwaukee B and synchronized interchanges trailed the diamond, Milwaukee A, and parclo B in travel times, but appear to offer relatively good pedestrian safety compared to other designs. Relatively poor pedestrian service is expected from the DDI since it got the worst ranking in all the MOEs. The diamond had the worst performance for vehicle travel time, and pedestrians affected the vehicle travel time at a diamond far more than at other designs, so a trade-off for the diamond between good pedestrian service and relatively poor vehicle service seems clear.

Turning volume ratio and traffic distribution were sometimes important variables in explaining travel time and the number of stops for pedestrians while the percentage of heavy vehicles was not found to be generally significant at the levels simulated.

At the levels simulated pedestrians did not impact the vehicle travel time significantly except in the case of the diamond interchange. The impact of pedestrians seemed to be higher as the traffic operations became more critical to handle. For example, pedestrians increased the vehicle travel time only about one second for the DDI during high-turning tests but the impact was raised to 13 seconds in the low-turning condition when the DDI should have a more difficult time handling the demand.

Future studies should analyze driver behavior and driver expectation and reaction to pedestrians in the new interchanges using a driving simulator. Simulator or field studies regarding pavement markings, traffic signs, and pedestrian signals would be helpful to maximize the safety and operation of pedestrians in the new designs. A useful future effort could focus on versions of these designs that

Table 5. Effects of Traffic Variables on the Pedestrians MOEs of the Existing Interchanges

		Travel time		Stops	
Type	Variable	F[a]	Sig[b]	F	Sig
DDI	Intercept	4240000	0.000	18600	0.000
	Turning case	314	0.000	11.0	0.003
	Traffic distribution	33.9	0.000	17.9	0.000
	Truck %	4.31	**0.064**	5.52	0.041
	Turning case–traffic distribution	15.1	0.000	9.66	0.001
	Turning case–truck %	0.81	**0.470**	1.01	**0.396**
	Traffic distribution–truck %	1.56	**0.254**	1.38	**0.307**
Diamond	Intercept	6050000	0.000	20400	0.000
	Turning case	43.2	0.000	16.8	0.001
	Traffic distribution	16.8	0.000	6.25	0.007
	Truck %	1.91	**0.197**	0.08	**0.775**
	Turning case–traffic distribution	7.09	0.002	6.17	0.004
	Turning case–truck %	6.71	0.014	0.01	**0.997**
	Traffic distribution–truck %	1.82	**0.196**	1.70	**0.221**
Milwaukee A	Intercept	19400000	0.000	11100	0.000
	Turning case	27.1	0.000	31.1	0.000
	Traffic distribution	11.8	0.001	2.61	**0.092**
	Truck %	1.15	**0.308**	0.72	**0.414**
	Turning case–traffic distribution	2.74	**0.064**	1.12	**0.429**
	Turning case–truck %	3.46	**0.072**	1.19	**0.343**
	Traffic distribution–truck %	0.53	**0.744**	1.09	**0.421**
Milwaukee B	Intercept	102000000	0.000	705000	0.000
	Turning case	1330	0.000	1560	0.000
	Traffic distribution	98.5	0.000	0.25	**0.930**
	Truck %	2.50	**0.145**	5.00	0.049
	Turning case–traffic distribution	62.5	0.000	676	0.000
	Turning case–truck %	2.50	**0.132**	5.00	0.031
	Traffic distribution–truck %	1.00	**0.465**	1.00	**0.465**
Parclo B	Intercept	20100000	0.000	3620000	0.000
	Turning case	693	0.000	3610	0.000
	Traffic distribution	478	0.000	4530	0.000
	Truck %	4.61	**0.057**	0.21	**0.651**
	Turning case–traffic distribution	27.6	0.000	1070	0.000
	Turning case–truck %	1.15	**0.354**	1.52	**0.265**
	Traffic distribution–truck %	0.30	**0.897**	1.26	**0.352**
Synchronized	Intercept	172000000	0.000	621000	0.000
	Turning case	13900	0.000	1850	0.000
	Traffic distribution	1550	0.000	7.64	0.003
	Truck %	1.00	**0.341**	0.93	**0.357**
	Turning case–traffic distribution	315	0.000	110	0.000
	Turning case–truck %	1.00	**0.402**	1.04	**0.387**
	Traffic distribution–truck %	1.00	**0.465**	1.02	**0.454**

Note: Bold represents the insignificant variables in the level of 0.05.
[a]Variation between sample means/variation within the samples.
[b]Sig determines whether any of the differences between the means are significant at the level of 0.05.

Table 6. Vehicle Travel Time in Different Conditions of Pedestrian Presence

	Overall			High turning			Moderate turning			Low turning		
Interchange type	With ped	No ped	Mean diff	With ped	No ped	Mean diff	With ped	No ped	Mean diff	With ped	No ped	Mean diff
DDI	139	133	5.41	122	121	0.96	131	129	1.78	164	151	12.9
Diamond	190	172	20.3[a]	188	172	15.6	208	174	34.8[a]	178	167	10.6
Milwaukee A	122	121	1.14	117	114	2.50	122	120	2.50	129	128	1.58
Milwaukee B	113	112	1.81	112	110	2.08	113	113	0.67	113	112	0.83
Parclo B	131	128	2.11	131	127	4.33	131	129	2.17	130	129	0.17
Synchronized	148	142	6.61	159	146	13.08	144	141	4.33	143	140	2.42

Note: Vehicle travel time unit is seconds.
[a]The mean difference is significant at the level of 0.05.

are the best possible for pedestrians; in other words, how narrow could the interchanges become while still providing reasonable service to vehicles. This study attempted to cover situations that were representative of those encountered by highway agencies at actual interchanges, but of course did not cover the entire range of possibilities, so more study of the new interchanges at macroscopic and microscopic levels under a wider range of conditions would of course be helpful.

Author Contributions

Both the authors confirm contribution to the paper as follows: study conception and design, Amirarsalan Mehrara Molan, Joseph E. Hummer; data collection, Amirarsalan Mehrara Molan, Joseph E. Hummer; analysis and interpretation of results, Amirarsalan Mehrara Molan, Joseph E. Hummer; draft manuscript preparation, Amirarsalan Mehrara Molan, Joseph E. Hummer. Both authors reviewed the results and approved the final version of the manuscript.

References

1. Moreno, L., P. Morency, and A. El-Geneidy. Pedestrian Crash Estimation Models for Signalized Intersections. *Accident Analysis and Prevention*, Vol. 43, No. 5, 2011, pp. 1624–1634.
2. Transport Canada. http://tc.gc.ca/, Retrieved October 2, 2017.
3. Zhang, W., K. Wang, L. Wang, Z. Feng, and Y. Du. Exploring factors affecting pedestrians' red-light running behaviors at intersections in China. *Accident Analysis and Prevention*, Vol. 96, 2016, pp. 71–78.
4. Oskarbski, J., L. Guminska, M. Miszewski, and I. Oskarbska. Analysis of Signalized Intersections in the Context of Pedestrian Traffic. *Proc., 6th Transport Research Arena*, Curran Associates, Inc., Warsaw, Poland, 2016.
5. Marisamynathan, S., and P. Vedagiri, P. Modeling Pedestrian Delay at Signalized Intersection Crosswalks under Mixed Traffic Condition. *Proc., 2nd Conference of Transportation Research Group of India, Elsevier Ltd.*, 2013, pp. 708–717.
6. Brosseau, M., S. Zangenehpour, and L. Moreno. The Impact of Waiting Time and Other Factors on Dangerous Pedestrian Crossings and Violations at Signalized Intersections: A Case Study in Montreal. *Transportation Research Part F*, Vol. 21, 2013, pp. 159–172.
7. Pulugurtha, S., and V. Sambhara. Pedestrian Crash Estimation Models for Signalized Intersections. *Accident Analysis and Prevention*, Vol. 43, 2011, pp. 439–446.
8. Vaughan, C., C. Cunningham, B. Schroeder, and J. Hummer. Empirical Study and Assessment of Operational Performance of Double Crossover Diamond Interchanges. Presented at 92nd Annual Meeting of the Transportation Research Board, Washington, D.C., 2013.
9. Transportation Research Board Website. http://apps.trb.org/. Accessed April 12, 2017.
10. Milazzo, S., N. Rouphail, J. Hummer, and P. Allen. Effect of Pedestrians on Capacity of Signalized Intersections. *Transportation Research Record: Journal of the Transportation Research Board*, 1998. 1646: 37–46.
11. Banerjee, I., S. Shladover, J. Misener, and D. Ragland. *Impact of Pedestrian Presence on Movement of Left-Turning Vehicles: Method, Preliminary Results & Possible Use in Intersection Decision Support*. Safe Transportation Research and Education Center, 2004.
12. Hubbard, S., D. Bullock, and F. Mannering. Right Turns on Green and Pedestrian Level of Service: Statistical Assessment. *Journal of Transportation Engineering*, Vol. 135, No. 4, 2009, pp. 153–159.
13. Mehrara Molan, A., and J. Hummer. Travel Time Evaluation of Milwaukee B and Synchronized as New Interchange Designs. *Journal of Transportation Engineering*, Vol. 144, No. 2, 2018. https://doi.org/10.1061/JTEPBS.0000107
14. Chlewicki, G. *Variations of the Diverging Diamond Interchange*. Second Urban Street Symposium of TRB, Anaheim, California, 2003.
15. Reid, J., L. Sutherland, P. Brinckerhoff, B. Ray, A. Daleiden, P. Jenior, and J. Knudsen. *Median U-Turn Informational Guide*. FHWA-SA-14-069. Federal Highway Administration (FHWA), Washington, D.C., 2014.
16. Hughes, W., R. Jagannathan, D. Sengupta, and J. Hummer. *Alternative Intersection and Interchanges: Informational Report (AIIR)*. FHWA-HRT-09-060. Federal Highway Administration (FHWA), Washington, D.C., 2010.
17. Hummer, J., B. Ray, A. Daleiden, P. Jenior, and J. Knudsen. *Restricted Crossing U-turn Informational Guide*. FHWA-SA-14-070. Federal Highway Administration (FHWA), Washington, D.C., 2014.
18. Jagannathan, R., and J. Bared. Design and Performance Analysis of Pedestrian Crossing Facilities for Continuous Flow

Intersections. *Transportation Research Record: Journal of the Transportation Research Board*, 2005. 1939: 133–144.

19. Hummer, J., A. Holzem, N. Rouphail, C. Cunningham, S. O'Brien, B. Schroeder, K. Salamati, and R. Foyle. *Pedestrian and Bicycle Accommodations on Superstreets*. North Carolina Department of Transportation, North Carolina, 2014.
20. Messer, C., J. Bonneson, and D. Anderson. *NCHRP Report 345: Single Point Urban Interchange Design and Operation Analysis*. Transportation Research Board of the National Academies, Washington D.C., 1991.
21. Diverging Diamond Interchange Website. http://divergingdiamond.com. Accessed February 8, 2017.
22. Bared, J., P. Edara, and R. Jagannathan. Design and Operational Performance of Double Crossover Intersection and Diverging Diamond Interchange. *Transportation Research Record: Journal of the Transportation Research Board*, 2005. 1912: 31–38.
23. Chlewicki, G. Learning from Eight Operational Diverging Diamond Interchanges in the United States. Presented at 92nd Annual Meeting of the Transportation Research Board, Washington, D.C., 2013.
24. Yeom, C., J. Hummer, B. Schroeder, C. Cunningham, and N. Rouphail. Empirical Before-After Comparison of the Operational Performance of Diverging and Conventional Diamond Interchanges. *Journal of Transportation*, Vol. 7, No. 1, pp. 35–56, 2015.
25. Khan, T., and M. Anderson. Evaluating the Application of Diverging Diamond Interchange in Athens, Alabama. *International Journal for Traffic and Transport Engineering*, Vol. 6, No. 1, pp. 38–50, 2016.
26. *Highway Capacity Manual 2010*. Transportation Research Board of the National Academies, Washington, D.C., 2010.
27. Ghasemzadeh, A., and M. Ahmed. Drivers Lane-Keeping Ability in Heavy Rain. *Transportation Research Record: Journal of the Transportation Research Board,* 2017. 2663: 99–108.
28. Mehrara Molan, A., and A. Kordani. Optimization of Speed Hump Profiles Based on Vehicle Dynamic Performance Modeling. *Journal of Transportation Engineering*, Vol. 140, No. 8, 2014.
29. Olya, M. H. Applying Dijkstra's Algorithm for General Shortest Path Problem with Normal Probability Distribution Arc Length. *International Journal of Operational Research*, Vol. 21, No. 2, 2014, 143.
30. Kordani, A., and A. Mehrara Molan. The Effect of Combined Horizontal Curve and Longitudinal Grade on Side Friction Factors. *Journal of Civil Engineering*, Vol. 19, No. 1, pp. 303–310, 2015.
31. Ishaque, M., and R. Noland. Pedestrian and Vehicle Flow Calibration in Multimodal Traffic Microsimulation. *Journal of Transportation Engineering*, Vol. 135, No. 6, 2009.
32. Schroeder, B., K. Salamati, and J. Hummer. Calibration and Field Validation of Four Double-Crossover Diamond Interchanges in VISSIM Microsimulation. *Transportation Research Record: Journal of the Transportation Research Board*, 2014. 2404: 49–58.
33. Eustace, D., and S. Ponnada. *L&D Manual Turn Lane Storage Validation/Update*. Report No. FHWA/OH-2012/13. Ohio Department of Transportation, 2012.
34. Mehrara Molan, A. *Evaluation of Milwaukee B and Synchronized as New Service Interchange Designs*. PhD dissertation. Wayne State University, Detroit, 2017.
35. Mehrara Molan, A., and J. Hummer. Safety Analysis of the New Synchronized and Milwaukee B Interchanges in Comparison to Existing Designs. *Journal of Accident Analysis and Prevention*, Vol. 109, 2017, pp. 29–35.
36. Fitzpatrick, K., W. Schneider, and E. Park. Predicting Speeds in an Urban Right-Turn Lane. *Journal of Transportation Engineering*, Vol. 132, No. 3, 2006.

The Standing Committee on Pedestrians (ANF10) peer-reviewed this paper (18-00146).

Sage Publications USA
sagepub.com

ISBN 978-1-5108-7836-5

9 781510 878365

CURRAN ASSOCIATES INC.
proceedings.com